CRUDE OIL EXPLORATION IN THE WORLD

Edited by **Mohamed Abdel-Aziz Younes**

Crude Oil Exploration in the World
Edited by Mohamed Abdel-Aziz Younes

Published by InTech
Janeza Trdine 9, 51000 Rijeka, Croatia

Publishing Process Manager Vana Persen

Technical Editor Teodora Smiljanic

Cover Designer InTech Design Team

Additional hard copies can be obtained from orders@intechopen.com

Crude Oil Exploration in the World, Edited by Mohamed Abdel-Aziz Younes
p. cm.
ISBN 978-953-51-0379-0

For EU product safety concerns:
IN TECH d.o.o., Prolaz Marije Krucifikse Kozulić 3, 51000 Rijeka, Croatia,
info@intechopen.com or visit our website at intechopen.com.

Meet the editor

Prof Dr Mohamed Abdel-Aziz Younes rejoined the staff of Geology Department, Faculty of Science, Alexandria University, Egypt as a Professor of Petroleum Geology and Geochemistry in 2007. Mohamed has B. Sc. Honors (1981) in Special Geology and M. Sc. (1985) in Petroleum Geology. He obtained his Ph. D. in Petroleum Geology and Geochemistry of the Gulf of Suez in (1991) from Alexandria University, Egypt. He is an active member of the American Association of Petroleum Geologists (AAPG), European Association of Organic Geochemists (EAOG), the American Association for the Advancement of Science (AAAS) and the Egyptian Geological Survey (EGS). He received the Alexandria University Scientific Encouragement Prize for the academic year 2003. Mohamed is the author of several international published research papers on Petroleum Geology and Geochemistry of the Gulf of Suez, Western Desert, Nile Delta and Offshore Mediterranean Sea and the Upper Egypt.

Contents

Preface

Exploring for oil and gas Traps is one of the main aim of the "Crude Oil Exploration in the World" which represents an important part of the Treatise of Petroleum Geology and Geochemistry in the world.

We have chosen eleven distinguished papers from the entire submitted researches around the world in the field of oil and gas exploration and environmental application from the world. These researches represent a guide to the petroleum geologists and geochemists all over the world.

The handbook of "Crude Oil Exploration in the World" is a professional exploration's guide to the petroleum technology, methodology to explore fields of oil and gas in the world.

Prof. Dr. Mohamed Abdel-Aziz Younes
General Coordinator of Petroleum Geology Program
Professor of Petroleum Geology and Geochemistry,
Geology Department, Moharrem Beck,
Faculty of Science,
Alexandria University,
Alexandria,
Egypt

1

Crude Oil Geochemistry Dependent Biomarker Distributions in the Gulf of Suez, Egypt

M. A. Younes
Geology Department, Moharrem Bek, Faculty of Science, Alexandria University, Alexandria, Egypt

1. Introduction

The Gulf of Suez occupies the northern end of the Red Sea rift (Said, 1962) Figure 1. It is a northwest-southeast fault-forming basin that provided adequate conditions for hydrocarbon generation, maturation and entrapment (Dolson et al., 2000). The Gulf of Suez province has been producing oil since 1908 and is reported to have 1.35 billion barrels of recoverable oil reserves. Intensive exploration has resulted in the discovery of more than 120 oil fields providing more than 50% of the overall daily oil production in Egypt (Egypt Country Analysis Briefs, 2009).

The Precambrian to Holocene lithostratigraphic succession of the Gulf reaches a total thickness of about 6,000 meters (Figure 2), which contributed to the development of different types of structural traps as well as different source, reservoir, and cap rocks (Khalil and Moustafa, 1995). It can be subdivided into three major lithostratigraphic sequences relative to the Miocene rifting of the Afro-Arabian Plate that led to the opening of the Suez rift and deposition of significant syn-rift facies from the Miocene Gharandal and Ras Malaab Groups (Evans, 1990). The pre-rift lithostratigraphic section, starting from the Nubia Sandstone to the Eocene Thebes Formation, rests unconformably on Precambrian basement. Rifting in the Gulf was associated with the upwelling of hot asthenosphere (Hammouda, 1992). Both crustal extension and tectonic subsidence reached their peaks between 19 and 15 Ma (Steckler, 1985; Steckler et al., 1988). Palaeozoic through Tertiary strata and major Precambrian basement blocks are exposed on both sides of the southern province which is characterized by structural and depositional complexity (Winn et al., 2001). The regional dip of strata is towards the SW (Meshref et al., 1988).

Previous geochemical studies throughout the Gulf of Suez have revealed that the oils are derived mainly from marine sources, which may be differentiated into three main groups (Mostafa, 1993 and Barakat et al., 1997). The distribution of these oil families are consistent with the geographic subdivisions of the Gulf of Suez provinces as northern, central and southern (Moustafa, 2002). Crude oil of the northern Gulf of Suez province is characterized by a C_{35}/C_{34} homohopane index <1 and a relatively heavy carbon isotope composition ($\delta^{13}C$ saturate -27‰) suggesting generation from a less reducing marine source rock environment at relatively low levels of thermal maturity. Meanwhile, crude oil of the central province is characterized by low API gravity, a predominance of pristane over phytane, a high C_{35}/C_{34}

homohopane index, and a lighter carbon isotopic composition ($\delta^{13}C$ saturate -29‰). oils of the southern province is characterized by a high API gravity, a low sulfur content and intermediate carbon isotopic composition values ($\delta^{13}C$ saturate -28 to -29 ‰). These two oil groups are believed to be derived from a marine source and exhibits compositional heterogenity suggesting a complex petroleum system may be present in the Gulf of Suez province.

In the present study saturate and aromatic biomarker distributions as well as stable carbon isotope compositions have been determined for a collection of crude oils of various ages and derived from different source rock types in the Gulf of Suez. These biomarker parameters have been used in an attempt to characterize the types of organofacies, and depositional environments, and to assess the thermal maturity of the source rocks responsible for oil generation.

2. Sampling and analytical techniques

Crude oils of various ages and derived from various source rock types were collected from the giant producing fields in the Gulf of Suez namely: July, Ramadan, Badri, El-Morgan, Sidki, Ras El-Bahar, East Zeit, Hilal, Zeit Bay and Shoab Ali (Fig. 1). These oil samples were collected from syn-rift (Miocene) and pre-rift (Palaeozoic, Lower and Upper Cretaceous) reservoirs (Fig. 2).

The crude oil samples were fractionated using high performance liquid chromatography (HPLC) into saturates, aromatics, and resins following the standard procedures outlined by Peters and Moldowan (1993). Saturate fractions were treated with a molecular sieve (silicate) to remove the *n*-alkanes. The saturate and aromatic fractions were analyzed on a Hewlett Packard 5890 Series-II gas chromatograph equipped with a Quadrex 50m fused silica capillary column. The gas chromatograph was programmed from 40°C to 340°C at 10 °C/min with a 2 min hold at 40° C and a 20 min hold at 340°C. The saturate and aromatic fractions were also analyzed by gas chromatography-mass spectrometry (GCMS) using a Hewlett Packard 5971A Mass Selective Detector (MSD) to determine terpane (*m/z* 191) and sterane (*m/z* 217) distributions. The aromatic steroid hydrocarbon fractions were analyzed to determine mono- and triaroaromatic (*m/z* 253 and *m/z* 231) steroid hydrocarbon distributions. Aromatic sulphur compounds were monitored to determine dibenzothiophene (*m/z* 184), methyldibenzothiophenes (*m/z* 198), dimethyl-dibenzothiophenes (*m/z* 212), methylnaphthalenes (*m/z* 142, 156 and 170) and phenanthrenes (*m/z* 178, 192 and 206). Stable carbon isotope values ($\delta^{13}C$) were determined for the whole oils, saturate and aromatic hydrocarbon fractions using a Micromass 602 D Mass-Spectrometer. Data are reported as $\delta^{13}C$ (‰) relative to the PDB standard. The organic geochemical analyses and stable carbon isotopes for the studied crude oil samples were conducted at the organic geochemical laboratories, Oklahoma State University, USA.

3. Results and discussions

Rohrback (1982) concluded that all the crude oils of the Gulf of Suez appear to be of the same genetic family. However, great variations in the biological marker distributions and stable carbon stable isotope compositions of the studied crude oils from this province suggest that this group should be subdivided into two subfamilies consistent vertically with

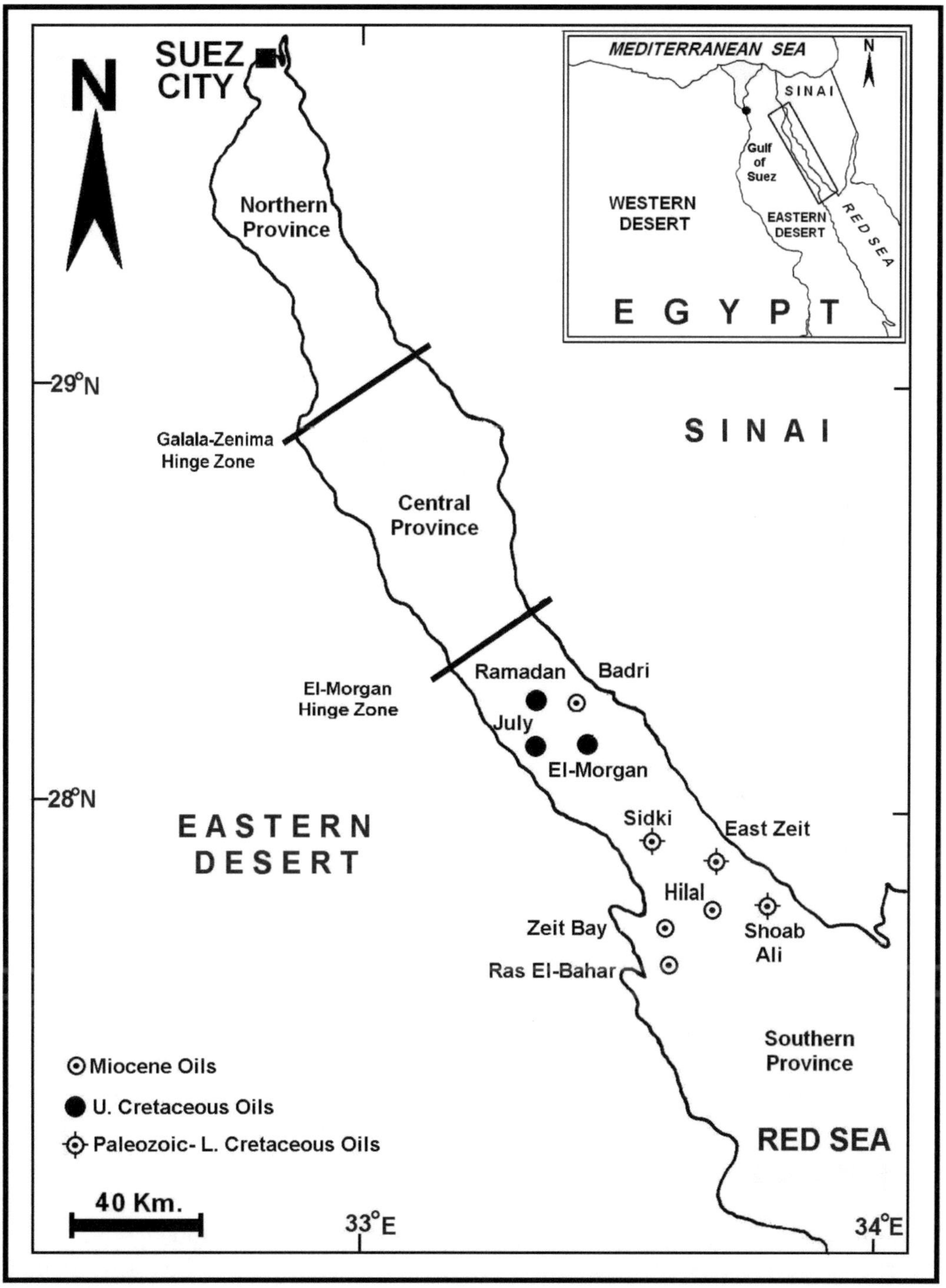

Fig. 1. Map showing the distribution of oil samples from the different fields of the southern Gulf of Suez province.

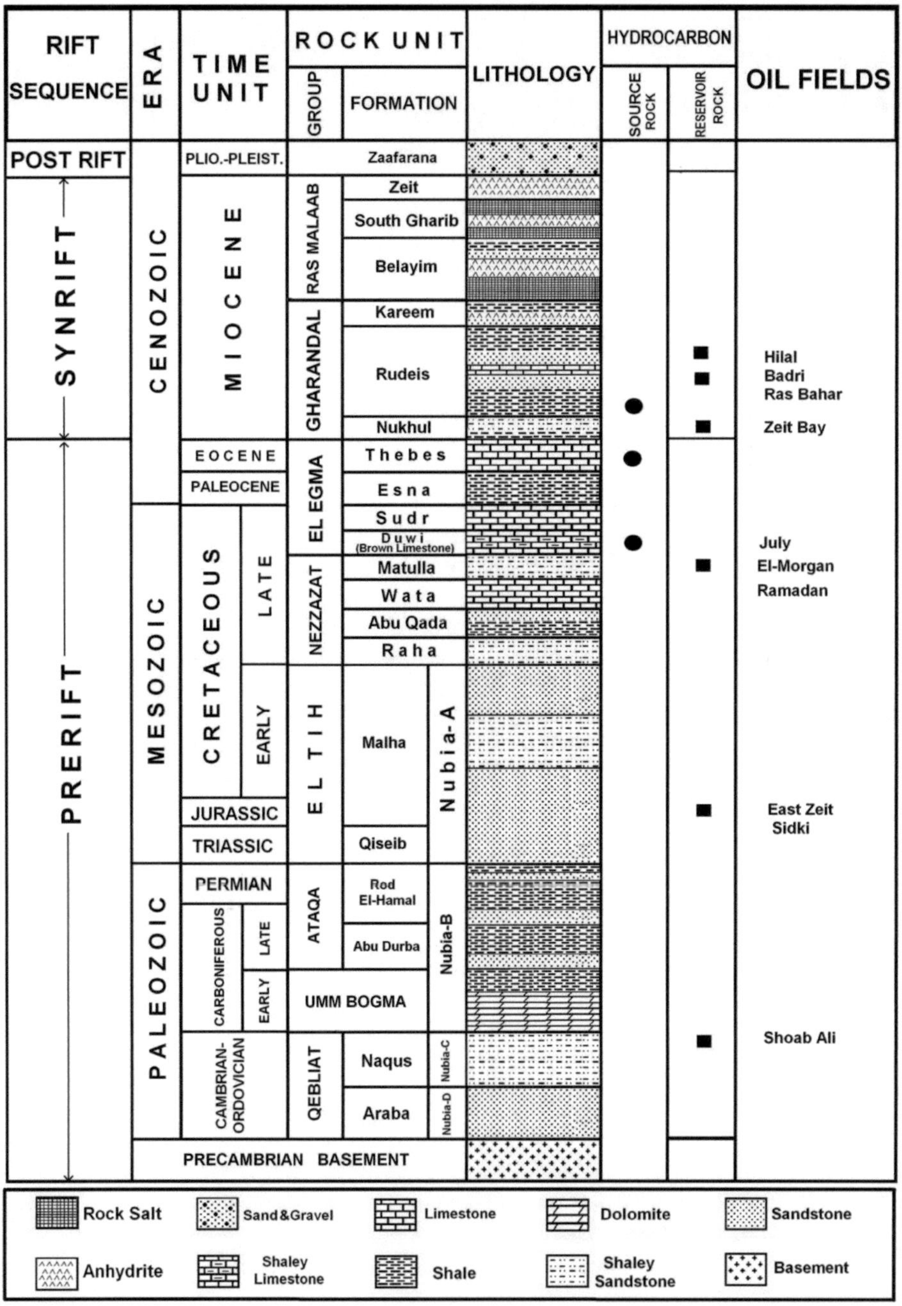

Fig. 2. Generalized lithostratigraphic succession illustrating the rifting sequences and hydrocarbon distributions in the southern Gulf of Suez modified after (Alsharhan, 2003).

the syn-rift and pre-rift tectonic sequences of the Gulf of Suez. Furthermore, the data from the present study suggests two oil families represent two distinct independent petroleum systems for hydrocarbon generation and maturation.

FIELD NAMES	HILAL	BADRI	R-BAHAR	ZEIT BAY	JULY	MORGAN	RAMADAN	E-Z E I T	SIDKI	SHOAB ALI
WELL NAMES	HA-11	BDR-5	RB-3	ZBC-1	J-26	M-208	R-55	EZA-18	GS 382 B-1	B-6
OIL TYPES	Type-I	Type-I	Type-I	Type-I	Type-II	Type-II	Type-II	Type-II	Type-II	Type-II
RESERVOIR AGE	Miocene	Miocene	Miocene	Miocene	U.Cretaceous	U.Cretaceous	U.Cretaceous	Paleoz.- L. Cretaceous	Paleoz.- L. Cretaceous	Paleoz.- L. Cretaceouse
BULK PARAMETERS										
API° gravity	32.2	27.9	34.9	32.2	35.8	34.8	36.8	41.3	37.2	43.2
Sulphur (wt.%)	0.97	0.89	0.98	0.88	1.36	1.32	1.23	1.28	1.33	1.39
GC PARAMETERS										
Pr / Ph	1.37	1.12	1.15	1.13	0.81	0.91	0.93	1.01	0.94	1.03
Pr / n-C17	0.72	0.57	0.52	0.54	0.37	0.37	0.52	0.49	0.33	0.53
Ph / n-C18	0.58	0.58	0.48	0.56	0.46	0.42	0.47	0.48	0.35	0.56
CPI	1.02	1.01	1.01	1	0.99	0.98	0.97	0.99	0.99	1.01
GC- MS (Saturated Hydrocarbons)										
1-Terpanes (m/z 191)										
Ts/(Ts+Tm)	0.8	0.56	0.78	0.67	1.01	1.07	1.1	1.03	1.08	1.13
Oleanane Index%	26.42	28.43	23.15	21.4	3.38	4.45	3.09	6.3	5.01	6.03
Gammacerane Index%	9.44	8.62	8.66	7.56	21.7	25.53	23.71	29.19	21.99	25
C_{35}/C_{34} Homohopanes	0.79	0.77	0.81	0.87	1.15	1.04	1.05	1.01	1.25	1.07
2- Steranes (m/z 217)										
C_{27} %	49.5	43.6	47.6	43.8	41.5	44.6	42.6	41.5	44.3	42.6
C_{28} %	25.6	29.8	24.8	28.5	26.7	25.1	23.3	23.4	26.1	27.9
C_{29} %	24.9	26.6	27.6	27.7	31.8	30.3	34.1	35.1	29.6	29.5
$C_{29}\alpha\alpha\alpha$ 20S/(S+R)	0.32	0.31	0.36	0.34	0.5	0.51	0.49	0.59	0.54	0.59
$C_{29}\alpha\beta\beta/(\alpha\beta\beta + \alpha\alpha\alpha)$	0.51	0.48	0.53	0.52	0.59	0.6	0.58	0.67	0.65	0.71
Diasterane Index	0.63	0.61	0.58	0.62	0.71	0.69	0.73	0.75	0.81	0.78
3- Aromatic Steranes										
TAS/(MAS+TAS)	0.46	0.38	0.41	0.45	0.64	0.68	0.62	0.71	0.73	0.76
TAS/MAS% (C_{27}/C_{28})	62.6	54.2	59.6	56.8	67.2	73.1	68.4	75.2	74.4	78.2
TAS/MAS% (all isomers)	66.1	58.6	63.2	64.2	73.8	76.2	72.3	76.5	77.2	83.2
GC- MS (Aromatic Sulfur Compounds)										
4-MDBT/1-MDBT (MDR)	1.87	1.88	1.81	1.85	2.31	2.27	2.34	2.85	2.91	2.87
2,4-DMDBT/1,4-DMDBT	2.13	2.25	2.3	2.46	4.2	4.1	4.56	5.23	5.98	5.45
4,6-DMDBT/1,4-DMDBT	1.98	2.45	2.29	2.84	3.98	4.15	4.38	4.96	5.23	5.58
DBT/P	0.31	0.34	0.37	0.38	0.57	0.62	0.69	0.73	0.94	0.81
Stable Carbon Isotope Composition										
$\delta^{13}C$ Saturate (‰ PDB)	-26.42	-27.26	-27.63	-28.36	-28.76	-28.46	-28.78	-26.31	-28.96	-28.16
$\delta^{13}C$ aromatic (‰ PDB)	-25.2	-26.92	-26.98	-28.04	-27.68	-28.43	-27.57	-26.25	-28.69	-28.06
$\delta^{13}C$ whole oil (‰ PDB)	-25.31	-26.96	-27.31	-27.56	-28.15	-27.78	-28.24	-26.98	-28.5	-27.76
Canonical Variable Parameters	-0.754	-1.407	-1.645	-2.148	-0.339	-2.76	-0.045	-3.365	-2.073	-2.698

Table 1. Bulk, biomarker properties and stable carbon isotope composition of crude oils from the Gulf of Suez.

Peak No.	Compound Name
A	C_{19} Tricyclic terpane
B	C_{20} Tricyclic terpane
C	C_{21} Tricyclic terpane
D	C_{22} Tricyclic terpane
E	C_{23} Tricyclic terpane
F	C_{24} Tricyclic terpane
G	C_{25} Tricyclic terpane (22*R*)
G	C_{25} Tricyclic terpane (22*S*)
H	C_{24} Tetracyclic terpane
I	C_{26} Tricyclic terpane (22*R*)
I	C_{26} Tricyclic terpane (22*S*)
J	C_{28} Tricyclic terpane (22*R*)
J	C_{28} Tricyclic terpane (22*S*)
K	C_{29} Tricyclic terpane (22*R*)
K	C_{29} Tricyclic terpane (22*S*)
L (Ts)	C_{27} 18α(*H*)-22, 29, 30- trisnorneohopane
M (Tm)	C_{27} 17α(*H*)-22, 29, 30- trisnorhopane
N	C_{30} Tricyclic terpane (22*R*)
N	C_{30} Tricyclic terpane (22*S*)
P	C_{31} Tricyclic terpane (22*R*)
P	C_{31} Tricyclic terpane (22*S*)
Q	C_{29} 18α(*H*)-norneohopane (29*Ts*)
R	C_{30} 18α(*H*)-oleanane
S	C_{30} 17α(*H*), 21β(*H*)-hopane
T	C_{30}17β(*H*), 21α(*H*)- moretane
U	C_{31} 17α(*H*), 21β(*H*)-30 homohopane (22*S*)
	C_{31} 17α(*H*), 21β(*H*)-30 homohopane (22*R*)
V	C_{30} Gammacerane
W	C_{32} 17 α(*H*), 21β(*H*)-30 bishomohopane (22*S*)
	C_{32} 17 α(*H*), 21β(*H*)-30 bishomohopane (22*R*)
X	C_{33} 17 α(*H*), 21β(*H*)-30 trishomohopane (22*S*)
	C_{33} 17 α(*H*), 21β(*H*)-30 trishomohopane (22*R*)
Y	C_{34} 17 α(*H*), 21β(*H*)-30 tetrakishomohopane (22*S*)
	C_{34} 17 α(*H*), 21β(*H*)-30 tetrakishomohopane (22*R*)
Z	C_{35} 17 α(*H*), 21β(*H*)-30 pentakishomohopane (22*S*)
	C_{35} 17 α(*H*), 21β(*H*)-30 pentakishomohopane (22*R*)

Table 2. Peak identifications in the *m/z* 191 mass fragmentograms.

Peak No.	Compound Name
	13β(H), 17α(H)- diacholestane (20*S*)
a	13β(H), 17α(H)- diacholestane (20*R*)
	13α(H), 17β(H)- diacholestane (20*S*)
b	13α(H), 17β(H)- diacholestane (20*R*) +
c	24- Methyl-13β(H), 17α(H)- diacholestane (20*S*)
	24- Methyl-13β(H), 17α(H)- diacholestane (20*R*)
d	5α(H), 14α(H), 17α(H) – cholestane (20*S*)
e	5α(H), 14β(H), 17β(H) – cholestane (20*R*) +
f	24-Ethyl-13β(H), 17α(H)- diacholestane (20*S*)
g	5α(H), 14β(H), 17β(H) – cholestane (20*S*) +
	24-Methyl-13β (H), 17α(H)- diacholestane (20*R*)
h	5α(H), 14α(H), 17α(H) – cholestane (20*R*)
i	24-Ethyl-13β(H), 17α(H)- diacholestane (20*R*)
j	24-Ethyl-13 α(H), 17β(H)- diacholestane (20*S*)
k	5α(H), 14α(H), 17β(H)– 24-methylcholestane (20*S*)
l	5α(H), 14α(H), 17β(H)– 24-methylcholestane (20*R*)+
m	24-Ethyl-13 α(H), 17β(H)- diacholestane (20*R*)
n	5α(H), 14β(H), 17β(H)– 24-methylcholestane (20*S*)
o	24-Propyl-13α(H), 17β(H)- diacholestane (20*S*)
p	5α(H), 14α(H), 17α(H)– 24-methylcholestane (20*R*)
q	5α(H), 14α(H), 17α(H)– 24-ethylcholestane (20*S*)
r	5α(H), 14β(H), 17β(H)– 24-ethylcholestane (20*R*)
s	5α(H), 14β(H), 17β(H)– 24-ethylcholestane (20*S*)
t	5α(H), 14α(H), 17α(H)– 24-ethylcholestane (20*R*)

Table 3. Peak identifications in the *m/z* 217 mass fragmentograms.

4. Gross geochemical characteristics

The syn-rift oil produced from (Miocene) reservoirs is a naphthenic, non-waxy crude with API gravity ranging from 27.9° to 34.9° and sulfur content between 0.78 to 0.98 wt.% (Table 1). Meanwhile, the second type, which occurs in the pre-rift lithostratigraphic units is paraffinic and waxy with API gravity ranging from 34° to 44° and sulfur content between 1.23 and 1.39 wt.%. The stratigraphic change in gross geochemical characteristics of the crude oils from a naphthenic to a paraffinic type is related probably to the change of source rock types from clastics to carbonate and environment of source rock deposition (Rohrback, 1982). High sulfur oils of the second oil type is indicative of carbonate evaporate source rocks, while the low sulfur concentrations are typical for siliciclastic source rocks (Gransch and Posthuma, 1974). The diversity of the gross geochemical characteristics of the crude oils is consistent vertically with a gradual change in API gravity and maturity variation (Matava et al., 2003).

5. Source-dependent biomarker distributions

Biomarkers are compounds that characterize certain biotic sources and retain their source information after burial in sediments (Meyers, 2003). It is used for oil-oil and oil-source rock correlations to assess the source of organofacies, kerogen types and the degree of thermal maturity (Philp and Gilbert, 1986; Waples and Machihara, 1991; Peters and Moldowan, 1993; Peters and Fowler, 2002). The great variability of saturate and aromatic biomarker indices, listed in Table 1, that enable subdivisions of the studied crude oil into two types referred as type-I and II as illustrated in (Figure 3). The predominance of *n*-alkanes and acyclic isoprenoids in the C_{11} to C_{35} region of the gas chromatograms is diagnostic of marine organofacies sources (Collister et al., 2004). A predominance of

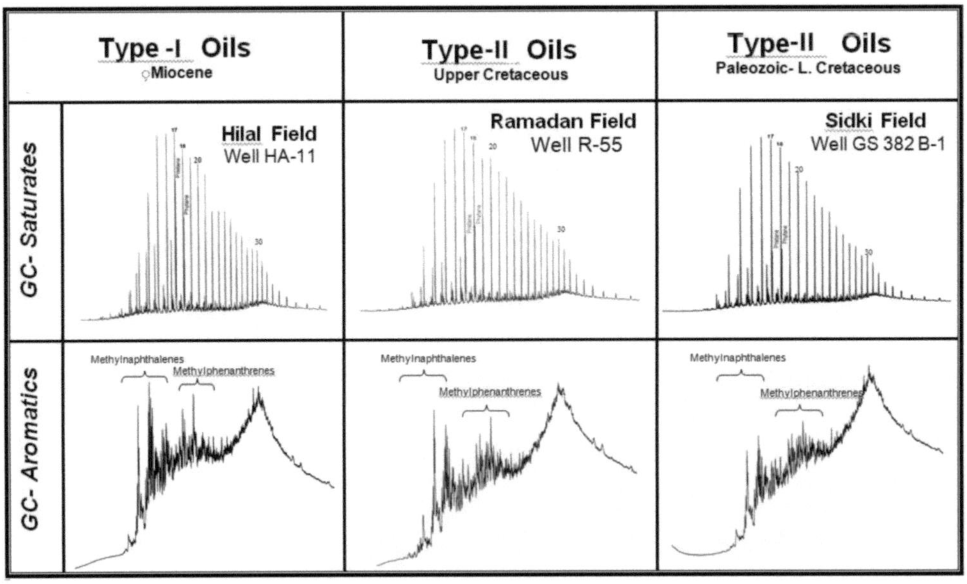

Fig. 3. Gas chromatograms of saturate and aromatic hydrocarbon fractions for representative crude oil types I and II.

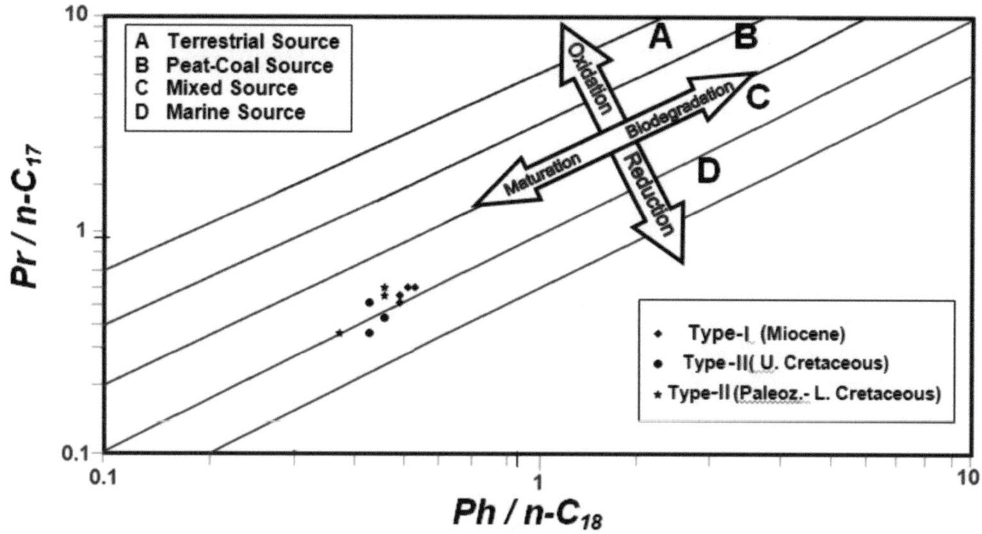

Fig. 4. Relationship between isoprenoids and *n*-alkanes showing source and depositional environments (Shanmugam, 1985) of the oil samples from the Gulf of Suez. All samples are located within the mixed to marine reducing depositional environments.

Peak No.	Compound Name
A	C_{20} Triaromatic Sterane
B	C_{21} Triaromatic Sterane
C	C_{26} Triaromatic Sterane (20*S*)
D	C_{26} Triaromatic Sterane (20*R*)+
	C_{27} Triaromatic Sterane (20*S*)
E	C_{28} Triaromatic Sterane (20*S*)
F	C_{27} Triaromatic Sterane (20*R*)
G	C_{28} Triaromatic Sterane (20*R*)

Table 4. Peak identifications in the *m/z* 231 mass fragmentograms.

pristane over phytane (Pr/Ph ratio >1) and the high odd-even carbon preference index (CPI>1) for the type-I oil is typical of crude oils generated from source facies containing terrigenous, wax-rich components (Peters et al., 2000). Type-II oil has lower Pr/Ph ratios (<1) and a slight even-carbon preference index (CPI<1) indicating algal/bacterial organic detritus in the kerogen (Collister et al., 2004), typical for a marine source rock deposited under less reducing conditions (Lijmbach, 1975). The nature of the source rock depositional environments can be further supported from the plotting of the isoprenoid ratios Pr/n-C_{17} versus Ph/n-C_{18} (Shanmugam,1985). It can be seen from Fig. 4 that both of the oil types plotted in the border region of marine-mixed organic matter with the source rocks being deposited under less reducing conditions and receiving significant clastic input (Bakr and Wilkes, 2002).

Peak No.	Compound Name
a	5 β- C_{27} Monoaromatic Sterane (20*S*)
b	dia- C_{27} Monoaromatic Sterane (20*S*)
c	5 β- C_{27} Monoaromatic Sterane (20*R*)+ dia- C_{27} Monoaromatic Sterane (20*R*)
d	5 α- C_{27} Monoaromatic Sterane (20*S*)
e	5 β- C_{28} Monoaromatic Sterane (20*S*)+ dia- C_{28} Monoaromatic Sterane (20*S*)
f	5 α- C_{27} Monoaromatic Sterane (20*R*)
g	5 α- C_{28} Monoaromatic Sterane (20*S*)
h	5 β- C_{28} Monoaromatic Sterane (20*R*)+ dia- C_{28} Monoaromatic Sterane (20*R*)
i	5 β- C_{29} Monoaromatic Sterane (20*S*)+ dia- C_{29} Monoaromatic Sterane (20*S*)
j	5 α- C_{29} Monoaromatic Sterane (20*S*)
k	5 α- C_{28} Monoaromatic Sterane (20*R*)
l	5 β- C_{29} Monoaromatic Sterane (20*R*)+ dia- C_{29} Monoaromatic Sterane (20*R*)
m	5 α- C_{29} Monoaromatic Sterane (20*R*)

Table 5. Peak identifications in the m/z 253 mass fragmentograms.

Terpane biomarker distributions derived from the *m/z* 191 mass chromatograms are shown in (Figure 5) and peak identifications are given in (Table 2). The ratio of Ts/(Ts+Tm) is considered as a facies and depositional environmental parameter of the relevant source rocks (Bakr and Wilkes, 2002). It is also considered a maturation parameter due to the greater thermal stability of Ts (18α(H)-22,29,30-trisnorneohopane) than its counterpart Tm (17α (H)-22,29,30-trisnorhopane) (Seifert and Moldowan, 1978; Cornford et al., 1988; Isaksen, 2004). Ts/(Ts+Tm) ratio for the crude oil is generally consistent with the carbon preference index CPI, indicating an anoxic marine depositional environment (Mello et al., 1988). The C_{35}/C_{34} homohopane ratio was found to be less than unity for type-I oil, suggesting a reducing marine environment. The Ts/(Ts+Tm) ratio is greater than unity for type-II oil suggesting a higher contribution of bacterial biomass to the sediments possibly reflecting a highly saline reducing environment (ten Haven et al., 1988; Mello et al., 1988).

Depositional environment biomarker parameters based on the terpanes (*m/z* 191), such as the oleanane index [oleanane/(oleanane+hopane)] and gammacerane index [gammacerane/(gammacerane+hopane)], illustrate that type-I oil is highly enriched in oleanane compared to the type-II oil. The oleanane ratio are 28.4% in some samples clearly demonstrating an enrichment of angiosperm higher land plant input to the source kerogen of Tertiary age (Ekweozor et al., 1979; Moldowan et al., 1994). Meanwhile, the low oleanane index in the type-II oil, ranging from 3.4 to 6.3%, suggesting generation from an Upper Cretaceous source rock or older (Moldowan et al., 1994). Higher values of the gammacerane index for type-II oil (21.7 to 25.5%) compared to type-I oil (7.6 to 9.4%) indicates a highly saline depositional environment associated with an evaporitic-carbonate deposition and low terrigenous input (Rohrback, 1982; Mello et al., 1988; Peters and Moldowan, 1994).

Sterane distributions for the two oil types (*m/z* 217) are shown in (Figures 5) and compound identifications are given in (Table 3). The predominance of C_{27} steranes (Table 1) and the presence of C_{30} *n*-propyl steranes (Figure 5) further support the idea of generation from bacterial-algal marine source rocks (Moldowan et al., 1985; Peters and Moldowan, 1991). Type-II oil is highly enriched in αββ sterane isomers relative to the type-I oil, which suggests that the type-II oil is probably generated from an evaporitic-carbonate source rock.

Cross plots of the Pr/Ph ratio for the two oil types against various depositional environment biomarker indices show an obvious separation of the two oil types, and a direct relationship of the Pr/Ph ratio with the oleanane index and an inverse relationship with gammacerane and the C_{35}/C_{34} homohopane ratio. An inverse relationship also exists between the oleanane and gammacerane indices for the two oil types (Figure 6). The separation of the two oil types is interpreted to indicate the presence of two independently sourced oils that consistent vertically with the pre-rift and syn-rift tectonic sequences of the Gulf of Suez.

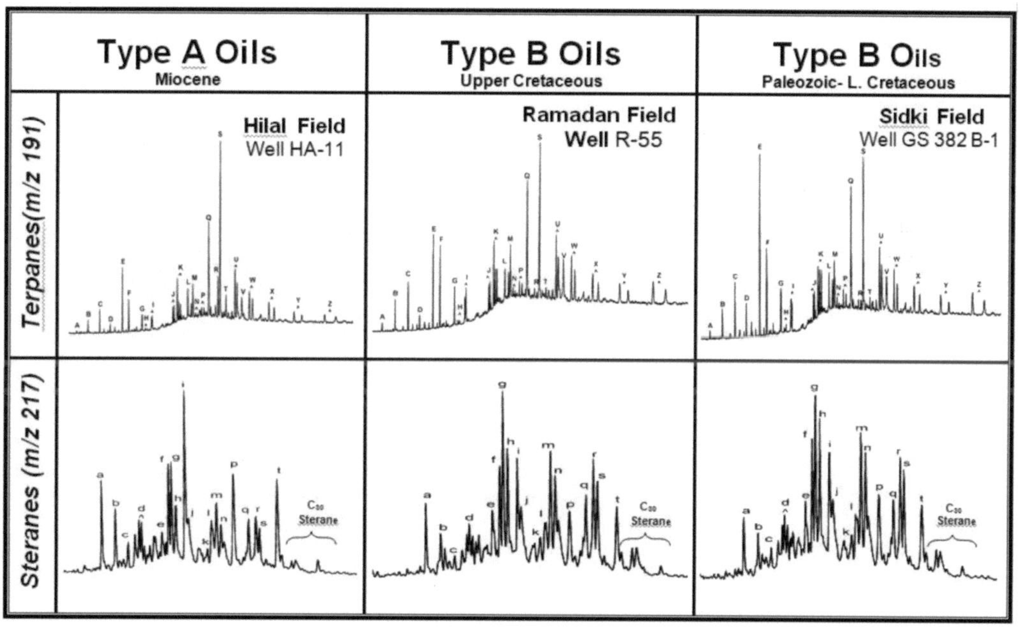

Fig. 5. Triterpane (*m/z* 191) and sterane (*m/z* 217) distribution patterns of the saturate hydrocarbon fractions from the two oil types in the Gulf of Suez. Labeled peaks are identified in Tables 2 and 3.

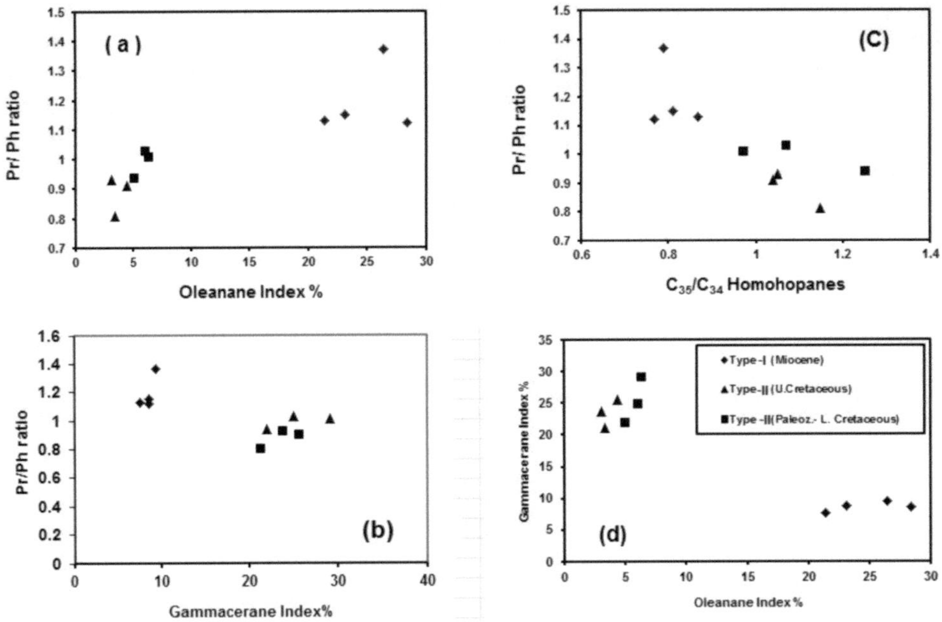

Fig. 6. A cross plot relation of source parameter Pr/Ph ratio for the studied crude oils that enable from differentiation of crude oils into two groups and show a direct relationship between Pr/Ph ratio with oleanane index and reverse relation with gammacerane and C_{35}/C_{34} homohopanes. A reverse relationship is shown on the basis of oleanane versus gammacerane indices.

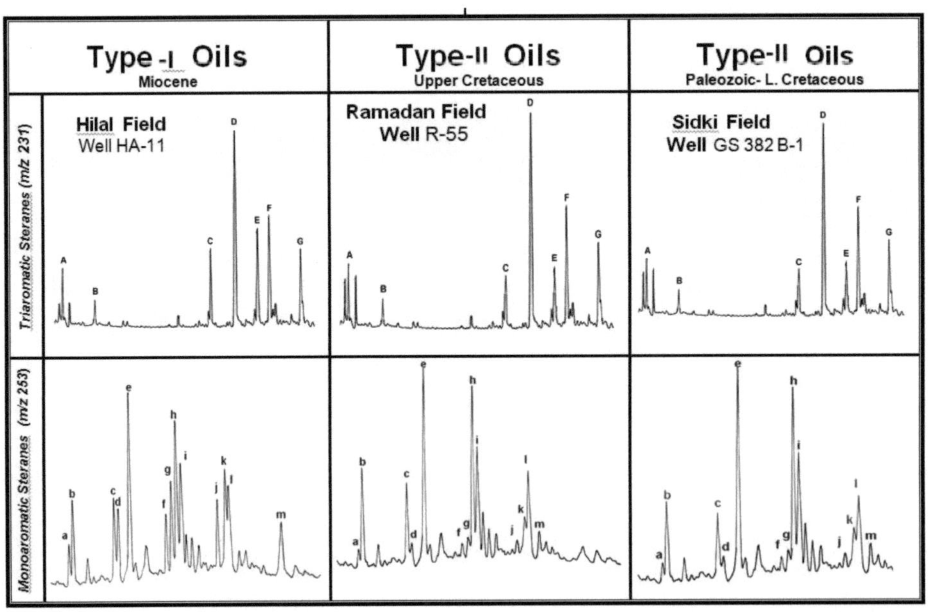

Fig. 7. Triaromatic (*m/z* 231) and monoaromatic (*m/z* 253) distribution patterns for two oil types from the Gulf of Suez. Labeled peaks are identified in Tables 4 and 5.

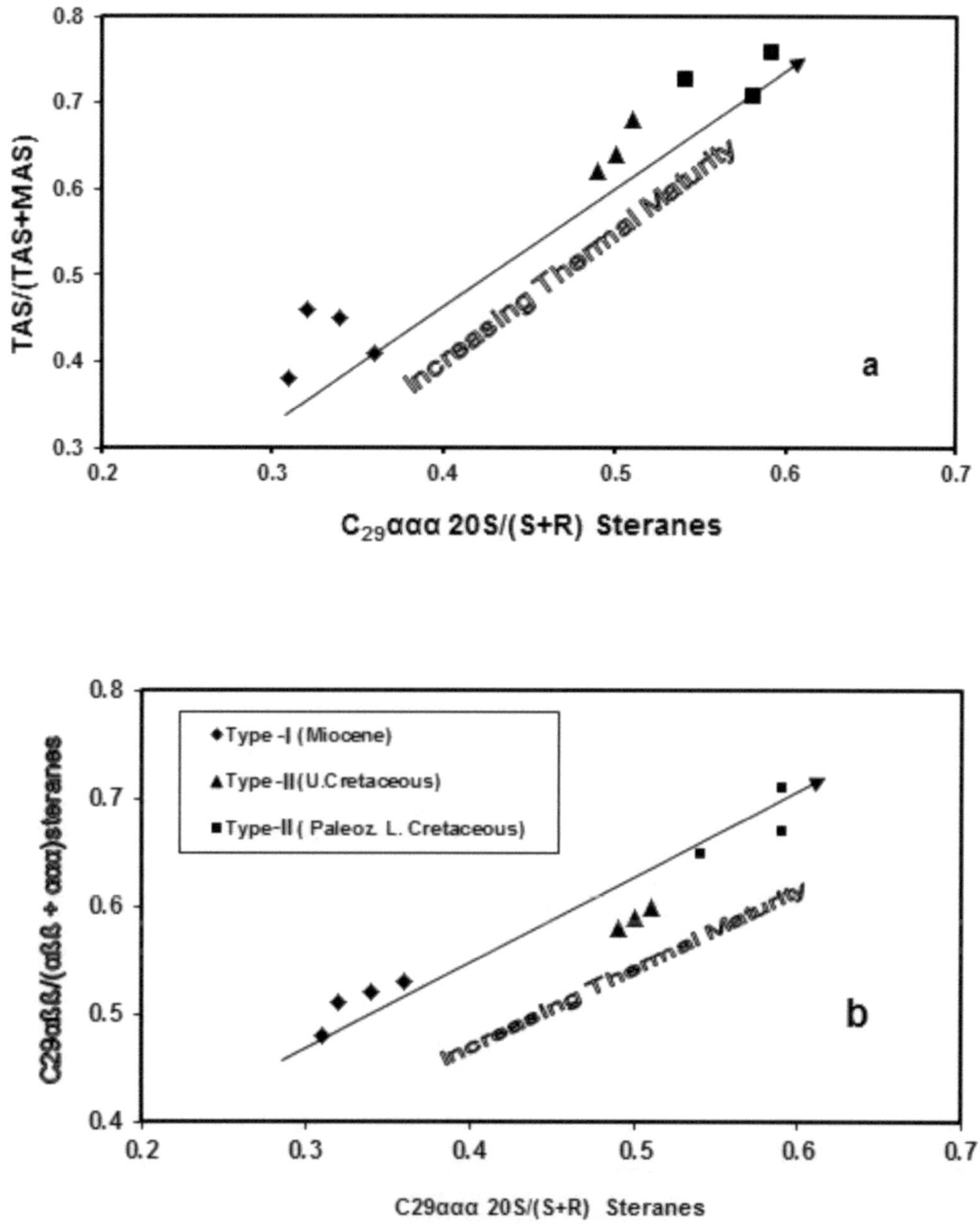

Fig. 8. Regular relationship between sterane maturity biomarkers C_{29} ααα 20S/(S+R) sterane with [(TAS/(TAS+MAS)] and C_{29}αßß/(αßß + ααα).

Peak No.	Compound Name
	Naphthalenes
A	2-Methylnaphthalene
B	1-Methylnaphthalene
C	2-Ethylnaphthalene
D	1-Ethylnaphthalene
E	2,6+2,7-Dimethylnaphthalene
F	1,3+1,7-Dimethylnaphthalene
G	1,6-Dimethylnaphthalene
H	1,4+2,3-Dimethylnaphthalene
I	1,5-Dimethylnaphthalene
J	1,2-Dimethylnaphthalene
K	1,3,7-Trimethylnaphthalene
L	1,3,6-Trimethylnaphthalene
M	1,3,5+1,4,6-Trimethylnaphthalene
N	2,3,6-Trimethylnaphthalene
O	1,2,4-Trimethylnaphthalene
P	1,2,5-Trimethylnaphthalene
	Phenanthrenes
a	Phenanthrene (P)
b	3-Methylphenanthrene
c	2-Methylphenanthrene
d	9-Methylphenanthrene
e	1-Methylphenanthrene
f	2,6+3,5-Dimethylphenanthrene
g	2,7-Dimethylphenanthrene
h	1,3+2,10+3,10+3,9-Dimethylphenanthrene
i	1,6-Dimethylphenanthrene
j	1,7-Dimethylphenanthrene
k	2,3-Dimethylphenanthrene
l	1,9-Dimethylphenanthrene
	Dibenzothiophenes
1	Dibenzothiophene
2	4-Methyldibenzothiophene
3	3,2-Methyldibenzothiophene
4	1-Methyldibenzothiophene
5	4-Ethyldibenzothiophene
6	4,6-Dimethyldibenzothiophene
7	2,4-Dimethyldibenzothiophene
8	1,4-Dimethyldibenzothiophene

Table 6. Peak identifications of the aromatic sulfur compound mass fragmentograms.

6. Maturation-dependent biomarker distributions

The Gulf of Suez province is characterized by local areas of higher heat flow due to the presence of hot spots in the southernmost Gulf and northern Red Sea (Alsharhan, 2003). Biomarker maturity parameters, including the sterane isomerization , C_{29} ααα20S/(S+R), and ratios based on the mono-and triaromatic steroidal hydrocarbon distributions (*m/z* 253 and 231) are shown in (Figure 7) with compound identifications in (Tables 4 and 5). These parameters also clearly distinguish the two different oil types on the basis of their different maturity levels consistent with the pre-rift and syn-rift tectonic sequences of the Gulf of Suez. Increasing source rock maturation from diagenesis to catagenesis is accompanied by an increase in the degree of aromaticity that converts monoaromatic steroids (MAS) to triaromatic steroids (TAS) lead to an increase thermal maturity through diagenetic/catagenetic processes results in the conversion of monoaromatic steroid to triaromatics (Seifert and Moldowan, 1978).

The triaromatic/monoaromatic maturity parameters (TAS/MAS) for all isomers and C_{27}/C_{28} ratios found to be 60% for type-I oil. For type-II B oil these ratios reaches more than 75%. Both of these ratios indicate a predominance of triaromatic relative to monoaromatic steroids for type-II oil compared to type-I oil which in turn reflect the higher maturity level for the type-II oil. Thus, it is proposed that type-II oil was generated from high mature source rock compared to type-I oil which are considered to be derived from a marginally mature source rock in the Gulf of Suez.

A plot showing the relationship between the sterane isomerization ratios C_{29}ααα 20*S*/(*S*+*R*) and C_{29}αßß/(αßß+ααα) and TAS/(MAS+TAS), that according to Seifert and Moldowan (1981), are genetically related to the effect of thermal maturity processes are shown in (Figure 8). It shows that there is a direct relationship between C_{29}ααα 20S/(S+R) and both TAS/(MAS+TAS) and C_{29}αßß/(αßß+ααα) increasing with burial depth of the source rocks (Matava et al., 2003). Type-II oil has a maximum value of 0.71 for the sterane isomerization ratio and 0.59 for the C_{29}ααα 20S/(S+R) ratio, while these ratios for type-I oil is 0.53 and 0.36 respectively. The API gravity is directly proportional to the maturity biomarker parameters as C_{29}ααα 20*S*/(*S*+*R*), C_{29}αßß/(αßß+ααα), TAS/(MAS+TAS) and C_{35}/C_{34} homohopanes as shown in (Figure 9). These relationships also support the high thermal maturity level of the type-II oil compared to the type-I oil in the Gulf of Suez province.

Diasterane/sterane ratios are highly dependent on both the nature of the source rock and level of thermal maturity. This ratio is commonly used to distinguish carbonate from clay rich source rocks and can be used to differentiate immature from the highly mature oils (Seifert and Moldowan, 1978). Type-I oil is slightly depleted in diasteranes relative to type-II oil, probably reflecting differences in their level of thermal maturity and also differing clastic input to their source rocks (Kennicutt et al., 1992). Aromatic sulfur compounds such as dibenzothiophene (DBT), methyldibenzothiophenes (MDBT) and dimethyldibenzothiophenes (DMDBT) can be used as maturity indicators of source rock and petroleum (Chakhmakhchev et al., 1997; Radke et al., 1997). Figure (10) displays representative partially expanded mass chromatograms of the aromatic sulfur hydrocarbons representing naphthalenes, phenanthrenes and dibenzothiophenes with compound identifications given in (Table 6). Previous studies (e.g. Radke et al., 1997) have demonstrated that the relative distributions of methylated aromatic compounds are thermodynamically controlled and, with increasing maturity, a decrease is observed in the amount of the less stable *a*-

substituted isomer (1-MDBT) compared with the amount of the more stable *ß*-substituted isomer (4-MDBT). A number of ratios are applicable for thermal maturity assessments on the basis of aromatic sulphur compounds. Logarithmic scale cross-plots of 4-MDBT/1-MDBT (MDR) parameter versus the three maturity parameters (4,6-/1,4-DMDBT; 2,4-/1,4-DMDBT; and DBT/Phenanthrene ratios) is presented in (Figure 11). An increase of MDR is accompanied by an increase of the 4,6-/1,4-DMDBT, 2,4-/1,4-DMDBT and DBT/Phenanthrene ratios, reflects the differences in aromatic sulfur compound maturity from the marginally mature type-I oil (syn-rift Miocene Rudeis Shale) to fully mature type-II oils (pre-rift Upper Cretaceous Brown Limestone and Middle Eocene Thebes Formation) in the Gulf of Suez.

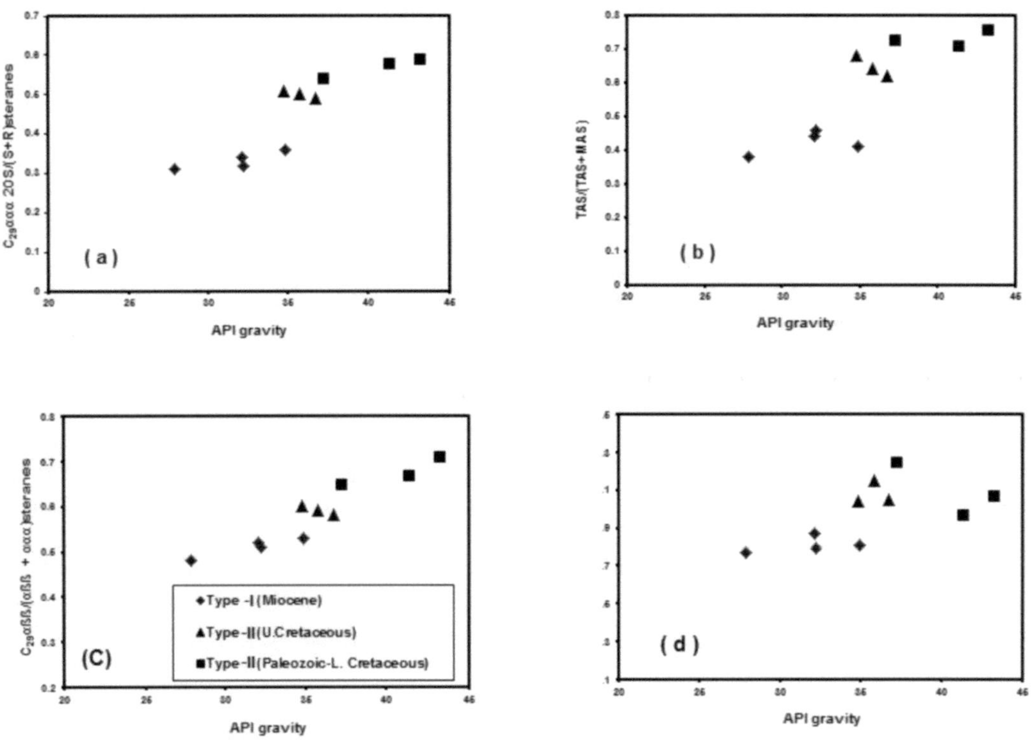

Fig. 9. Illustrates the direct relationship between gross geochemical attribute API gravity of crude oils and the sterane and triterpane maturity biomarkers C_{29} ααα 20S/(S+R), [(TAS/(MAS+TAS)], C_{29}αßß/(αßß + ααα) and C_{35}/C_{34} homohopanes.

6.1 Stable carbon isotopic compositions

The stable carbon isotopic composition of organic matter is an important tool in differentiating algal from land plant source materials and marine from continental depositional environments (Meyers, 2003). Rohrback (1982) and Zein El-Din and Shaltout (1987) found that the crude oils of the Gulf of Suez were relatively light with δ^{13} C values for the saturate fractions between -27‰ to -29‰. They concluded that the stable carbon isotope values of crude oils are dependent mainly on the depositional environment of the source rock and the degree of thermal maturity at which the oil was expelled.

Sofer (1984) distinguished oils derived from marine and non-marine sources from different parts of the world, including Egypt on the basis of the δ^{13} C composition of the saturate and aromatic hydrocarbon fractions.

Using the canonical variable relationship CV= $-2.53\delta^{13}C_{sat.} + 2.22\delta^{13}C_{arom.} - 11.65$, postulated by Sofer (1984), the Gulf of Suez province oil yield canonical variable values between -3.365 and -0.045. These values are generally lower than 0.47 indicating typical marine (non-waxy) oils. Stable carbon isotope data of the saturate and aromatic hydrocarbons and whole oils are shown in (Table 1) and plotted in (Figure 11). The stable carbon isotope composition of the saturate fraction ranges between -28.96 and -26.42‰, while the aromatic fraction has a range of -28.69 to –25.2‰. The results show an almost complete separation of the type-I and II oils. The results of the stable carbon isotope values are consistent with the results obtained by Rohrback (1982) and Alsharhan (2003), who concluded that all the Gulf of Suez crude oils were derived from marine source rocks. Type-I oil is generally exhibit heavier isotopic values than type-II oil, which is consistent with source rock variations. Miocene oil from the Zeit Bay well has a stable carbon isotope composition, which is more consistent with Type-II oil. Paleozoic-Lowe Cretaceous oil from the well East Zeit A-18 has a stable carbon isotope composition which is more like type-I oil.

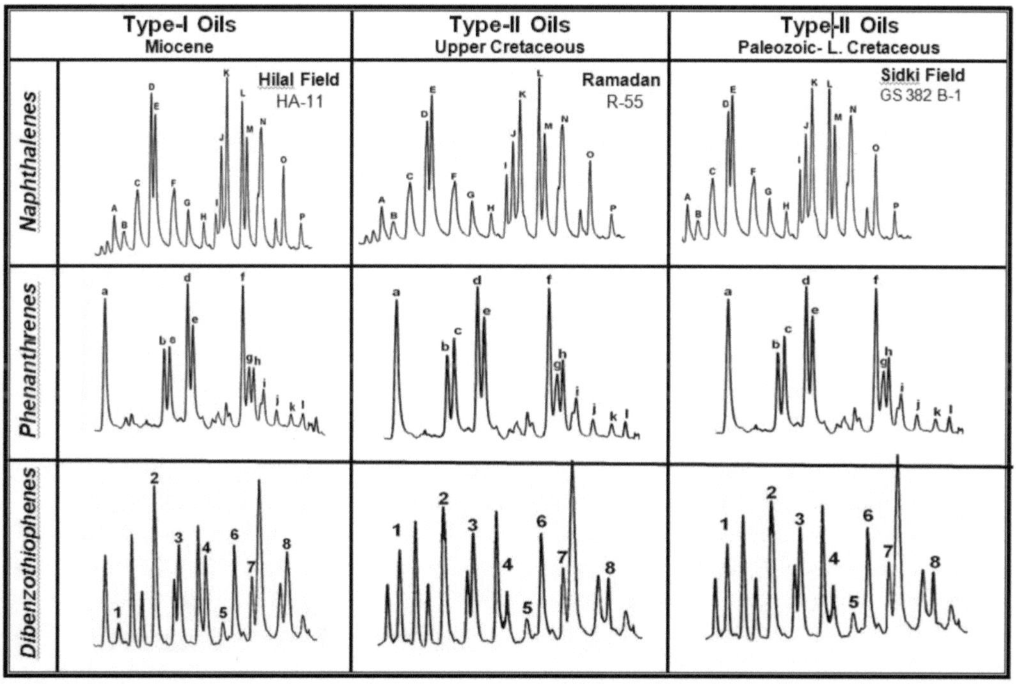

Fig. 10. Representative partial expanded gas chromatograms-mass spectrometry of the aromatic fractions to the naphthalenes (*m/z* 142, 156 and 170), phenanthrenes (*m/z* 178, 192 and 206) dibenzothiophenes, methyldibenzothiophenes, and dimethyldibenzothiophenes (*m/z* 184, 198 and 212) with peak identifications in Table 6.

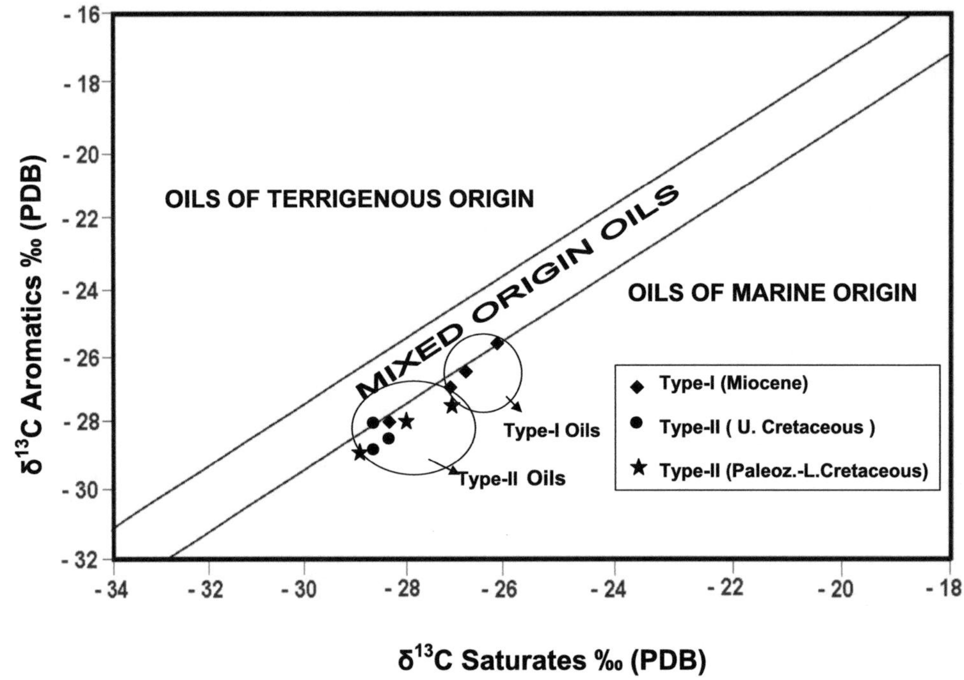

Fig. 11. Relationship between the carbon stable isotopic composition of the saturate and aromatic fractions for crude oils from the southern Gulf of Suez province (Sofer,1984).

7. Inferred oil to source rock correlation

Comprehensive studies published on the source rock potential in the Gulf of Suez were by Shahin and Shehab, 1984; Chowdhary and Taha, 1987; Alsharhan and Salah,1995; Barakat et al., 1997; Lindquist, 1998; Weaver, 2000; Younes, 2001; Younes, 2003 a and b; Alsharhan, 2003; and El-Ghamri and Mostafa, 2004. They found that the Black Shale of the Nubia-B, Upper Cretaceous Brown Limestone, Middle Eocene Thebes Formation and the Lower Miocene Rudeis Shale all appear to have good source rock potential in the Gulf of Suez.

As mentioned above, detailed biomarker distributions in conjunction with stable carbon isotopic composition distinguished the studied crude oils into two types referred to as type-I and II consistent vertically with the pre-rift and syn-rift tectonic rift sequences of the Gulf of Suez province. High oleanane, low gammacerane and marginally mature type-I oil possess organic geochemical characteristics with close similarities to the Tertiary Lower Miocene Rudeis Shale source rock. This formation reached the oil generation window at vitrinite reflectance measurements Ro% between 0.60 and 0.85 at 3-4 Ma and began to generate oils at a depth of 3000 meters in the deeper basin of the Gulf of Suez. Meanwhile, type-II oil, characterized by low oleanane, high gammacerane indices and high level of thermal maturity are fully mature with more advanced level of aromatization and complete sterane isomerisation ratios. Type-II oil has been generated at a depth of approximately 5000 meters in a deeper kitchen within the pre-rift source rocks

(Upper Cretaceous Brown limestone and Middle Eocene Thebes Formation) that entered the oil generation window at vitrinite reflectance measurements Ro% between 0.85-1.35 at 8-10 Ma (Younes, 2003a).

8. Conclusions

Two independent petroleum systems for oil generation, maturation and entrapment consistent vertically with the pre-rift and syn-rift tectonic sequences of the Gulf of Suez province were revealed from biomarker distributions in conjunction with stable carbon isotopic compositions of crude oils. Biomarker variations in crude oils of various ages and source rock types dividing the Gulf of Suez crude oils into two oil types referred as type-I and II that were generated from two types of source rocks of different levels of thermal maturation. Type-I oil is characterized by a predominance of oleanane and low gammacerane indices suggesting an angiosperm higher land plants input of terrigenous organofacies source rock within the marginally mature syn-rift Lower Miocene Rudeis Shale. By contrast, type-II oil is distinguished by a relatively high gammacerane content and low oleanane indices, and may be generated from fully mature marine carbonate source rocks within the Upper Cretaceous Brown Limestone to Middle Eocene Thebes Formation. The higher sterane isomerization as well as aromatic sulfur compound further support the higher thermal maturation level for the type II oils rather than type I.

9. Acknowledgements

The author wishes to express the deepest gratitude to the management of the Gulf of Suez Petroleum Company (GUPCO), Suez Oil Company (SUCO) and Suez Esso Petroleum Company (SUESSO) for providing crude oil samples to accomplish this work. Sincere thanks and gratitudes are also due to the Egyptian General Petroleum Corporation (EGPC) for granting the permission to publish the paper.

10. References

Alsharhan, A. S. 2003. Petroleum geology and potential hydrocarbon plays in the Gulf of Suez rift basin, Egypt. American Association of Petroleum Geologist Bulletin,87, 1, 143-180.

Alsharhan, A. S. and Salah, M. G., 1995 Geology and hydrocarbon habitat in the rift setting: northern and central Gulf of Suez, Egypt. Bulletin of Canadian Petroleum Geology, 43, 156-176.

Bakr, M. Y, and Wilkes, H. 2002, The influence of facies and depositional environment on the occurrence and distribution of carbazoles and benzocarbazoles in crude oils: A case study from the Gulf of Suez, Egypt: Organic Geochemistry , 33, 561-580.

Barakat, A.O., Mostafa, A., El-Gayar, M. S., and Rullkőtter, J. 1997. Source-dependent biomarker properties of five crude oils from the Gulf of Suez, Egypt. Organic Geochemistry , 26, 441-450.

Chakhmakhchev, A., Suzuki, M. and Takayama, K. 1997. Distribution of alkylated dibenzothiophenes in petroleum as tool for maturity assessments. Organic Geochemistry , 26, 483-490.

Chowdhary, L.R. and Taha, S. 1987. Geology and habitat of oil in Ras Budran Field, Gulf of Suez, Egypt: American Association of Petroleum Geologist Bulletin,71, 1274-1293.

Collister, J., Ehrlich, R., Mango, F. and Johnson, G. 2004. Modification of the petroleum system concepts: Origin of alkanes and isoprenoids in crude oils. American Association of Petroleum Geologist Bulletin, 88, 5, 587-611.

Cornford, C., Christie U., Enderesen, P., Jensen and Myhr. M. B. 1988. Source rock and seep maturity in Dorset , Southern England: Organic Geochemistry, 13, 399-409.

Dolson, J. C. Shann, M. V. Matbouly, S. Hammouda, H. and Rashed, R. 2000. Egypt in the next millenium: petroleum potential in offshore trends. Proceeding of the Mediterranean Offshore Conference, Alexandria, 109-131.

Egypt Country Analysis Briefs 2009. Country analysis briefs (Oil and natural gas energy annual report, Egypt. February, 2009 (internet report).

Egyptian General Petroleum Corporation 1996. Gulf of Suez oil fields (A comprehensive Overview). Cairo, 736 pp.

El-Ghamri , M. A. and Mostafa, A. R. 2004. Geochemical evaluation of the possible source rocks in the October Field, Gulf of Suez, Egypt. Sedimentological Society of Egypt, v. 12, p. 55-67.

Ekweozor, C. M., Okogun, J. I., Ekong, D.E.U. and Maxwell, J. R. 1979. Preliminary organic geochemical studies of samples from the Niger Delta (Nigeria). Analyses of crude oils for triterpanes. Chemical Geology, 27, 11-29.

Evans, A. I., 1990. Miocene sandstone provenance relations in the Gulf of Suez. Insight into synrift unroofing and uplift history. American Association of Petroleum Geologist Bulletin, 74,1386-1400.

Gransch, J. A. and Posthuma, J. 1974. On the origin of sulfur in crude. In Advance in Organic Geochemistry.1973, eds. B. Tissot and F. Bienner, 727-739.

Hammouda, H., 1992. Rift tectonics in the southern Gulf of Suez, 11th Petroleum Exploration and Production Conference, Cairo, pp.18-19.

Hunt, J. M., 1996. Petroleum Geochemistry and Geology: New York, Freeman, 743 p.

Isaksen, G. H. 2004, Central North Sea hydrocarbon systems: Generation, migration, entrapment and thermal degradation of oil and gas. American Association of Petroleum Geologist Bulletin, 68, 1545-1572.

Khalil, M. and Moustafa, A. R.1995. Tectonic framework of northeast Egypt and its bearing on hydrocarbon exploration. American Association of Petroleum Geologist Bulletin, 79, 8, 1409-1423.

Kennicutt, H. M.C., Mcdonald, T. J. Comet, P. A. Denoux, G. J. and Brooks, J. M. 1992. The origin of petroleum in the northern Gulf of Mexico. Geochemica et Cosmochemica Acta 56, 1259-1280.

Lijambach, G.W., 1975. On the origin of petroleum. Proceeding of the 9th World Petroleum Congress, Applied Science Publisher, London. 2, 357-369.

Lindquist, S. J. 1998. The Red Sea basin province: Sudr-Nubia and Maqna petroleum systems. Open file report 99-50-A. U. S. Geological Survey, Denver, Colorado. 21p.

Matava, T., Rooney, M. A., Chung, H. M., Nwankwo, B. and Unomah, G. 2003. Migration effect on the composition of hydrocarbon accumulations in the OML 67-70 areas of the Niger Delta. American Association of Petroleum Geologist Bulletin, 87, 1193-1206.

Meshref, W. M., Abu Karamat, M. S. and Gindi, M., 1988. Exploration Concepts for oil in the Gulf of Suez. 9th Petroleum Exploration and Production Conference, Cairo. P. 1-24.

Mello, M.R., Gaglianone, P.C., Brassell, S.C. and Maxwell, J. R., 1988. Geochemical and biological marker assessment of depositional environments using Brazilian offshore oils. Marine and Petroleum Geology , 5 , 205-233 .

Meyers, P. A. 2003. Application of organic geochemistry to paleolimnological reconstructions: a summary of examples from Laurentian Great Lakes. Organic Geochemistry, 34, 2, 261-289.

Moldowan, J.M., Dahl, J., Huizinga, B. and Fago, F., 1994. The molecular fossil record of oleanane and its relation to angiosperms. Science 265, 768-771.

Moldowan, J.M., Seifert, W.K. and Gallegos, E. J., 1985. Relationship between petroleum composition and depositional environment of petroleum source rocks. American Association of Petroleum Geologist Bulletin, 69, 1255-1268.

Mostafa, A. R., 1993. Organic geochemistry of source rocks and related crude oils in the Gulf of Suez, Egypt. Berlin.Geowiss.Abh.A147, 163pp. Berlin.

Moustafa, A. R., 2002. Control on the geometry of transfer zones in the Suez rift and northwest Red Sea: Implications for the structural geometry of rift systems. American Association of Petroleum Geologist Bulletin, 86, 979-1002.

Peters, K.E. and Moldowan, J. M. 1991. Effect of source, thermal maturity and biodegradation on the distribution and isomerization of homohopanes. Organic Geochemistry , 17, 46-61.

Peters, K.E. and Moldowan, J. M. 1993. The biomarker guide interpreting molecular fossils in petroleum and ancient sediments, Prentice Hall, Engelwood Cliffs, NJ,363 pp.

Peters, K. E. and Fowler, M. G. 2002. Application of petroleum geochemistry to exploration and reservoir management. Organic Geochemistry, 33, 5-36.

Peters, K. E. Snedden, J. W., Sulaeman, A., Sarg, J. F. and Enrico, R. J. 2000. A new geochemical sequence stratigraphic model for the Mahakam delta and Makassar slope, Kalimantan, Indonesia: American Association of Petroleum Geologist Bulletin, 84, 12-44.

Philp, R. P. and Gilbert, T. D. 1986. Biomarker distributions in oils predominantly derived from terrigenous source materials. In advances in Organic Geochemistry 1985, eds. D. Leythaeuser and R. Rullkőtter. Organic Geochemistry 10, 73-84, Pergamon, Oxford.

Radke, M., Horsfield, B., Littke, R. and Rullkőtter, J. 1997. Maturation and petroleum generation. In migration of hydrocarbons in sedimentary basins, ed. B. Doligez. pp.649-665.

Rohrback, B.G. 1982. Crude oil geochemistry of the Gulf of Suez. 6th EGPC Exploration and Production Conference, v. 1, p. 212-224.

Said, R. 1962. The Geology of Egypt: Amesterdam, Elsevier, 317p.

Steckler, M. S., 1985. Uplift and extension of the Gulf of Suez. Nature. 317, 135-139.

Steckler, M. S., F. Berthelot, N. Lyberris and Le Pichon, 1988. Subsidence in the Gulf of Suez: Implication for rifting and plate kinematics: Tectonophysics, 153, 249-270.

Seifert, W.K. and Moldowan, J. M. 1978. Applications of steranes, terpanes and mono-aromatics to the maturation , migration and source of crude oils. Geochimica et Cosmochimica Acta, 42, 77-95.

Seifert, W.K. and Moldowan, J. M. 1981. Paleoreconstruction by biological markers. Geochimica et Cosmochimica Acta , 45, 783-795.

Shahin, A.N. and Shehab, M. M. 1984. Petroleum generation, migration and occurrences in the Gulf off Shore of South Sinai, Proceeding of the 7 th Petroleum Exploration Seminar, 126-151, Cairo.

Shanmugam, G. 1985. Significance of coniferous rain forests and related oil , Gippsland Basin , Australia. American Association of Petroleum Geologists Bulletin, 69, 1241-1254 .

Sofer, Z. 1984. Stable carbon isotope compositions of crude oils: Applications to source depositional environments and petroleum alteration. American Association of Petroleum Geologists Bulletin, 68, 31-49.

Ten Haven, H.L., De Leeuw, J.W., Sinninghe Damasté, J.S., Schenck, P.A., Palmer, S.E. and Zumberge, J.E. 1988. Application of biological marker in the recognition of paleo-hypersaline environments: In Kelts K., Fleet A. and Talbot M.(eds) Lacustrine petroleum source rocks, pp.123-130.

Younes, M. A. 2003a. Hydrocarbon seepage generation and migration in the southern Gulf of Suez, Egypt: insight from biomarker characteristics and source rock modeling. International Journal of Petroleum Geology, 26 , 211-224.

Younes, M. A. 2003b. Organic and carbon isotope geochemistry of crude oils from Ashrafi field, southern Gulf of Suez province, Egypt: implications for the processes of hydrocarbon generation and maturation. Petroleum Science and Technology Journal, 21, 971-995.

Younes, M. A., 2001. Application of Biomarkers and Stable Carbon Isotopic Composition to assess the Depositional Environment of Source Rock and the Maturation of Crude Oils from East Zeit Field, Southern Gulf of Suez, Egypt . Petroleum Science and Technology Journal.19, 1039-1061.

Waples, D. and Machihara, T. 1991. Biomarkers for geologists. America Association of Petroleum Geologists. Methods in Exploration Series, no. 9, pp.91.

Wever, H. 2000. Petroleum and Source Rock Characterization Based on C7 Star Plot Results: Examples from Egypt. American Association of Petroleum Geologist Bulletin, 84, 1041–1054.

Winn, R. D., Crevello, P. D. and Bosworth, W., 2001. Lower Miocene Nukhul Formation, Gebel El-Zeit, Egypt: Model for structural control on early synrift strata and reservoirs, Gulf of Suez. AAPG Bull., 85,10,1871-1890.

Zein El-Din, M.Y. and Shaltout, E.M. 1987. Application of isotope type curves to multiple source rock correlation in the Gulf of Suez, Egypt. Abstract, American Association of Petroleum Geologist Bulletin, 61, 631.

2

Hydrocarbon Potentials in the Northern Western Desert of Egypt

M. A. Younes
Geology Department, Moharrem Bek, Faculty of Science, Alexandria University, Alexandria, Egypt

1. Introduction

The Western Desert of Egypt covers two thirds of the whole area of Egypt. The coastal basins (Matruh, Shushan, Alamein and Natrun) located in the northern half of the Western Desert 75 kilometers to the southwest of Matruh City, covering an area of about 3800 Km^2 which forms the major part of the unstable shelf as defined by Said (1990). It is located northeast-southwest trending basin. This basin characterizes by its high oil and gas accumulations and its oil productivity about 45,000 BOPD from 150 producing wells in 16 oilfields, which represents more than one third of the oil production from the northern Western Desert of Egypt (EGPC, 1992).

Khalda was the first discovered field in 1970 by Conco Egypt Inc. and Phoenix Resources and after that followed the discovery of Kahraman, Meleiha, Tut, Salam, Yasser, Shrouk, Safir, Hayat and Kenz oilfields (Figure1).

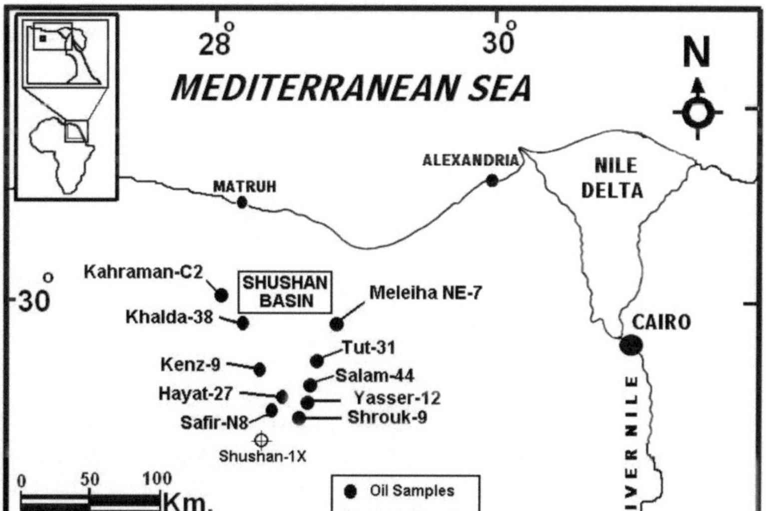

Fig. 1. Location map to the Shushan Basin oilfields.

The sedimentary cover within the northern coastal basins reaches about 14,000 ft. The stratigraphic column includes most of the sedimentary succession from Pre-Middle Jurassic to Recent (Figure 2).

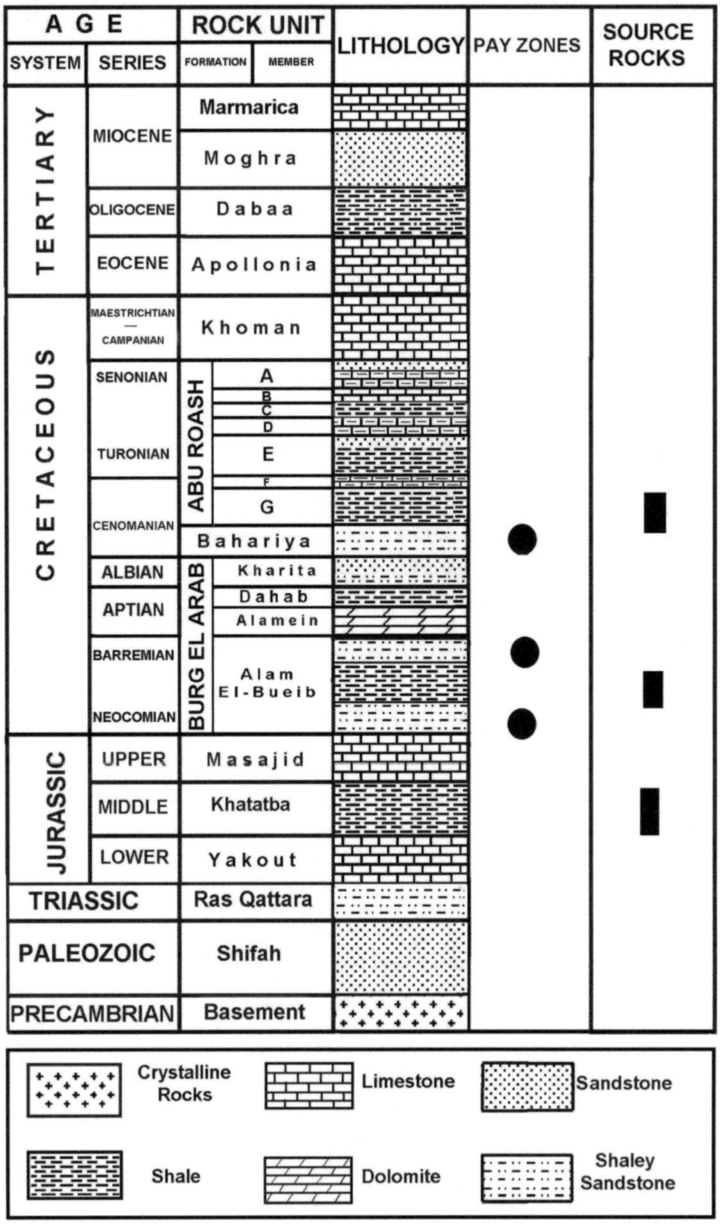

Fig. 2. Generalized lithostratigraphic column of the north Western Desert of Egypt, modified after Abdou (1998).

The northern coastal basins have potential hydrocarbon traps within the sandstones of Lower Cretaceous (Alam El-Bueib Member) and Upper Cretaceous (Bahariya Formation).

The main objectives of this paper are to evaluate the hydrocarbon generation potentials using the Rock-Eval pyrolysis technique applied to shale rock samples representing the succession of Khatatba, Alam El-Bueib and Abu Roash-G formation in addition to burial history modeling to evaluate the degree of thermal maturation in the well Shushan-1X.

The biomarker characteristics as well as stable carbon isotope composition applied on crude oils produced from Bahariya and Alam El-Bueib reservoirs from the different fields of Shushan Basin. This technique was applied on the shale source rock extracts for correlation between them to conclude the depositional environmental conditions prevailing during the hydrocarbon generation.

2. Sampling and analytical techniques

Source rock potential was evaluated by measuring the amount of hydrocarbons generated through thermal cracking of the contained kerogen by Rock-Eval pyrolysis technique. This method was applied on fifteen selected core shale rock samples from Khatatba, Alam El-Bueib and Abu Roash-G lithostratigraphic succession of the well Shushan-1X. These samples were analyzed for total organic carbon, Rock-Eval pyrolysis and vitrinite reflectance measurements. Ten crude oil samples were collected from different pay zones (Alam El-Bueib and Bahariya reservoirs) of the all fields located within Shushan Basin and were analyzed for the biomarker properties and stable carbon isotopes. Meanwhile, three extracts from core shale representing the source rocks of (Khatatba, Alam El-Bueib and Abu Roash-G formations) were used for the same purposes. The crude oils and the extracts from shale source rocks were fractionated by column chromatography where asphaltenes were precipitated with hexane, and the soluble fraction was separated into saturates, aromatics and resins (NSO compounds) on a silica-alumina column by successive elution with hexane, benzene and benzene-methanol. The solvents were evaporated and the weight percent of each component was determined. Gas chromatography (GC) was carried out on Perkin-Elmer 9600 for the saturate fractions equipped with a capillary column (30m x 0.32mm i.d) and the gas chromatograph was programmed from 40°C to 340°C at10 °C/min with a 2 min. hold at 40°C and a 20 minutes hold at 340°C. The saturate fractions were analyzed using an automated Gas Chromatograph-Mass Spectrometer (GC-MS); the fractions were injected into a Finnigan-MAT

SSQ-7000 operated at 70 ev with a scan range of m/z (50-600), fitted with DB-5 (J&W) fused silica capillary column (60 m × 0.32 mm i.d) with helium as carrier gas. The temperature was programmed from 60°C (1 min. isothermal) to 300°C (50 min. isothermal) at 3°C/min. GC-MS analysis of the saturate fraction targeted: terpanes (m/z 191) and steranes (m/z 217). Stable carbon isotope analyses were performed on the saturate and aromatic fractions of crude oils and extracts using a Micromass 602 D Mass-Spectrometer. Data are reported as $\delta^{13}C$ relative to the PDB (‰) standard. Both the Rock-Eval Pyrolysis and the biomarker fingerprints were conducted by StratoChem Services, New Maadi, Cairo.

3. Source rock evaluation

A potential source rock has the capability of generation and expulsion thermally mature oil and gas accumulations (Peters and Cassa, 1994). Source rock evaluation includes quantity and quality of organic matter in addition to thermal maturity or burial heating of organic matter buried in sedimentary succession (Waples, 1994).

The source rock potential and the hydrocarbon generation of the northern Western Desert of Egypt were studied by many authors among them, Parker (1982), Shahin and Shehab (1988), Taher *et al.*, (1988), Zein El-Din *et al.*, (1990), Abdel-Gawad *et al.*, (1996), Abdou (1998), McCain (1998), Abdel-Aziz and Hassan (1998), Khaled (1999), Ghanem *et al.*, (1999), Sharaf *et al.*, (1999), Wever (2000), Waly *et al.*, (2001), Al-Sharhan and Abdel-Gawad (2002), Shahin and El-Lebbudy (2002), Metwally and Pigott (2002), El-Gayar *et al.*, (2002), Younes (2002), El-Nadi *et al.*, (2003) and Harb *et al.*, (2003).

Accordingly, these studies concluded that the stratigraphic section of the northern Western Desert contains multiple source rocks of different degrees of thermal maturation. The dark shale of Khatatba Formation that considered mature source rock with an excellent capability for both oil and gas generation. Shale rocks of Alam El-Bueib and Abu Roash-G formations considered a marginally to good mature source rock for oil generation during the Late Cretaceous.

Source rock evaluation were applied on fifteen core shale rock samples representing the lithostratigraphic section of the Khatatba, Alam El-Bueib and Abu Roash-G formations of the well Shushan-1X including total organic carbon (wt.%), pyrolysis parameters (S1 and S2 values) and vitrinite reflectance measurements (Ro%) to evaluate their organic richness, kerogen types, and the degree of thermal maturity in the northern Western Desert of Egypt.

4. Quantity of organic matter

The available Rock-Eval pyrolysis data of the studied rock units from the well Shushan-1X are summarized in (Table 1) and graphically represented in (Figure 3). The results show that organic-rich intervals are present at three stratigraphic intervals starting with the oldest.

Khatatba Formation:

It consists of dark shale contains TOC ranges between 3.60 and 4.20 wt.% indicating an excellent source rock (Peters and Cassa, 1994). The pyrolysis yield S1+S2 varies between 8.00 and 10.65 kg HC/ton rock and the productivity index (S1/S1+S2) of these rocks ranges between 1.35 and 1.70 therefore the shale rocks of the Khatatba Formation has an excellent source rock potential.

Alam El-Bueib Member:

The shale section of Alam El-Bueib Member contains TOC varies from 1.85 to 2.40 wt.% indicating a good source rock. The pyrolysis yield S1+S2 ranges between 3.60 and 4.50 kg HC/ton rock and the productivity index (S1/S1+S2) of these rocks are generally less than unity, therefore the shale rocks of the Alam El-Bueib Member has a good source rock generating potential.

Abu Roash-G Member:

The organic richness of Abu Roash-G Member varies from 1.10 to 1.50 TOC (wt.%) reflect a medium to good source rock. The pyrolysis yield S1+S2 ranges between 0.85 and 1.10 kg HC/ton rock and the productivity index (S1/S1+S2) of these rocks are generally less than unity, therefore the shale rocks of Abu Roash-G Member indicating fair source rock generating potential.

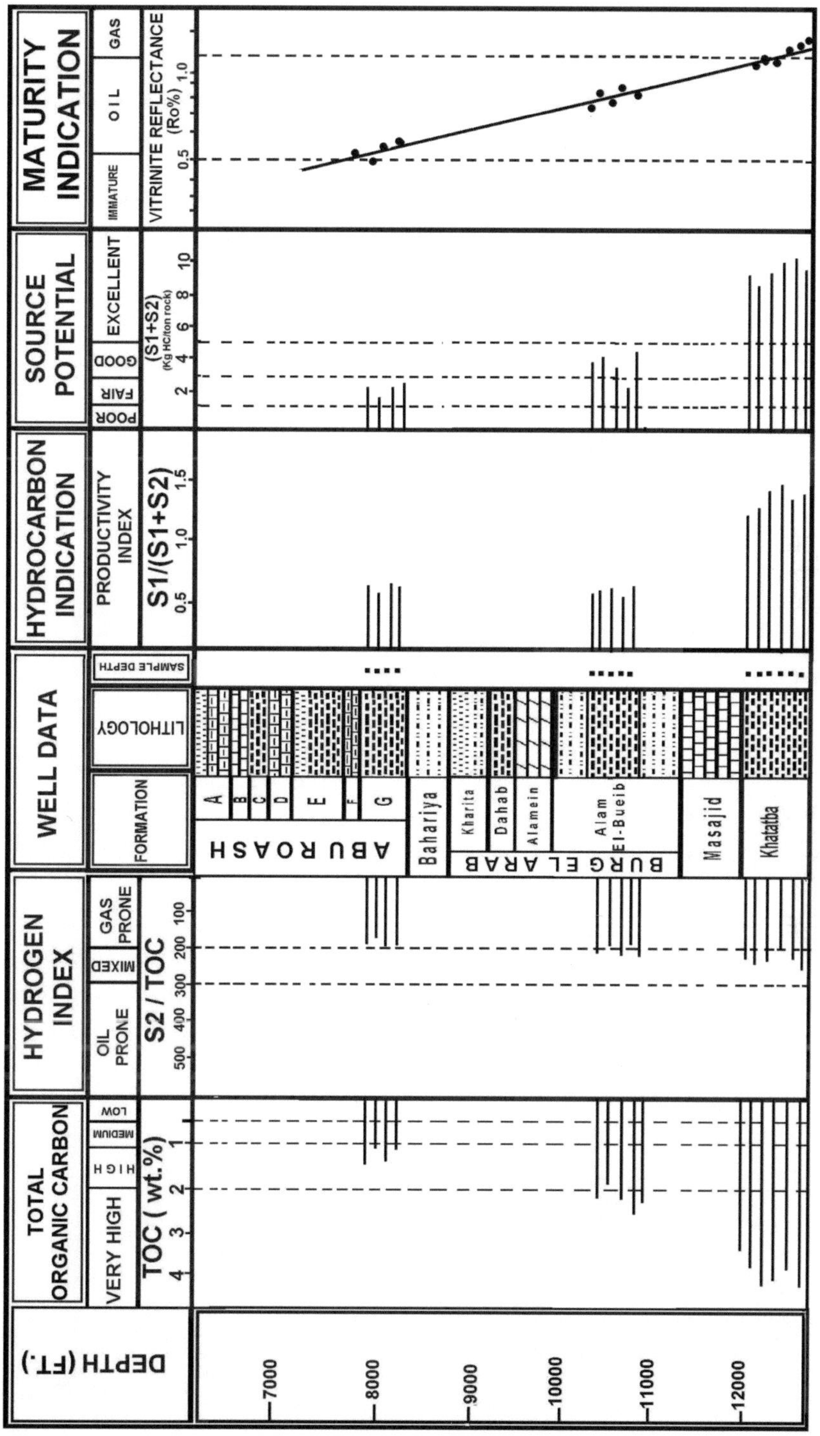

Fig. 3. Idealized geochemical log to the well Shushan-1X, showing Rock-Eval pyrolysis data, total organic carbon and vitrinite reflectance measurements.

5. Type of organic matters (kerogen types)

Kerogen types are distinguished using the Hydrogen Index (HI) versus Oxygen Index (OI) on Van Krevelen Diagram originally developed to characterize kerogen types (Van Krevelen, 1961 and modified by Tissot *et al.,* 1974). Figure (4) shows a plot of hydrogen index (HI) versus oxygen index (OI) on Van Krevelen diagram for the studied shale source rock intervals of Khatatba, Alam El-Bueib and Abu Roash-G from well Shushan-1X. The figure shows that Khatatba Alam El-Bueib and Abu Roash shales contain mixed kerogen types II-III. This kerogen type of mixed vitrinite-inertinite derived from land plants and preserved remains of algae (Peters *et al.,*1994). Mixed kerogen type characterizes mixed environment containing admixture of continental and marginal marine organic matter have the ability to generate oil and gas accumulations (Hunt, 1996).

6. Thermal maturity of organic matters

Thermal maturation of organic material is a process controlled by both temperature and time (Waples, 1994). The vitrinite reflectance is used to predict the hydrocarbon generation and maturation.

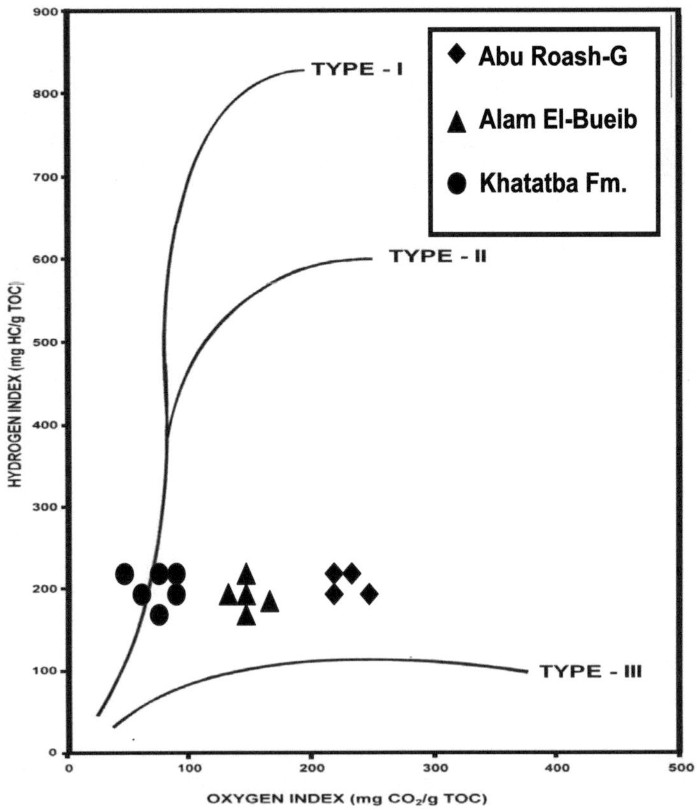

Fig. 4. Hydrogen index (HI) versus oxygen index (OI) and the locations of source rock kerogen types after Van Krevelen (1961).

The data of vitrinite reflectance measurements (Ro%) for the well Shushan-1X were plotted against depth (Figure 3) to indicate the phases of hydrocarbon generation and expulsion. Based on the maturity profile in the burial history model of the well Shushan-1X (Fig.5). The burial history model of the different hydrocarbon bearing rock units indicate that the shale source rock of Khatatba Formation entered the late mature stage of oil and gas generation window between vitrinite reflectance measurements between 1.0-1.3 Ro% during the Late Cretaceous. The shale source rock of Alam El-Bueib Member entered the mid mature stage of oil generation window between vitrinite reflectance measurements between 0.7-1.0 Ro% during the Late Cretaceous while shale source rock of Abu Roash-G Member entered the early mature stage of oil generation at vitrinite reflectance values between 0.5-0.7 Ro% at time varying from Late Cretaceous to Late Eocene.

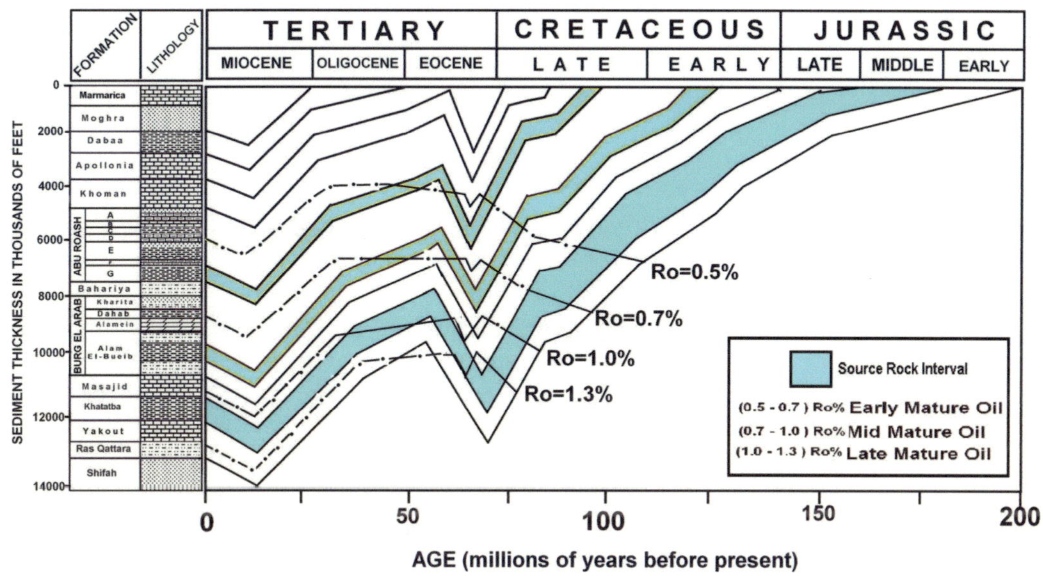

Fig. 5. Burial history model of the well Shushan-1X and stages of hydrocarbon generation windows.

The relationship between Hydrogen Index (HI) with Maximum Temperature (Tmax) and Total Organic Carbon (TOC) to the studied shale source rocks of the Khatatba, Alam El-Bueib and Abu Roash-G succession (Figures 6 & 7) indicate that the shale source rocks of Khatatba Formation located within the oil and gas generation window and considered excellent source rock potential. Meanwhile, the shale rocks of Alam El-Bueib and Abu Roash-G members are considered good source rock for oil generation having a less degree of thermal maturation in comparable with the shale source rock of Khatatba Formation.

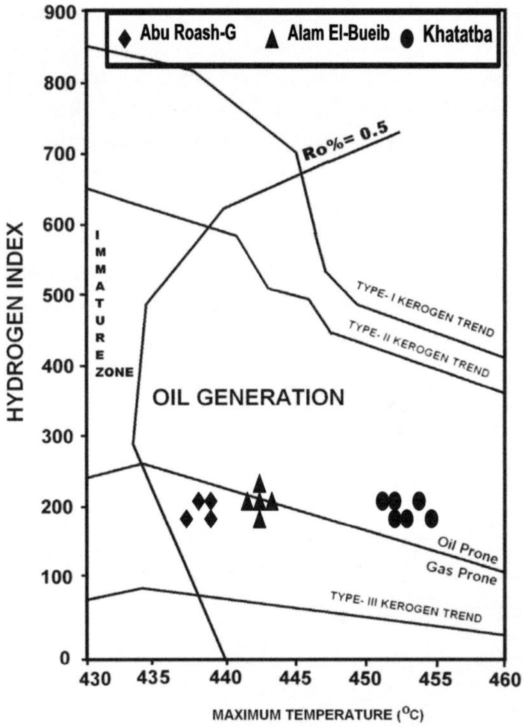

Fig. 6. Relation between HI and T_{max}. and the locations of Khatatba, Alam El-Bueib and Abu Roash-G source rocks.

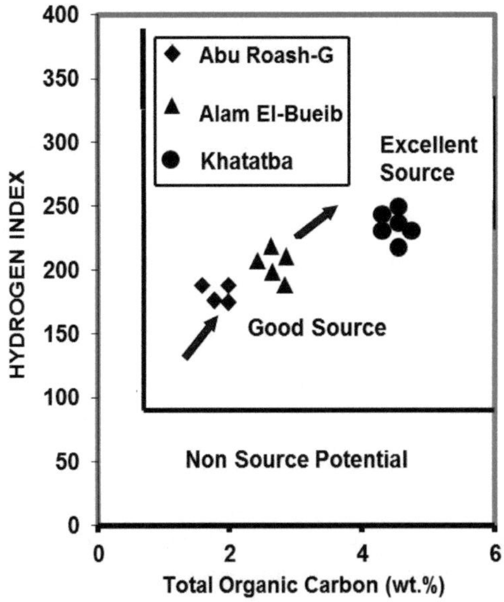

Fig. 7. Relation between HI and TOC and the locations of Khatatba, Alam El-Bueib and Abu Roash-G source rocks.

7. Source rock extracts

Two types of extracts can be identified in this study on the basis of the saturate/aromatic and the pristine/phytane ratios. Type (A) includes Abu Roash-G and Alam El-Bueib while, type (B) represents Khatatba Formation (Table 2).

GC and GC-MS chromatograms (Figure 8) with peak identification in Tables (3&4) of type (A) extracts have a predominance of saturate compounds rather than aromatic with the ratios of saturate/aromatic of about 2.5, pristane/phytane around 0.6 and Ts/Tm 0.5. The plotting of isoprenoids/n-alkanes on the diagram (Figure 9) after Shanmugam (1985) shows that the organic matters in the shale source rocks of Alam el-Bueib and Abu Roash-G are of mixed sources with significant terrestrial contribution as indicated from the low ratio of C_{30} moretane less than 10%. The predominance of C_{27} regular steranes (Figure 10) and diasteranes indicates the greater input of terrestrial organic matters. The low ratio of [20S/20S+20R] $C_{29}\alpha\alpha\alpha$ steranes of about 0.4, which in turn low maturity level of hydrocarbon generation of Alam El-Bueib and Abu Roash-G members shale source rocks.

Type (B) extract of Khatatba Formation of waxy type with the ratio of saturate/aromatic of 1.2, pristane/phytane of 0.7 and Ts/Tm 0.7. The plotting of isoprenoids/n-alkanes on Fig. 9 suggests that the organic matters derivation from terrestrial sources (Moldowan et al., 1985). This conclusion is further supported from the relatively high ratio of C_{30} moretane of 14%. The predominance of C_{27} regular steranes (Figure 10) indicates higher land plants input of terrestrial of sources (Huang and Meinschein, 1979). The high ratio of [20S/20S+20R] $C_{29}\alpha\alpha\alpha$ steranes 0.59,which in turn high maturity level of hydrocarbon generation of Khatatba shale source rocks than the shale source rocks of Alam El-Bueib and Abu Roash-G members.

Depth (ft.)	Formation	TOC wt.%	HI	OI	S1/(S1+S2)	S1+S2	Vitrinite Reflectance Ro%
7850	Abu Roash-G	1.50	195	215	0.55	0.90	0.55
7900	Abu Roash-G	1.10	188	210	0.60	0.85	0.50
8000	Abu Roash-G	1.40	200	207	0.65	0.85	0.57
8050	Abu Roash-G	1.20	200	213	0.60	1.10	0.60
10500	Alam El-Bueib	2.15	205	196	0.71	4.00	0.85
10650	Alam El-Bueib	1.85	200	185	0.72	4.20	0.83
10780	Alam El-Bueib	2.10	210	188	0.65	4.10	0.82
10820	Alam El-Bueib	2.40	195	168	0.62	3.60	0.91
10900	Alam El-Bueib	2.20	205	150	0.61	4.50	0.95
11920	Khatatba	3.60	220	100	1.35	8.00	1.10
11980	Khatatba	3.85	240	89	1.45	9.00	1.22
12000	Khatatba	4.20	235	94	1.62	10.10	1.25
12150	Khatatba	4.10	200	77	1.70	10.00	1.30
12340	Khatatba	3.95	215	87	1.58	10.65	1.35
12560	Khatatba	4.15	245	58	1.69	9.85	1.36

S1: mg HC/g rock., HI (Hydrogen Index): mg HC/g TOC, S1 + S2: Source Potential (Kgm HC/ton rock)
S2: mg CO_2/g rock., OI (Oxygen Index): mg CO_2 / g TOC, S1/(S1+S2): Productivity index

Table 1. Rock-Eval pyrolysis data and vitrinite reflectance (R_o%) data of the shale rock samples representing Abu Roash-G, Alam El-Bueib and Khatatba formations from the well Shushan-1X.

Biomarker Characteristics	Abu Roash-G Extract	Alam El-Bueib Extract	Khatatba Extract
Liguid Chromatography			
Saturates %	59.20	62.18	34.50
Aromatics %	23.50	24.45	26.80
NSO (Polars) %	17.30	13.37	38.70
Sat./Arom.	2.52	2.54	1.28
GC Parameters			
Pr/Ph	0.67	0.68	0.76
Pr/n-C17	0.49	0.44	0.31
Ph/n-C18	0.45	0.40	0.40
CPI	1.03	1.09	1.06
GC- MS Parameters			
1-Terpanes (m/z 191)		0.07	0.14
C_{30} Moretane	0.08	0.58	0.76
Ts/Tm	0.54	1.48	0.89
Homohopane Index	1.52		
2- Steranes (m/z 217)		48	54
C_{27}%	50	25	24
C_{28}%	23	27	22
C_{29}%	27	0.46	0.59
[20S/(20S+20R)] C_{29} $\alpha\alpha\alpha$ steranes	0.48		
StableCarbon Isotopes		-24.8	-26.2
δ^{13}C Saturates ‰ (PDB)	-24.7	-22.6	-24.3
δ^{13}C Aromatics ‰ (PDB)	-22.4	0.92	0.69
Canonical Variable Parameter	1.11		

Table 2. Mathematical classification of crude oils is dependent on the stable carbon isotopic composition of saturate and aromatic fractions of crude oils using the canonical variable parameter postulated by Sofer (1984) that equals ($CV= -2.53\delta^{13}C_{sat.} + 2.22\delta^{13}C_{arom.} - 11.65$).

Peak No.	Compound Name
1	C_{19} Tricyclic terpane
2	C_{20} Tricyclic terpane
3	C_{21} Tricyclic terpane
4	C_{22} Tricyclic terpane
5	C_{23} Tricyclic terpane
6	C_{24} Tricyclic terpane
Ts	C_{27} 18 α (*H*)-22, 29, 30- trisnorneohopane
Tm	C_{27} 17 α (*H*)-22, 29, 30- trisnorhopane
9	C_{29} 17 β (*H*), 21 α (*H*)- 30-normoretane
10	C_{30} Moretane
11	C_{30}17 β (*H*), 21 α (*H*)- moretane
12	C_{31} 17 α (*H*), 21β (*H*)-30 homohopane (22S)
13	C_{31} 17 α (*H*), 21β (*H*)-30 homohopane (22R) C_{32} 17 α (*H*), 21β (*H*)-30 bishomohopane (22S)
14	C_{32} 17 α (*H*), 21β (*H*)-30 bishomohopane (22R) C_{33} 17 α (*H*), 21β (*H*)-30 trishomohopane (22S)
15	C_{33} 17 α (*H*), 21β (*H*)-30 trishomohopane (22R) C_{34} 17 α (*H*), 21β (*H*)-30 tetrakishomohopane (22S)
16	C_{34} 17 α (*H*), 21β (*H*)-30 tetrakishomohopane (22R) C_{35} 17 α (*H*), 21β (*H*)-30 pentakishomohopane (22S)
	C_{35} 17 α (*H*), 21β (*H*)-30 pentakishomohopane (22R)

Table 3. Peak identification in the m/z 191 mass fragmentogram (Terpanes).

Peak No.	Compound Name
A	13β (H), 17α(H)- diacholestane (20S)
B	13β (H), 17α(H)- diacholestane (20R)
C	13 α (H), 17 β (H)- diacholestane (20S)
D	13 α (H), 17 β (H)- diacholestane (20R) + 24- Methyl-13β (H), 17α(H)- diacholestane (20S) 24- Methyl-13β (H), 17α(H)- diacholestane (20R)
E	5 α (H), 14 α (H), 17 α (H) - cholestane (20S)
F	5 α (H), 14 β (H), 17 β (H) - cholestane (20R) +
G	24- Ethyl-13β (H), 17α(H)- diacholestane (20S)
H	5 α (H), 14 β (H), 17 β (H) - cholestane (20S) + 24- Methyl-13β (H), 17α(H)- diacholestane (20R) 5 α (H), 14 α (H), 17 α (H) - cholestane (20R)
I	24- Ethyl-13β (H), 17α(H)- diacholestane (20R)
J	24- Ethyl-13 α (H), 17 β (H)- diacholestane (20S)
K	5 α (H), 14 α (H), 17 β (H)– 24-methylcholestane (20S)
L	5 α (H), 14 α (H), 17 β (H)– 24-methylcholestane (20R)+
M	24- Ethyl-13 α (H), 17 β (H)- diacholestane (20R)
N	5 α (H), 14 β (H), 17 β (H)– 24-methylcholestane (20S)
O	24- Propyl-13 α (H), 17 β (H)- diacholestane (20S)
P	5 α (H), 14 α (H), 17 α (H)– 24-methylcholestane (20R)
Q	5 α (H), 14 α (H), 17 α (H)– 24-ethylcholestane (20S)
R	5 α (H), 14 β (H), 17 β (H)– 24-ethylcholestane (20R)
S	5 α (H), 14 β (H), 17 β (H)– 24-ethylcholestane (20S)
T	5 α (H), 14 α (H), 17 α (H)– 24-ethylcholestane (20R)

Table 4. Peak identification in the m/z 217 mass fragmentogram (Steranes).

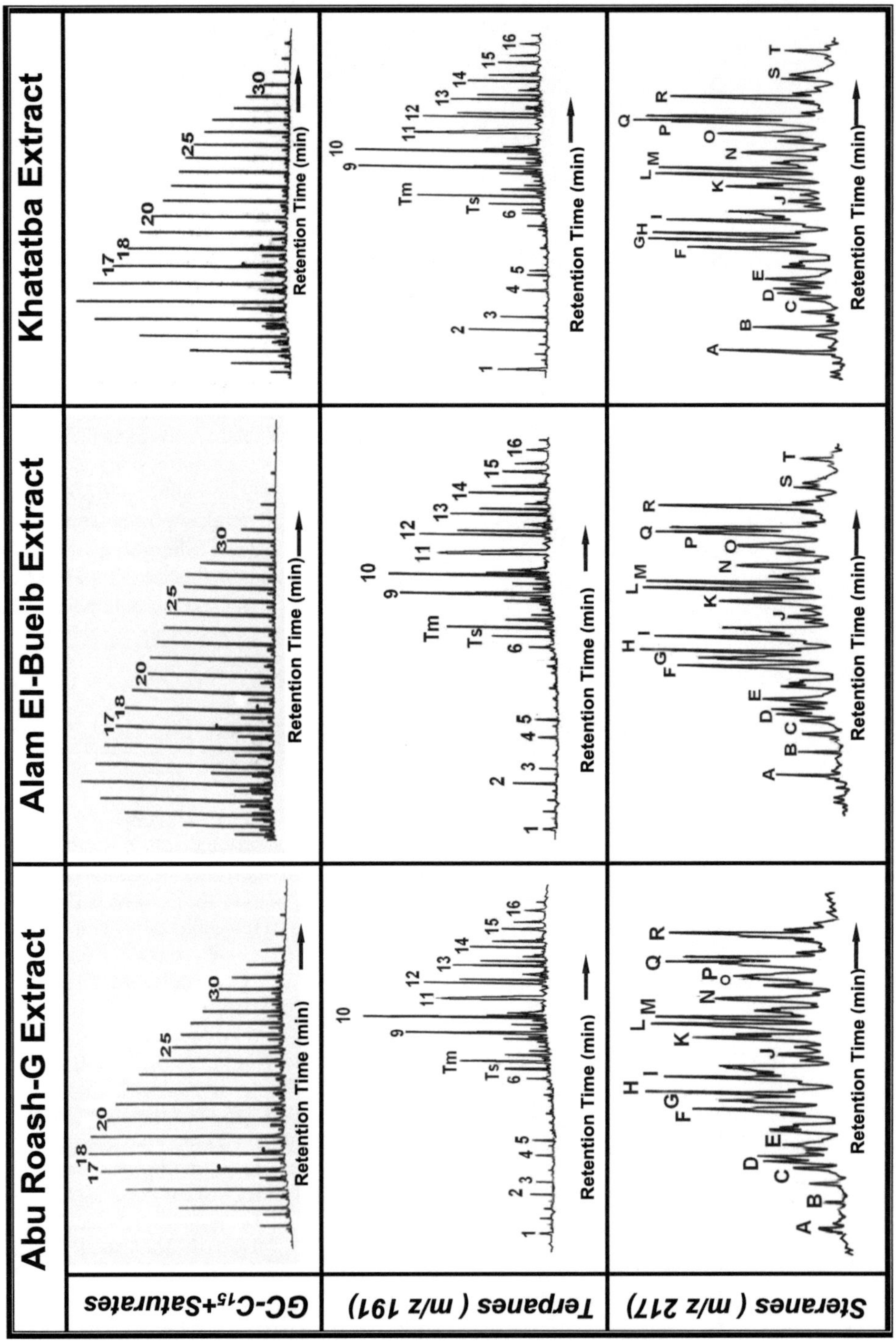

Fig. 8. C_{15}+ saturated hydrocarbons gas chromatograms, Triterpanes (m/z 191) and Steranes (m/z 217) mass fragmentograms for the Khatatba, Alam El-Bueib and Abu Roash-G extracts with peaks identification in tables (3 &4).

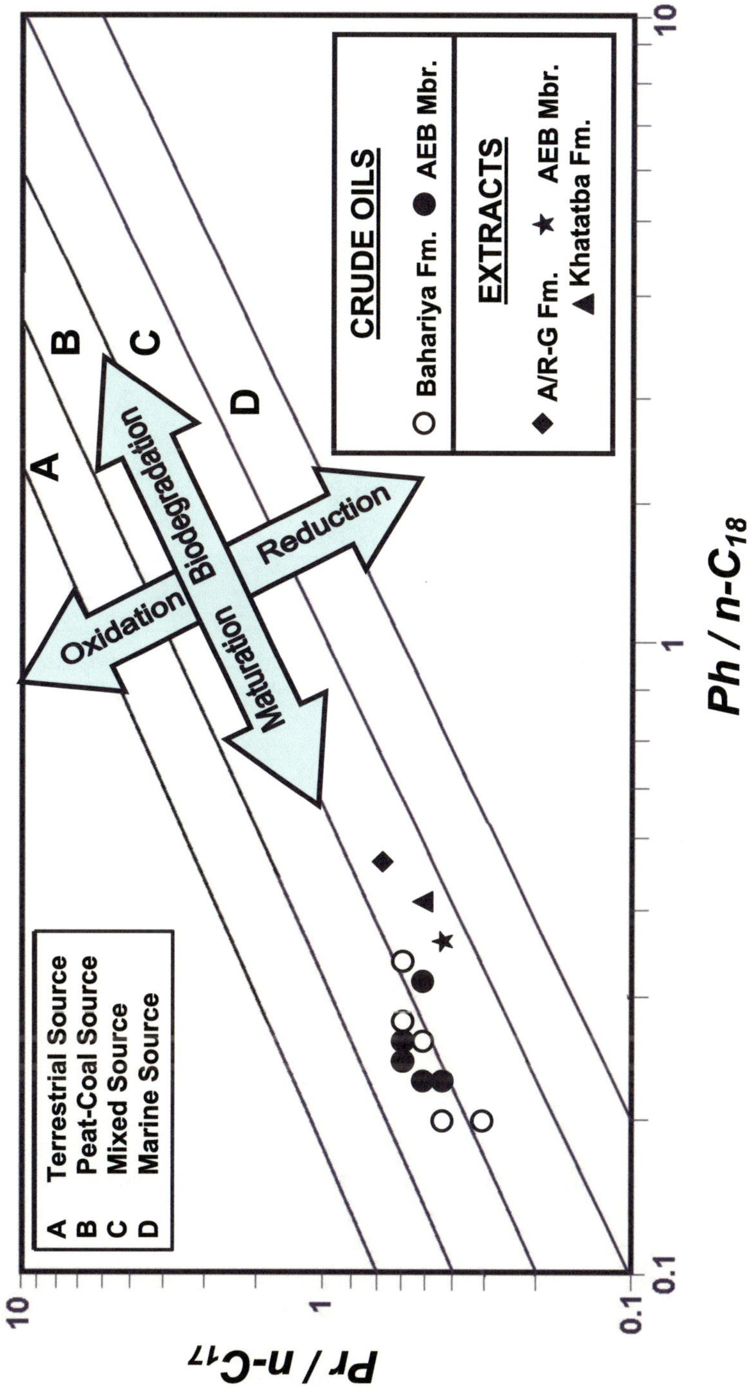

Fig. 9. Relationship between isoprenoid n-alkanes showing source and depositional environments (Shanmugam, 1985) and the locations of crude oils and extracts. from Shushan oilfields.

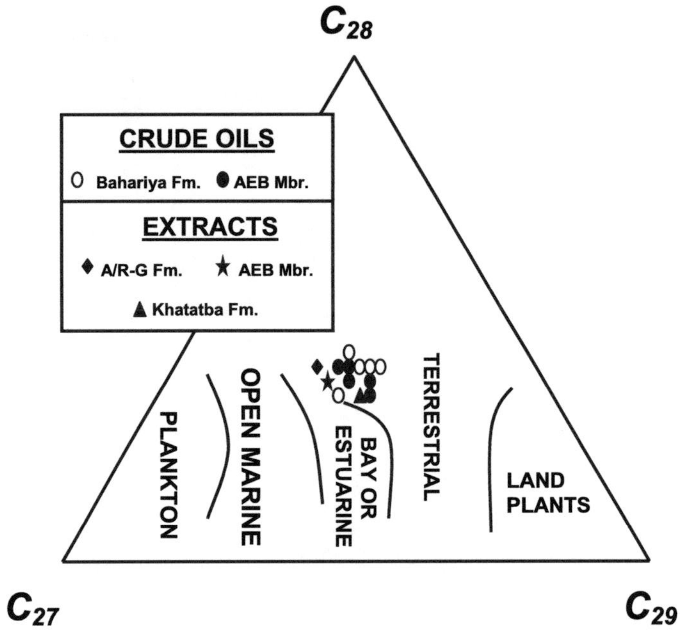

Fig. 10. Distribution of C_{27}, C_{28} and C_{29} steranes illustrating the depositional environments of the studied crude oils and extracts of Shushan oilfields (Huang and Meinschein, 1979).

8. Geochemical relations of fluids

8.1 Crude oils characteristics

Taher *et al.*, (1998), Zein El-Din *et al.*, (1990), Younes (2002 and 2003), El-Nadi *et al.*, (2003), Harb et al., (2003) and El-Gayar (2003) used the geochemical fingerprints of crude oils produced from different basins of the northern Western Desert to assess the genetic relationship between hydrocarbon generation and their source rock depositional environments. Ten crude oil samples from several wells recovered from the different producing zones of the Bahariya and Alam El-Bueib reservoirs (Table 5).

8.2 Formational water characteristics

The chemical composition, properties and the calculated reaction statements are given in Table 6. According to Palmer׳s system of formational water characterization (Collins, 1975), the Bahariya and Alam El-Bueib waters (Fig.13) are assigned to "class 6", whose reaction statement is "S_1, S_2, S_3, A_2 and A_3" (i.e. Primary, secondary and tertiary salinity in addition to secondary and tertiary alkalinity). According to Sulin׳s method of water characterization the ratio of Na/Cl for all the studied samples are generally less than 1 ranging between 0.65 to 0.50 characterized by complete isolation of the hydrocarbon accumulations and the basin of accumulation is considered a good zone for hydrocarbon preservation (El-Zarka and Ahmed, 1983 and Hunt, 1996).

Figure 13, denotes that the Bahariya and Alam El-Bueib water types as chloride-calcium type where this type of waters are the most likely type to be associated with a hydrostatic environment which promotes the accumulation of hydrocarbon in the Jurassic-Cretaceous rocks. Waters of the chloride- calcium type characterized by Cl-Na/Mg >1 indicate an increase of chloride with respect to Na and Mg and $CaCl_2$ is yielded.

The index of base exchange (Cl-Na/Cl) for the studied formational waters equals about 0.2 indicates that the metal ions dissolved in the water have been exchanged with the alkali metals on the clays constituting the source rocks of Khatatba, Alam el-Bueib and Abu Roash-G. The predominant cation sequence is Na >Ca >Mg >Fe, while the predominant anion sequence is Cl >SO_4 >HCO_3.

The salinity of formational water decreases stratigraphically upwards from Alam El-Bueib to Bahariya this related most probably to the mutual reaction between formation water with oil and source rocks during upward migration process (El-Zarka and Younes, 1987).

8.3 Family (I): Bahariya crude oils

Family (I) represents Bahariya crude oils which have a wide range of API gravities between 32.6° and 43.3° with correspond to a high variation of sulfur content, which was found to be ranges between 0.05 and 0.13 wt.%. Liquid chromatograph data indicate a predominant composition of saturates to become more than 65%, with saturate/aromatic ratio more than 2.30. The plot of Pr/n-C_{17} versus Ph/n-C_{18} (Figure 9) suggests that these oils were derived from peat coal source environment of terrestrial origin (Shanmugam, 1985).

Mass fragmentograms of triterpane (m/z 191) and sterane (m/z 217) are shown in Figure 11, with peaks identification in (Tables 3&4). The ratio of C_{30} moretane was found to be <10% further suggests that the Bahariya crude oils may be derived from mixed source rocks dominated by terrestrial organic matters (Moldowan et al., 1985 and Zumberge, 1987). The regular C_{27} sterane distribution (Figure 10) further suggests their derivation from higher land plants input of terrestrial and estuarine environments (Huang and Meinschein, 1979). The ratio of [20S/(20S+20R)] $C_{29}\alpha\alpha\alpha$ sterane was found to be around 0.4, which in turn that these crude oils were generated at low level of thermal maturation (Peters and Fowler, 2002).

8.4 Family (II): Alam El-Bueib crude oils

Family (II) represents Alam El-Bueib crude oils which have low range of API gravities between 40.0° and 42.9° with correspond to a low variation of sulfur content, which was found to be ranged between 0.03 and 0.07 wt.%. The plotting of isoprenoids/n-alkanes on the diagram (Figure 9) suggests that these oils were derived from peat coal source environment derived from terrestrial sources (Shanmugam, 1985).

Mass fragmentograms of triterpane (m/z 191) and sterane (m/z 217) are shown in Figure 11, with peaks identification in Tables 4&5. The ratio of C_{30} moretane was found to be >10% further suggests that Alam El-Bueib crude oils may be derived from source rocks of higher input from terrestrial organic matters. The plotting of C_{27}, C_{28} and C_{29} on a triangular diagram (Figure 10) further suggests their derivation from higher land plants input of terrestrial and estuarine environments (Huang and Meinschein, 1979). The ratio of [20S/(20S+20R)] $C_{29}\alpha\alpha\alpha$ sterane was found to be >0.5, which in turn that the Alam El-Bueib crude oils were generated at relatively high level of thermal maturation rather than Bahariya crudes.

Crude Oil Characteristics	WELL NAMES									
	Kah-C2	Kh-38	Kenz-9	Hay.-27	Safir-N8	Tut-31	Sal.-44	Yass.-12	Shrouk-9	Mel. NE-7
Reservoir Age	Bahariya	Bahariya	Bahariya	Bahariya	Bahariya	AEB	AEB	AEB	AEB	AEB
API°	32.60	38.00	37.60	43.30	42.00	42.70	40.00	41.60	43.20	42.90
Sulfur (wt.%)	0.13	0.12	0.07	0.05	0.05	0.05	0.05	0.07	0.06	0.03
Liguid Chromatography										
Saturates %	64.10	65.20	65.10	66.30	67.00	64.10	66.30	67.20	71.30	74.30
Aromatics %	27.65	26.40	27.50	28.40	27.20	28.50	27.40	26.10	22.10	20.20
NSO (Polars) %	8.25	8.40	7.40	5.30	5.80	7.40	6.30	6.70	6.60	5.50
Sat./Arom.	2.31	2.47	2.37	2.33	2.46	2.25	2.42	2.57	3.23	3.68
GC Parameters										
Pr/Ph	1.51	1.58	1.47	1.13	1.49	1.68	1.97	2.10	2.15	3.71
Pr/n-C17	0.41	0.37	0.17	0.28	0.35	0.43	0.28	0.43	0.62	0.35
Ph/n-C18	0.31	0.28	0.18	0.19	0.29	0.32	0.21	0.18	0.24	0.12
CPI	1.01	1.01	0.98	1.02	1.00	1.05	1.01	0.97	1.01	1.08
GC- MS Parameters										
1-Terpanes (m/z 191)										
C_{30} Moretane	0.06	0.05	0.04	0.07	0.05	0.13	0.10	0.11	0.12	0.11
Ts/Tm	0.61	0.62	0.57	0.49	0.55	0.71	0.66	0.68	0.69	0.71
Homohopane Index	1.85	1.65	1.48	1.70	1.50	0.90	0.85	0.78	0.69	0.88
2- Steranes (m/z 217)										
C_{27}%	45	44	46	43	47	52	51	55	48	51
C_{28}%	20	23	26	22	25	21	24	20	26	28
C_{29}%	35	33	28	35	28	27	25	25	26	21
[20S/(20S+20R)] C_{29} ααα steranes	0.45	0.46	0.43	0.47	0.46	0.57	0.55	0.54	0.55	0.57
StableCarbon Isotopes										
δ^{13}C Saturates ‰ (PDB)	-24.8	-24.9	-25.9	-24.8	-24.7	-24.6	-25.3	-24.9	-25.2	-25.4
δ^{13}C Aromatics ‰ (PDB)	-22.3	-22.7	-23.3	-22.6	-21.9	-22.3	-21.5	-22.9	-22.6	-23.1
Canonical Variable Parameter	1.59	1.95	2.15	0.92	2.22	1.08	4.62	0.51	1.93	1.32

Table 5. A summary of biomaker characteristics and stable carbon isotope composition of the Bhariya and Alam El-Bueib crude oils from Shushan oilfields.

Formational Water Characteristics	WELL NAMES									
	Kah-C10	Sal.-21	Yasser-3	Hayat-2	Shrouk-1	Sal.-11	Sal.-22	Tut-9	Tut-10	Tut-11
Reservoir Age	**Bahariya**	**Bahariya**	**Bahariya**	**Bahariya**	**Bahariya**	**Bahariya**	**Bahariya**	**AEB**	**AEB**	**AEB**
Specific Gravity (gm/cm^3)	1.09	1.1	1.07	1.08	1.05	1.1	1.1	1.03	1.1	1.14
PH	6	6.3	6.5	6.9	6.6	5.3	5.5	6.0	6.5	6.6
TDS (ppm)	139200	179400	124000	112240	101200	150889	146699	226845	226200	219533
Composition %epm										
CATIONS: **Na**	83.6	79.1	83.9	82.1	86.2	76.2	77.5	74.0	72.5	72.7
CATIONS: **Ca**	12.3	16.2	12.5	14.3	11.1	18.1	17.1	20.0	21.3	19.9
CATIONS: **Mg**	4	4.6	3.4	3.5	2.6	5.2	5.2	5.8	5.9	7.1
CATIONS: **Fe**	0.1	0.1	0.2	0.1	0.1	0.3	0.2	0.2	0.3	0.3
ANIONS: **Cl**	99.5	99.1	98.4	99.5	97.1	99.4	99.1	99.5	99.6	99.6
ANIONS: **SO_4**	0.3	0.7	1.1	0.2	1.9	0.5	0.7	0.4	0.2	0.2
ANIONS: **HCO_3**	0.2	0.2	0.5	0.3	0.5	0.1	0.2	0.1	0.2	0.2
Reaction Statements										
S_1	83.6	79.1	83.9	82.1	86.2	76.4	77.5	74.0	72.5	72.7
S_2	16.2	20.7	15.6	17.6	13.3	23.2	22.3	25.8	27.2	27.0
S_3	-----	-----	------	-----	-----	-----	-----	0.1	0.1	0.1
A_2	0.1	0.1	0.3	0.2	0.4	0.2	0.1	------	-----	-----
A_3	0.1	0.1	0.2	0.1	0.1	0.1	0.1	0.1	0.2	0.2
Reaction Parameters										
SULIN: Na/Cl	0.58	0.50	0.59	0.58	0.61	0.55	0.55	0.54	0.53	0.53
SULIN: Na-Cl/SO_4	-53.3	-28.5	-13.2	-87.0	-6.0	-46.0	-30.8	-63.8	-135.5	-134.5
SULIN: Cl-Na	15.9	20	14.5	17.4	15	23	21.6	25.5	27.1	26.9
SULIN: Cl-Na/Mg	3.9	4.3	4.26	4.97	4.38	4.42	4.15	4.39	5.42	3.8
SULIN: Cl-Na/Cl	0.16	0.2	0.15	0.17	0.15	0.23	0.22	0.26	0.27	0.27
SULIN: Cl/Na	1.19	1.19	1.17	1.21	1.13	1.30	1.28	1.28	1.37	1.37
WATER TYPE	(Cl-Ca)	(Cl-Ca)	(Cl-Ca)	(Cl-Ca)	(Cl-Ca)	(Cl-Ca)	(Cl-Ca)	(Cl-Ca)	(Cl-Ca)	(Cl-Ca)
TDS (Total Dissolved Salts)	C H L O R I D E – C A L C I U M									

Table 6.

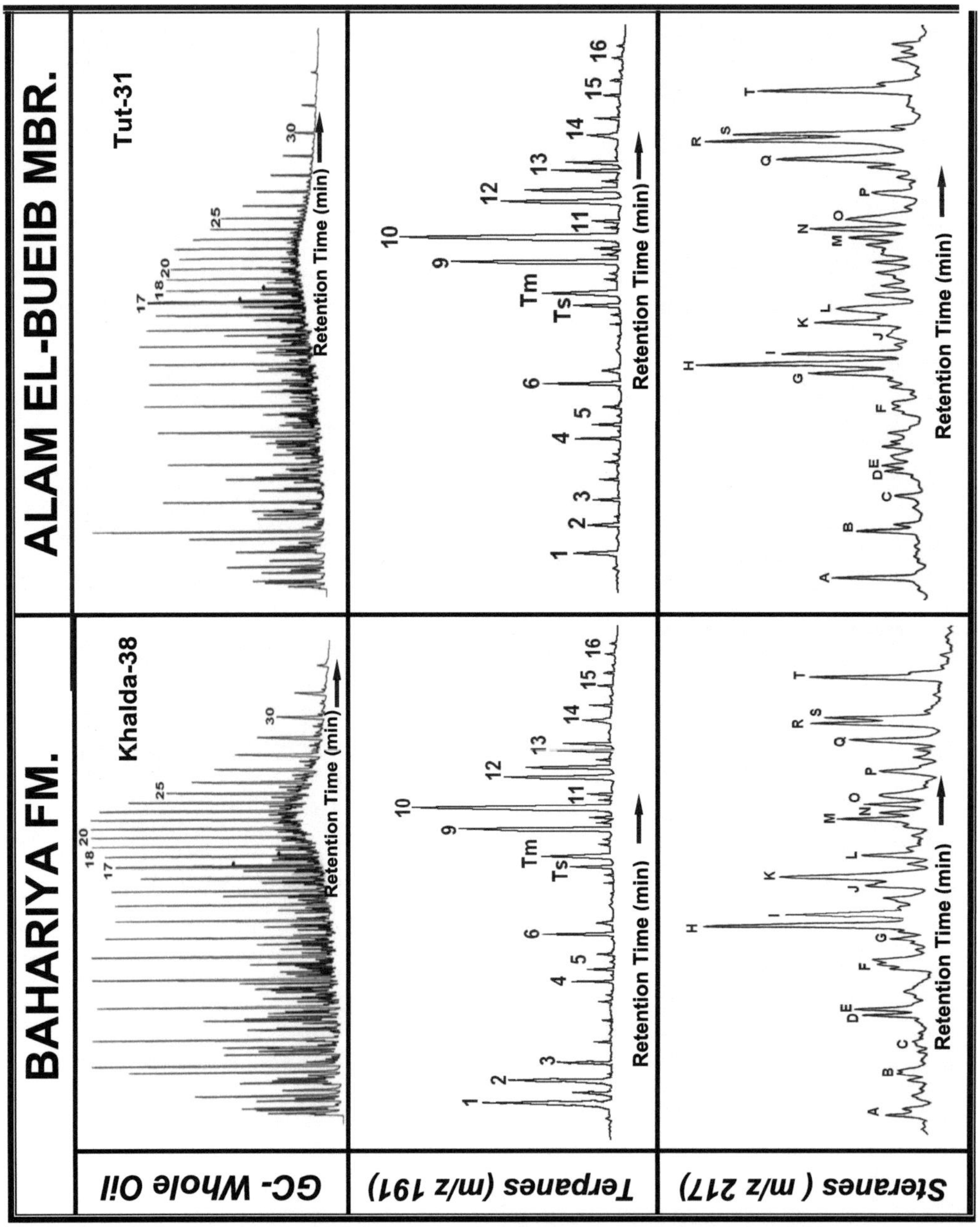

Fig. 11. Gas chromatograms and mass fragmentograms of triterpane m/z 191 and sterane m/z 217, with peaks identification in tables (3 &4) of crude oils from different wells of Shushan oilfields.

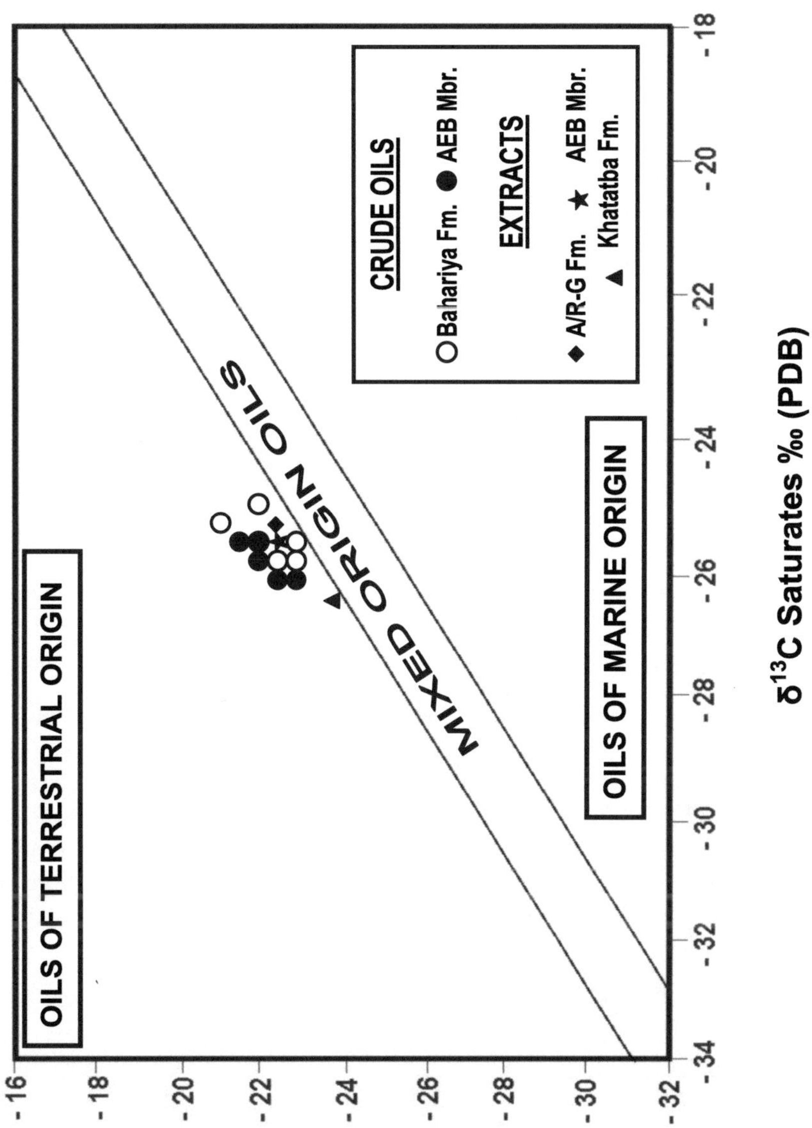

Fig. 12. Relationship between the stable carbon isotope composition of the saturate and aromatic fractions to the crude oils and extracts from Shushan oilfields (Sofer,1984).

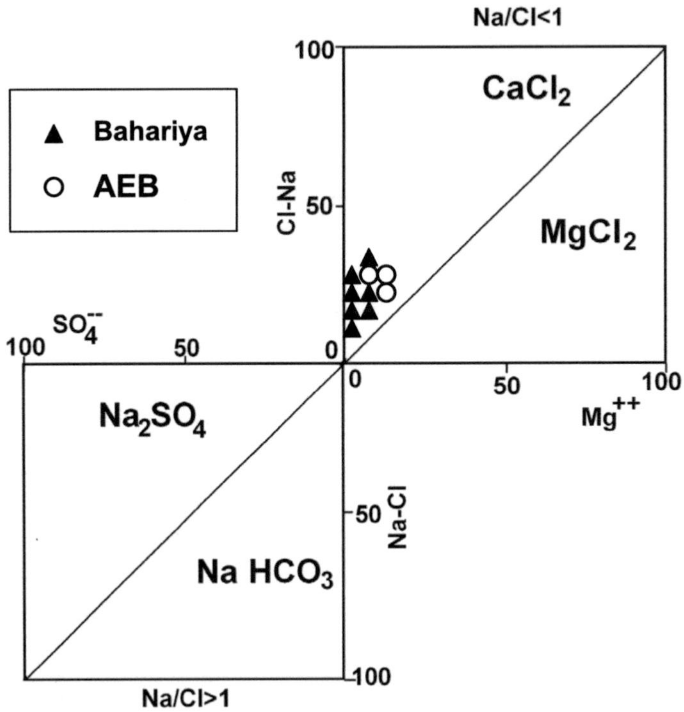

Fig. 13. Location of Bahariya and Alam El-Bueib formation waters on Sulin's diagram.

9. Stable carbon isotope composition

Taher *et al.*, (1988), Zein El-Din *et al.*, (1990) Ghanem *et al.*, (1999) Sharaf et al., (1999) and Younes (2002) used the stable carbon isotope composition to the aromatic and saturate fractions of the Western Desert crude oils and extracts to characterize waxy from non-waxy oil sources. Sofer (1984) distinguished the crude oils derived from marine and non-marine sources for crude oils from different areas of the world including Egypt depending on the stable carbon isotope δ^{13}C compositions to the saturate and aromatic fractions. He applied a mathematical relation to conclude the canonical variable parameter that differentiates between the source of crude oils and their depositional environments.

Despite there are two types of extracts and two types of crude oils but they are isotopically similar and genetically related which may be attributed to slight differences in the degree of thermal maturity. The plotting of the stable carbon isotopic composition to the saturate and aromatic fractions of the studied crude oils in addition to the source rock extracts are shown in (Figure 12). The stable carbon isotopes of the saturate fraction to the extracts range between -26.2 and –24.7 ‰PDB); while for the aromatic fraction ranges between -24.3 and –22.6 ‰PDB. The stable carbon isotopes of the saturate fraction to the crude oils range between -25.4 and –24.7 ‰PDB); while for the aromatic fraction ranges between -23.1 and –21.5 ‰PDB.

The figure reveals that the studied crude oils are of terrestrial origin and the organic matter responsible for hydrocarbon generation in shale source rock of Khatatba, Alam El-Bueib and

Abu Roash-G were probably originated from terrestrial sources. This conclusion is also supported from the calculated canonical variable parameter which was found to be >0.47 for all the studied crude oils and source rock extracts indicate waxy oils type rich in terrigenous organic matters. This is confirm with the results achieved by Zein El-Din *et al.*, (1990), Ghanem *et al.*, (1999) and Younes (2002), who concluded that the Western Desert crude oils are characterized by waxy nature, high maturity level and less negative carbon isotope values derived from terrestrial origin.

10. Inferred oil - source correlation

An oil-source rock correlation is defined as a relationship between geochemical characteristics of crude oils and its original source facies (Curiale, 1994). Many attempts were made to correlate source rock extracts with the biomarker characteristics of crude oils in the Western Desert of Egypt by Taher *et al.*, (1988), Abdel-Gawad *et al.*, (1996), Sharaf *et al.*, (1999) El-Nadi *et al.*, (2003) Harb *et al.*, (2003) and Younes (2003).

10.1 Bahariya oils-type (A) extracts

The organic geochemical characteristics of the type (A) extract has a close similarities with the crude oils reservoired from Bahariya Formation. It is referred to similar biomarker characteristics for both the oils and extracts. They both show a similar C_{30} moretane ratio which found to be <10%, suggest terrestrial land plants influence and similar ratio of [20S/(20S+20R)] $C_{29}\alpha\alpha\alpha$ sterane, which found to be <0.5, reflect that these crude oils were generated at low level of thermal maturation.

10.2 Alam El-Bueib oils-type (B) extracts

Gas chromatograms show that both crude oils of Alam El-Bueib Member and type (B) extracts are of waxy type. They show identical C_{27} regular sterane distributions and similar C_{30} moretane ratio which found to be >10%, further suggest a higher terrestrial land plants input. The identical ratios of [20S/(20S+20R)] $C_{29}\alpha\alpha\alpha$ sterane, which found to be >0.5, reflect that these crude oils were generated from shale source rocks of Khatatba Formation at higher level of thermal maturation than those of Alam El-Bueib and Abu Roash-G source rocks.

11. Conclusions

The organic geochemical characteristics of crude oils and related source rock extracts in Shushan Basin of the northern Western Desert of Egypt revealed two types of extracts (A) and (B) and two families of crude oils. Fair correlation can be seen between type (A) extracts of Alam El-Bueib and Abu Roash-G source rocks and Bahariya crude oils, where the similar biomarker properties as C_{30} moretane ratio <10% and [20S/(20S+20R)] $C_{29}\alpha\alpha\alpha$ sterane <0.5, suggest that the Bahariya crude oils were derived from terrestrial land plants influence at low thermal maturity level. Alam El-Bueib crude oils and type (B) extracts of Khatatba Formation are genetically related and bear the same terrestrial source input but generated at higher thermal maturity level than those of Alam El-Bueib and Abu Roash –G source rocks as indicated from higher C_{30} moretane ratio >10% and [20S/(20S+20R)] $C_{29}\alpha\alpha\alpha$ sterane >0.5.

Organic rich rocks with excellent potential to generate mainly oil are present in the Middle Jurassic Khatatba Formation that entered the late mature stage of oil and gas generation window at vitrinite reflectance measurements between 1.0-1.3 Ro% during the Late Cretaceous. Meanwhile, a good to fair source rocks of Alam El-Bueib and Abu Roash-G Member that located within the early to mid and mature stages of oil generation window between vitrinite reflectance measurements 0.5 and 1.0 Ro% at time varying from Late Cretaceous to Late Eocene.

12. Acknowlegements

The author is grateful to the authorities of the Egyptian General Petroleum Corporation (EGPC) for granting the permission to publish this paper. Also, extend my deepest gratitude to Khalda and Agiba petroleum companies for providing the crude oil and core rock samples and the data required to accomplish this work. Deepest thanks to the authorities of both Stratochem and Corex Laboratories.

13. References

Abdel-Aziz, A. L. and Hassan, Z. A. 1998. Kenz Field: A new play concept in Khalda Concession. 14th EGPC Petroleum Conference, Cairo.v.1, pp.118-138.

Abdel-Gawad, E. A., Philip, R. P. and Zein El-Din, M. Y. 1996. Evaluation of possible source rocks in Faghur-Siwa Basin, Western Desert, Egypt. Proceeding of the EGPC 13th Petroleum Exploration and Production Conference, Cairo. v.1, pp. 417-432.

Abdou, A., 1998. Deep wells in Khalda West: A brief review. 14th EGPC Petroleum Conference, Cairo.v.2, pp.517-533.

Al-Sharhan, A. S. and Abdel-Gawad, E. A. 2002. Integrating kerogen and bitumen analyses to enhance characterization of source rocks in northwestern Desert of Egypt. AAPG International Petroleum Conference and Exhibition, Cairo. Abstract Accepted.

Collins, A. G. 1975. Geochemistry of oil field water. Elsevier, Amsterdam, pp. 496.

Curiale, J. 1994. Correlation of oils and source rocks: A conceptual and historical prespective. In Magoon, L. B. and Dow, W. G. eds., 1994, The petroleum system from source to trap: AAPG Memoir 60, pp. 251-260.

EGPC (Egyptian General Petroleum Corporation), 1992. Western Desert, oil and gas fields (A comprehensive Overview).EGPC, Cairo, Egypt, 431 pp.

El-Gayar, M. Sh. 2003. Utilization of trace metals and sulfur contents in correlating crude oils and petroleum heavy ends. Petroleum Science and Technology, v.21(5&6), pp.719-726.

El-Gayar, M. Sh., Abdel-Fattah, A. E. and Barakat, A. O. 2002. Maturity dependent geochemical markers of crude petroleums from Egypt. Petroleum Science and Technology Journal, v.20 (9&10), pp.1057-1070.

El-Nadi, M., Harb, F. and Basta, J. 2003, Crude oil geochemistry and its relation to the potential source beds of some Meleiha oil fields in the north Western Desert, Egypt. Petroleum Science and Technology Journal, v.21 (1&2), pp.1-28.

El-Zarka, M.H. and Younes, M.A. 1987. Generation, Migration and Accumulation of oil of El-Ayun Field, Gulf of Suez, Egypt : Marine and Petroleum geology Journal, v.4 , pp.320-333.

El-Zarka, M. H. and Ahmed, W. A. 1983. Formational water characteristics as an indicator for the process of oil migration and accumulation at the Ain Zalah Fields, Northern Iraq. Journal of petroleum geology, v.6 (2), pp. 165-178.

Ghanem, M., Sharaf, L., Hussein, S. and El-Nadi, M. 1999. Crude oil characteristics and source correlation of Jurassic and Cretaceous oils in some fields, north Western Desert, Egypt. Bulletin of Egyptian Society of Sedimentology, v.7,pp. 85-98.

Harb, F., El-Nadi, M., and Basta, J. 2003, Oil : oil correlation for some oil fields in the north western part of the Western Desert, Egypt. Petroleum Science and Technology Journal, v.21 (9&10), pp.1583-1600.

Halim, M. A. Said, M. and El-Azhary, T. 1996. The geochemical characteristics of the Mesozoic and Tertiary hydrocarbons in the Western Desert and Nile delta Basins, Egypt: 13 th EGPC Petroleum Conference, Cairo.v.1, p.401-416.

Huang, W.Y. and Meinschein, W.G., 1979. Sterols and ecological indicators: Geochimica et Cosmochimica Acta, v. 43 , p.739-745.

Hunt, J. M. 1996. Petroleum geochemistry and geology: second edition, W.H. Freeman and company, New York.

Khaled , K. A. 1999. Cretaceous source rocks at Abu Gharadig oil and gas field, North Western Desert, Egypt. Journal of Petroleum Geology, 22(24) pp. 377-395.

McCain, W. 1998. Preparation of reservoir rock and fluid properties for simulation, Lower Bahariya Formation, Hayat-Yasser Field, Western Desert, Egypt. 14th EGPC Petroleum Conference, Cairo.v.2, p.406-421.

Moldowan, J.M. Dahl, J. Huizinga, B. and Fago, F. 1994. The molecular fossil record of oleanane and its relation to angiosperms: Science, v. 265, p.768-771.

Parker, J. R. 1982. Hydrocarbon Habitat of the Western Desert, Egypt. 6th Exploration Seminar, Cairo.

Peters, K. E., 1986: Guidelines for evaluating petroleum source rocks using programmed pyrolysis. AAPG Bulletin, v. 70, no. 3, p. 315-329.

Peters, K. and Cassa, M. 1994. Applied source rock geochemistry. In Magoon, L. B. and Dow, W. G. eds., 1994, The petroleum system from source to trap: AAPG Memoir 60, pp. 93-117.

Peters, K. E. and Fowler, M. G. 2002, Application of petroleum geochemistry to exploration and reservoir management. Organic Geochemistry, v. 33, pp. 5-36.

Shahin, A. N. and Lebbudy, M. M. 2002. Geohistory analysis, mapping of source beds, timming of hydrocarbon generated and undiscovered reserves in East Abu Gharadig Basin, Western Desert, Egypt. AAPG International Petroleum Conference and Exhibition, Cairo. Accepted Abstract.

Shahin, A.N. and Shehab, M.M.,1988. Undiscovered hydrocarbon reserves and their preservation time basin, Western Desert, Egypt. EGPC 9thPetroleum Exploration and Production Conference, Cairo.pp.134-163.

Shanmugam, G. 1985. Significance of coniferous rain forests and related oil, Gippsland Basin , Australia: AAPG Bulletin, v.69 , p.1241-1254.

Sharaf, L., Ghanem, M., Hussein, S. and El-Nadi, M. 1999. Contribution to petroleum source rocks and thermal maturation of Jurassic - Cretaceous sequence, south Matruh, north Western Desert, Egypt. Bulletin of Egyptian Society of Sedimentology, v.7,pp. 71-83.

Taher, M., Said, M. and El-Azhary, T.1988. Organic geochemical study in Meleiha area, Western Desert, Egypt. EGPC 9th Petroleum Exploration and Production Conference, Cairo.pp.190-212.

Tissot, B. P., Durand, B., Espitalie, J., and Combaz, A., 1974: Influence of nature and diagenesis of organic matter in formation of petroleum.AAPG Bulletin, v. 58, p. 499-506.

Van Krevelen, D. W. 1961. Coal. New York, Elsevier Science, pp. 514.

Waly, M., Allard, A. and Abdel-Razek, M. 2001. Alamein basin hydrocarbon expulsion models, Proceeding of the 5th Conference on Geochemistry, V. II, pp. 293-302.

Waples, D. W. 1994. Modeling of sedimentary basins and petroleum systems. In Magoon, L. B. and Dow, W. G. eds., 1994, The petroleum system from source to trap: AAPG Memoir 60, pp. 307-322.

Wever, H. 2000. Petroleum and Source Rock Characterization Based on C7 Star Plot Results: Examples from Egypt. AAPG Bulletin, v. 84 (7), pp. 1041–1054.

Younes, M. A. 2002. Alamein basin hydrocarbon potential of the Jurassic-Cretaceous source rocks, north Western Desert, Egypt. OilGas European Magazine , v.28, (3), pp. 22-28.

Younes, M. A. 2003. Geochemical fingerprints of crude oils and related source rock potentials in the Gindi Basin, Northern Western Desert of Egypt: OilGas European Magazine, v.29, (4), pp. 1-6.

Zein El-Din, M. Y., Matbouly, S., Moussa, S. and Abdel-Khalek, M. 1990. Geochemistry and oil-oil correlation in the Western Desert, Egypt. EGPC 10th Petroleum Exploration and Production Conference, Cairo.

Zumberge, J. E., 1987. Terpenoid biomarker distributions in low maturity crude oils: Organic geochemistry, v. 11(6), pp. 479-496.

3

Crude Oil and Fractional Spillages Resulting from Exploration and Exploitation in Niger-Delta Region of Nigeria: A Review About the Environmental and Public Health Impact

John Kanayochukwu Nduka, Fabian Onyeka Obumselu and Ngozi Lilian Umedum
Environmental Chemistry and Toxicological Research Unit, Pure and Industrial Chemistry Department, Nnamdi Azikiwe University, Awka, Nigeria

1. Introduction

The Niger Delta basin of Nigeria situates on the continental margin of the Gulf of Guinea in the equatorial West Africa between latitudes 3º and 6º North and longitudes 5º and 8º East. The word "Delta" is derived from a Greek alphabet in shape of a pyramidal triangle. A delta is a geographical feature formed when a River diversifies into numerous streams that sometimes inter-connect into an intricate web of Rivers, lagoons, swamps and wet land. The point where the main rivers divides is the top of the pyramid while the base is where the division ends or the River enters the sea or larger water course such as a lake. The Niger Delta on the Atlantic coast is one of the moist important in the world. Other worlds notable Deltas include the Nile, Mississippi, Orinoco, Ganges and Mekong (Ibru, 2001). Deltas are usually fertile, contain diverse resources and are therefore noted for large human settlements and civilizations. The Niger Delta complex is one of the most prominent basins in West Africa and actually the largest delta in Africa. It includes the Imo River and Cross River deltas and extends into the continental margins of Cameroun and Equatorial Guinea (Reijers et al 1996).

A more rigorous scientific definition of the territory locates it between Aboh to the North, the Benin River to the west and the Imo River to the East. It is located in the Atlantic coast of Southern Nigeria where River Niger divides into numerous tributaries. The area is the second largest delta in the world spanning a coastline of about 450kms foreclosing at Imo River (Awosika, 1995). It is the largest wetland in Africa, spanning over 20,000 square kilometer and among the three (3) largest in the world. It is estimated that about 2,400sq.km of the area consist of rivers, creeks and estuaries while stagnant swamp covers about 8,600sq.km. Its mangrove swamp of 1,900sq.km is the largest in Africa (Awosika, 1995). The area falls within the tropical rain forest zone of Nigeria. The ecosystem is highly diverse and supportive of several species of terrestrial and aquatic flora, fauna and human life. It ranks

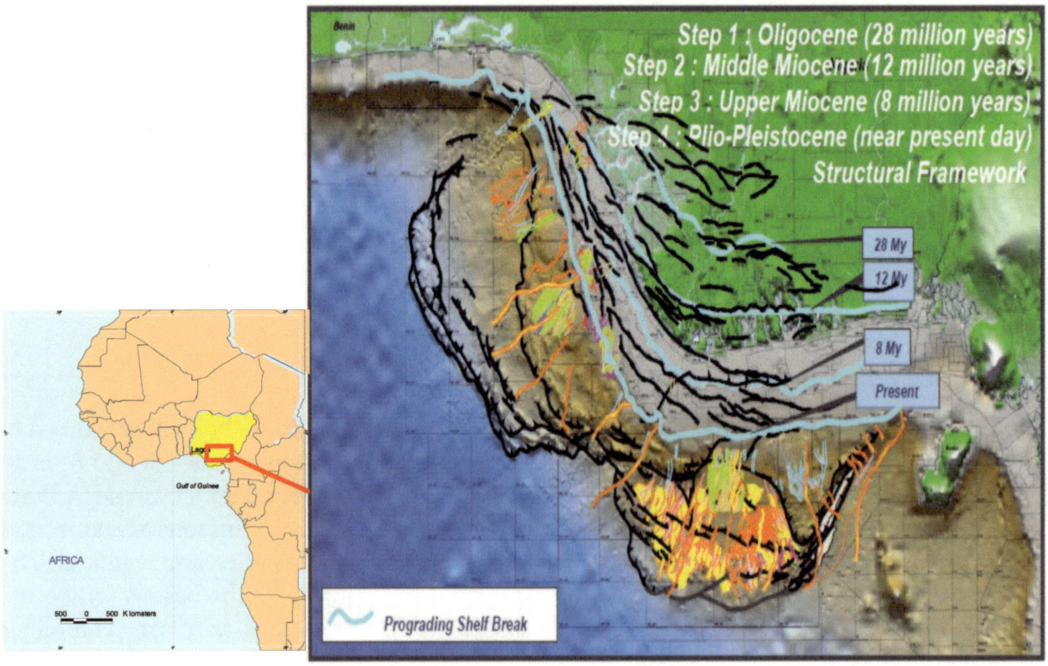

Fig. 1a. Location map of Niger Delta (Reijers et al, 1996).

amongst the world's most prolific petroleum tertiary delta that together accounts for about 5% of the world's oil and gas reserves and for about 2:55% of the present day basin area on earth. Enormous petroleum reserves in the Niger Delta is estimated at about 30 billion barrels of oil and 260 trillion cubic feet of natural gas ranks the basin 6th in the world production (Reigers et al, 1996). The structural features and petrophysical properties account for the hydrocarbon occurrence. The region is divided into four zones namely coastal inland zone, mangrove swamp zone, fresh water zone and lowland rain forest zone. It has been reported that delta sediments are layered in structure, but layers of alternating sands, silts and clays may not be homogenous (Abam and Okagbue, 1997). Three (3) major formations exist in the area, namely Benin, Agbada and Akata.

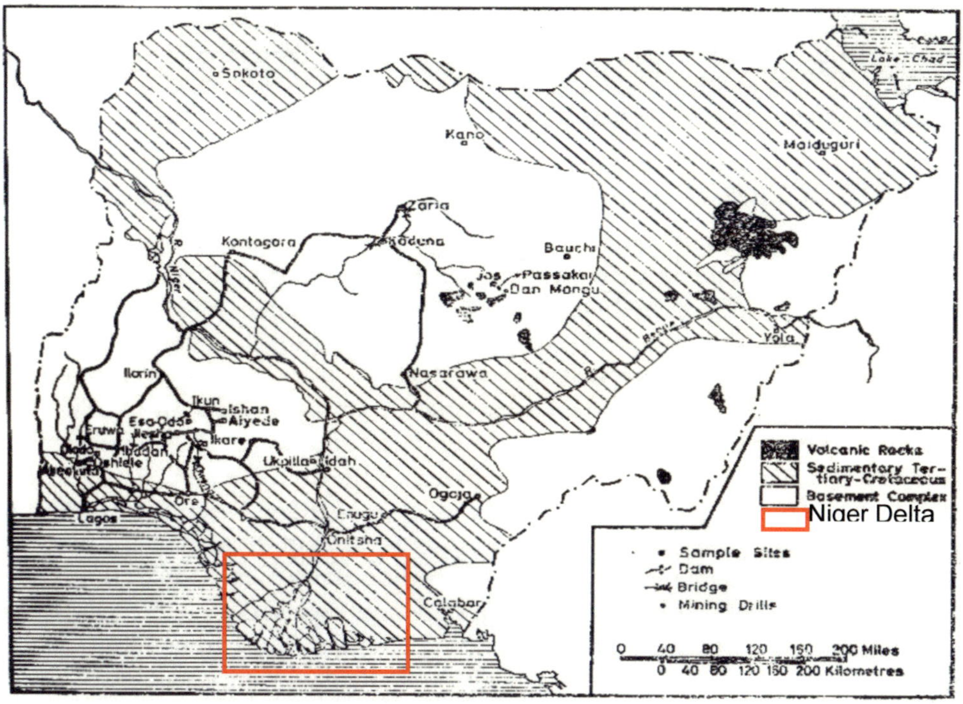

Fig. 1b. Location map of Niger Delta (Reijers et al 1996).

The Benin formation is mostly continental units of sand mixed with gravel and restricted clays, sequence of sandstone/shale exist in Agbada formation, which is the oil-reservoir in the Niger Delta basin (i.e. oil and gas in the area mainly occur in sandstone reservoirs throughout the Agbada formation). The Akata formation is regarded as over pressured shale (Lambert-Alklionbare et al, 1990). The soil of the area is mostly made up of Entisols and inceptisols, with traces of Alfisols (Ekundayo and Obuekwe, 2001). The dominant fresh water aquifer is found in the Benin formation, though there is shallow aquifer, the thickness of Benin formation generally exceeds 2000m with high consolidation at deep levels (Amadi and Amadi, 1990). Accessible fresh water could occur in the first 100 - 200m (Oteze 1983). Water levels can be located at less than 1m near the coast to more than 10m further inland (Amadi, 1986). The whole of delta region receives more than 5000mm of rainfall annually (National Atlas, 1987). Evapo-transpiration is estimated at over 1000mm per year, so there is adequate rainfall to recharge surface and ground water system, although organic matter and clay matrix that consolidate the soil may interfere with infiltration and seapage into the aquifer (Ekundayo and Obuekwe, 2001). The rapid urbanization and industrialization of the Niger delta region of Nigeria occasioned by huge crude oil and gas reserves has had its tole on the environment. Being the most naturally endowed by housing the oil and gas reserves that derive the nation's economy to the vast network of interwoven freshwater aquifers, extensive low lands, tropical and fresh water forest and aquatic ecosystems to its biodiversity with temperature, sunlight and rainfall in an amount and combination that support cultivation and bountiful harvest of rice, sugar cane, yams, plantains, cassava, oil palm, rubber and timber (Nduka et al, 2008). The exploitation of the huge reserve of crude oil and gas deposits has resulted in several million barrels of crude and fractional spillages and several billion cubic feet of gas flaring leading to environmental damage. The US Department of Energy estimates that since 1960, there have been more than four thousand (4,000) oil spills, discharging several million barrels of crude oil into the ponds, ditches, creeks, beaches, streams and rivers of the Niger Delta (Amaize, 2007).

2. Composition of crude oil

Crude or petroleum based oil refers to a wide range of natural hydrocarbon substances and refined products, each having a different chemical composition. Crude oil is a mixture of highly variable proportion of hydrocarbon but differs from lighter oils to heavier oils and bitumen. It is believed to have been formed several million years ago from decayed remains of animals and plants. Under the effects of heat and pressure, there is breaking down of decayed matter into liquids and gases, both collect into pools under the earth's surface. Definite molecular composition varies widely from formation to formation but the proportion of chemical elements differs little (Speight, 1979). Crude oil is the basic mineral product obtained from the geological strata. Though it is referred to as a single (uniform) mineral substance, in reality crude oil is a complex mixture of thousands of hydrocarbons and non hydrocarbons compounds. Of these, the hydrocarbons (Neumann et al, 1981) are by far the major components linked together with inter atom bonds, these hydrocarbons form a variety of kinds of molecules of many different shapes and sizes. Although crude oil are mixtures of some compounds, but the quantity of individual components vary widely in crude oils from different locations. The physical characteristics of oil, such as density, viscosity, flash point, pour point etc are determined by the characteristics of individual component and their relative quantities within the petroleum specific gravity determines whether a crude oil is classified as light or heavy (Wilson and LeBlanc, 1999/2000).

Elements	**% Range**
Carbon	83 to 87
Hydrogen	10 to 14
Nitrogen	0.1 to 2
Oxygen	0,1 to 1.5
Sulfur	0.5 to 6

Table 1. Major Composition of crude Oil.

2.1 Other constituents of crude oil

2.1.1 Polycyclic aromatic hydrocarbons (PAHs): PAHs is a group of about 100 chemicals that are formed during the incomplete burning of coal, oil, gas, garbage, tobacco and other organic substances. They are also present in crude oil, plastics and pesticides (ATSDR, 1995).

2.1.2 Volatile organic compounds (VOCs): The most common volatile organic compounds in crude oil are the benzene, toluene, ethylbenzene and xylene. They are also mainly found in household products (paints, paint strippers, solvents, aerosol).

2.1.3 Hydrogen sulphide gas: Some crude oils release high concentrations of hydrogen sulphide gas.

2.1.4 Alkanes (paraffin) and cycloalkanes: These are major constituents of crude oil, lower fractions are volatile at ordinary temperature.

2.1.5 Naturally occurring radon materials (NORM): the occurrence of natural radioactivity in oil and gas fields is well recognized worldwide and has been reported in Nigeria's Niger Delta (Jibiri and Emelue 2008). This radioactivity can result from the occurrence in both rocks and specific ores, of isotopes from the uranium and thorium decay series, normally with alpha and gamma radiation activity (Hamlat et al, 2001).

2.1.6 Metals and heavy metals. Crude oil contains trace metals present as minor constituents. More than sixty metals have been established to be in crude oils with Vanadium (V) and Nickel (Ni) as the most abundant (Tiratsoo, 1973). Metal components of crude oil occur as metalloporphyrin chelates, transition metal complexes, organometallic compounds, carbonylacids salts of polar functional groups and colloidal minerals (William and Robert, 1967, Lewis and Sani, 1981). Metal components of crude oil constitute serious problem to petroleum refining by causing corrosion or poisoning the cracking catalysts which reduces efficiency (Tiratsoo, 1973; William and Robert, 1967; Lewis and Sani, 1981). Trace metals such as arsenic (AS) decreases hydrogenation and isomerisation activity of catalysts, acting as permanent poison (Hettinger et al, 1955). Gas and coke formation in crack stock is enhanced by copper (Cu) which promotes low gasoline production. Iron (fe) poisons catalysts and accelerates oxidation, which may result in unstable products (Milner, 1963, Ming and Bott 1956, Karchmer and Gunn 1952), Nickel deposits contaminate cracking catalysts more than any common metal leading to low gasoline yield. Vanadium during cracking processes is oxidized, yielding low melting pentoxide which deposits readily and contributes to catalyst poisoning, metal embrittlement and pitting of refining equipment and combustion deposits ash known to be toxic. Sodium (Na) on its own causes loss in catalytic activity, it also imparts corrosive properties and thereby reduces the life of furnace tubes, turbine blade and metal wearing (Karchmer and Gunn, 1952; Kawchan 1955; Bowman and Wills, 1967 and Garner et al., 1953). Presence of certain trace metals particularly nickel and vanadium in crude oils is geochemically significant and provides information on the origin of crude oil. The work of Achi and Shide, 2004 revealed that Nigerian Bonny light crude and Bonny medium crude contains varying quantities of calcium, magnesium, zinc, iron, nickel, sodium and potassium in the following order: Na> Ca> Fe> Ni> Mg> >K >Zn in Bonny light crude while Bonny medium crude is of the order: Na> Ca> Fe> Ni> Mg> >K > Zn. Sundry constituents of crude oils or those that may be synergized by certain or natural processes include nitrate, nitrite, polychlorinated-n-alkanes, polychlorinated biphenyls and several others.

3. Activities involved in oil extraction

3.1 Exploration/drilling operations

This include drilling and work over activities which takes place on land, swamp and offshore. Oil spill during drilling results from blow outs, equipment failure, waste pits, overflow and sometimes human error.

3.2 Production operations

Production facilities include producing wells, flow stations, gas plants, compressors station, gas and water injection stations and numerous pipelines that connect these facilities. In production operations there are several potential sources of spill which ranges from equipment failure, valve and seals failure, operational errors and sabotage. Sometimes oil/chemical spills can be due to corrosion.

3.3 Terminal operations

This involves the filling of tanks, barges, vessels, dehydration of crude, crude storage, effluent water disposal and loading of tankers. Crude oil/fractional distillates spill can be due to hose and valve failure, tanker collision and grounding, ship to ship transfer, improper drainage of tanks, storage tank and pipeline failure.

3.4 Engineering operations

This include dredging, flow line replacement, flow station upgrade etc.

3.5 Sabotage/theft

It involves vandalization of manifold, pipelines delivery lines, cutting or removal of pipelines.

3.6 Others

Include falling trees, lightning and mystery (unexplainable) spills.

4. Classification of oil spill

Crude oil/chemical spills are classified according to a combination of factors notably real or potential impact on environment and the resources required for effective response. The Department of Petroleum Resources (DPR) has classified the magnitude of oil spillage into minor, medium and major spills.

4.1 Minor spills

The spills that are less than 25 barrels of crude oil discharged on inland water or less than 250 barrels discharged on land, coastal/offshore water.

4.2 Medium spill

This releases between 25 - 250 barrels on inland water or 250 – 2500 barrels discharged on land, coastal/offshore water.

4.3 Major spill

It releases greater than 250 barrels discharged on inland water or greater than 2,500 barrels discharged on land, coastal/offshore waters.

5. Fate and behavior of spilled oil

When oil is spilled, over a period it undergoes a number of physical and chemical changes which affect its behavior. These changes start from the spread and drift of the oil through its evaporation, dissolution, dispersion and emulsification to its sedimentation, photo-oxidation and biodegradation. Biotic factors, like bacteria, yeast and filamentous fungi play an important role in the degradation of petroleum and may be the dominant factor controlling the fate of petroleum hydrocarbons in marine environment (Delaune et al, 1990).

5.1 Spreading

The spreading rate of oil will be affected by its viscosity, pour point, wax content, marine state and weather conditions. Slicks formed will move in the same direction and in the same speed as the current and will move in the same direction as the wind at approximately 3% of the wind speed. Less viscous oil spread faster than heavy oils (Clark, 1992), but all oils spread faster on warmer waters. Spreading also induces a change in the composition of the oil by promoting the dissolution and evaporation of certain components, also as hydrocarbon dissolves in water, they alter the water-air interfacial tension (National Research Council, 1985). The force of gravity acting downwards through the thickness of a considerable oil spill tends to spread it out sideways. This movement compares with that of oil on solid surface but limited when the viscosity of the oil counter balances the spreading force. When compared with effect of surface tension, this force becomes operational if there is a difference between the oil/air and oil/water interface tension, while it remains positive, the oil will spread out. Fay (1969), gave a theoretical treatment of the spreading of oils into three (3) separate phases.

- Where the spreading rate is controlled by the difference in density between the oil and the water and the speed of spreading controlled by inertia resistance.
- Where spreading rate is controlled by gravity but viscous drag between the oil and water limits the spreading rate.
- When the spreading is controlled by the surface tension difference between the oil and the water.

Here we consider the third or surface tension type, earlier work on this by Blokker (1964) considered the dynamics of the spreading and proposed an empirical formula based on the assumption that oil spreads in a uniform way where the instantaneous rate of spreading is proportional to the slick thickness.

Blokker's formula thus simplified:

$$D^3 - D_o^3 = \frac{24k}{2}(dw - do)\frac{do}{dw}Vo \cdot t \cdot g^2 \quad (1)$$

where;

K = Blokker's constant (depending on oil type)
D = Diameter of the oil spread in meters
t = time
do = D at time (t = o)
do = density of oil

dw = density of water
Vo = initial volume of oil

An extensive experiment on spreading of oil and its disappearance was carried out by Warren Spring laboratory in 1972 and was described by Jeffrey (1973), here 120 tons of light Arabian crude oil was discharged at sea and its appearance and dimension studied for four (4) days on a uniform water surface, as might be found on the open sea, oil spread evenly in all directions but it is not practicable because of effect of wind and waves.

5.2 Evaporation

In evaporation, the rate and extent of evaporation is determined primarily by the volatility of oil, spreading rate, marine and wind conditions and temperature. Evaporation can be responsible for the loss of up to 40% spilled oil in the first day (Jordan and Payne, 1980). The amount of evaporation differs from about 10% in very heavy crude and refined product to as much as 75% in very light crude and refined products (Albers, 1995). From the study of the fate of oil in the *Amoco Cadiz* spill in 1978, off Brittany, France, it was established that lower molecular weight alkanes and single-ringed aromatics (benzenes) were rapidly depleted through evaporation (Gundlach et al, 1983).

5.3 Dissolution

The extent of dissolution is influenced by the oils aqueous solubility which for crude oil is put at 30mg/L (National Research Council, 1985). Low molecular weight aromatics such as benzene, toluene and xylenes which are also among the most volatile has the highest solubility. Actual dissolution of the slick would be expected to account for only around 1% of the mass balance (Mackay and McAuliffe, 1989).

5.4 Dispersion

Oil-in-water emulsion or dispersion is due to the incorporation of small globules of oil into the water column. Dispersions are considered beneficial because they increasingly disperse over time and dramatically increase the surface area of the oil available for degradation (Jordan and Payne, 1980). These emulsions are inherently unstable and larger particles tend to rise and coalesce but small droplets can be conveyed with water eddies to become a part of the water column. Oil-in-water dispersions can also be stabilized by suspended particulates (Huang and Elliot, 1987).

5.5 Emulsification

Low viscosity oils tend to form emulsion very quickly (2 to 3 hours) and can be up to 80% content. Oils that have asphaltenes content, 70.5% are likely to form stable emulsion. These emulsions are problematic because they slow microbial degradation and are resistant to dispersion (Payne and Philips, 1985).

5.6 Sedimentation

Sedimentation of oil is facilitated by the sorption of hydrocarbons to particulate matter suspended in the water column. Since coastal environment of the Niger Delta contains large amount of suspended particulate matter (Nduka and Orisakwe, 2011), adsorption of oil onto suspended particulate matter results in a high specific gravity mixture more than twice that of sea water alone (1.025g/cm^3). The high specific gravity of these particles causes deposition of the sorbed oil (Kennish, 1997). Increasing density of oil due to weathering can also promote its movement below the surface into the water column and eventually into sediments (National Research Council, 1985).

5.7 Photo-oxidation

UV radiation has enough energy to transform many petroleum hydrocarbons into compounds possessing significant chemical and biological activity (Jordan and Payne, 1980). The mechanism is known as auto catalytic free-radical chain reaction, which results in the formation of hydroxyl compounds, aldehydes, ketones and low molecular weight carboxylic acids (Burrwood and Speers, 1974, Jordan and Payne, 1980). Photo-oxidation of hydrocarbons derived from crude oil can occur from evaporated components in the gaseous state (Baek et al, 1991), from the dissolved fraction of petroleum (Neff, 1985) and from non dissolved oil like slicks and colloidally dispersed oil (Jordan an Payne, 1980). Organo-sulphur compounds reduce complete oxidation to carboxylic acids by leading to termination of free radical chain reaction (Jordan and Payne, 1980). Photo-oxidation rates can be increased by the presence of photo-sensitizing compounds such as xanthone-1-naphthol and other naphthalene derivatives and by the effect of dissolved ions of variable oxidation state such as vanadium that acts as catalysts (Jordan and Payne, 1980). Obviously, throughout the life of an oil slick, it continues to drift on the sea surface. The wind induced effect is normally taken as 3% of the wind velocity and the current effect is taken as 100% of the current velocity.

5.8 Time scale for different process in weathering

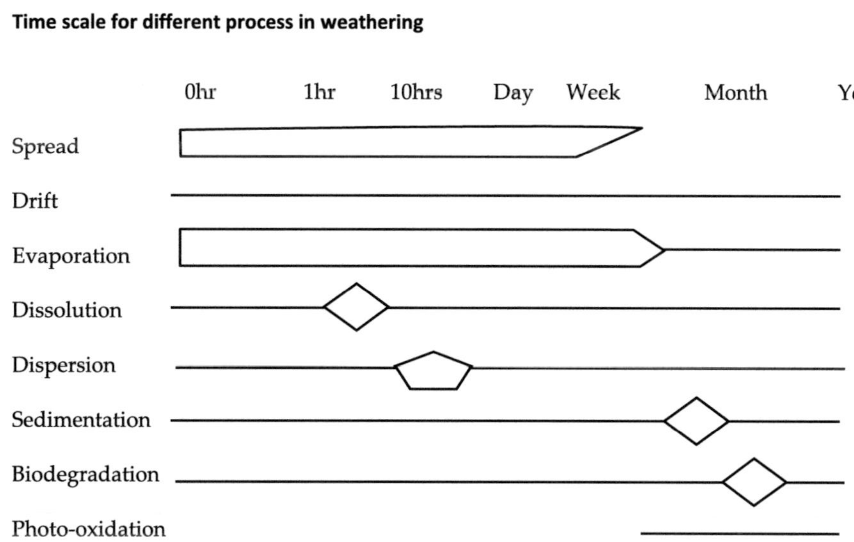

Scheme 1.

5.9 Micro-bial degradation

Micro-organism indigenous to soil, groundwater and marine ecosystems degrade a wide range of compounds like aromatic and aliphatic hydrocarbons, chlorinated solvents and pesticides released into natural environment (BICNEWS, 2005). Since empirical evidence that aromatic hydrocarbons exist in the Niger Delta marine ecosystem due to extensive crude oil and gas exploitation, fire explosions, leachlets from decomposing refuse and industrial effluents (Okoro and Ikolo, 2005; Anyakora et al 2004; Anyakora and Coker 2006; Olajire et al, 2005), native microbes could be assisting Nigerians in the Niger Delta region in cleaning water system unnoticeable and free of charge. From our work (Nduka and Orisakwe, 2011), fewer or no counts of *achromobacter* and *aspergillus* (polyaromatic hydrocarbon (PAHs) degraders) and *proteus* (straight chain hydrocarbon degraders) were found in most water samples while *bacillus* and *pseudomonas*, both PAHs and straight chain hydrocarbon degraders (Rosenberg, 1993), were found in highest counts and in almost all the water samples. It was evident that water samples in which PAHs and straight chain hydrocarbon were highest, had highest microbial count per specie (figs 2a, 2b, 2c and Tables 2 and 3) (Nduka and Orisakwe, 2011). Therefore the low levels or non detection of PAHs and straight chain hydrocarbons could mean degradation by microbes (utilization of carbon as food by microbes for sustenance). This observation agrees with previous report (Atlas 1991, Young and Cerninglia 1995).

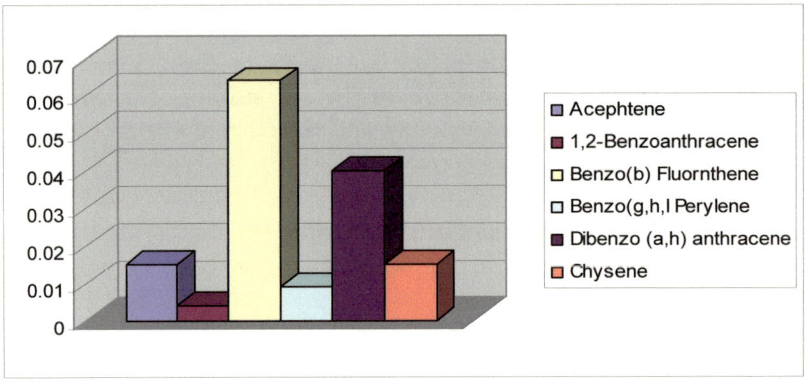

Fig. 2a. Some PAHs (mg/l) in water sample from Anieze river in River State.

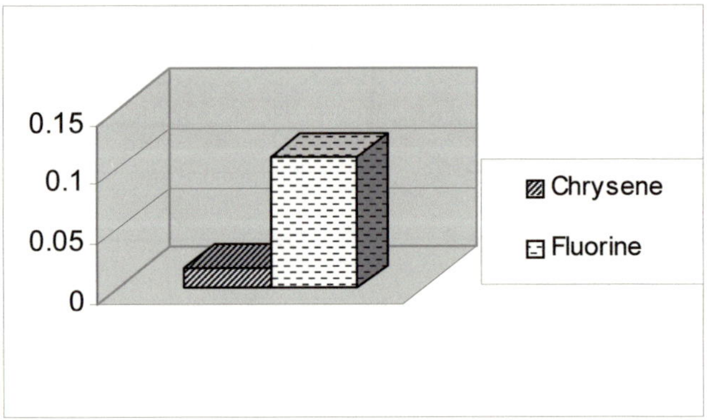

Fig. 2b. Some PAHs (mg/l) in water sample from Orashi River in Rivers State.

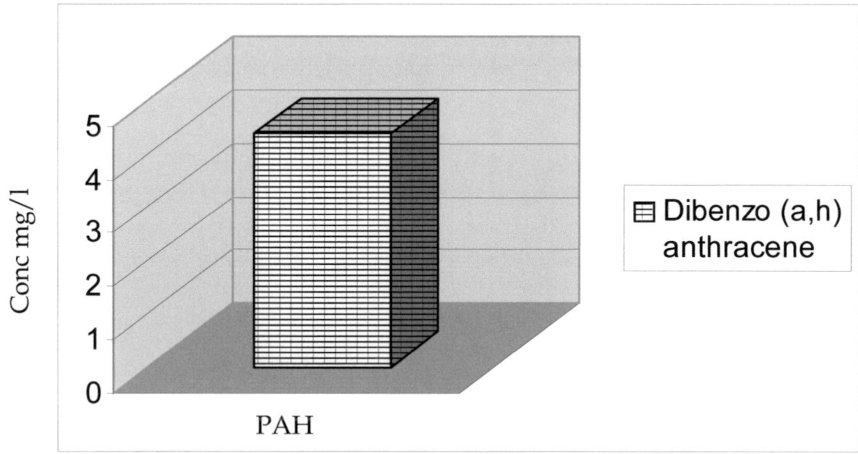

Fig. 2c. Dibenzo (a,h) anthracene (mg/l) in water sample from Ifie-Kporo River in Delta State (Nduka and Orisakwe, 2011).

Heavy and complex compounds are more resistant to microbial degradation, but many are degradable by micro organism to a large extent. Some of the heavy and more complex compounds will eventually settle to the marine floor (Neilson, 1994). Put together, 40 to 80% of crude oil can be degraded due to microbial action (Albers, 1995). The most significant factors that affect crude oil (hydrocarbon) degradation by microbes include microbe type, water temperature, nutrient availability, amount of oxygen, salinity and bioavaible surface of the oil (Atlas and Bartha, 1973, Kennish, 1997). Microbial degradation is inhibited by mouse formation (which reduces the surface area of the oil available to biodegradation) and by cool water temperatures (Galt et al, 1991). Anaerobic conditions severely restrict microbial degradation (Gundlach et al, 1983). Oil deposited in the bottom sediments of sheltered estuaries and wetlands can persist for more than ten(10) years (Teal, 1993).

6. Crude oil toxicity

Toxicity of crude oil largely depends on the concentration of oil in water, dispersed into the water at low concentration (<1ppm), it is very little problem and will quickly get broken down. At higher concentration, it is toxic to most life forms and floating on the surface of water, it will physically smoother on everything it touches. The majority of most types of oil will float on water surface when spilled, but some will dissolve into water and with the help of wave action, some will be dispersed into small droplets. The most likely impact of oil floating in the sea is on the sea birds. Any oil sinking down into the water, both dissolved and dispersed could affect marine life particularly if the water is shallow and the oil toxic.

It is unusual, however for significant concentration of oil to go deeper than 5 metres below the surface, so there is not usually great concern for marine life below this depth. (SPDC, Nigeria, 1997).

S/no	State	Surface water	C-8	C-9	C-10	C-11	C-12	C-13	C-14	C-15	C-16	C-17	C-18	C-19	C-20	C-22	C-24	C-26	C-28	C-30	C-32	C-34	C-36	C-38	C-40
1	Bayelsa	Akiplai Wellhead stream	-	3.560	-	-	-	-	-	-	2.014	2.245	3.247	5.210	3.250	3.064	3.064	3.015	2.423	2.033	-	-	-	-	-
		Nembe Creek	-	-	-	-	-	-	-	-	-	3.900	1.891	3.232	2.263	2.681	2.993	2.639	2.398	2.738	2.408	-	-	-	-
		Otensigha Wellhead Stream	-	-	-	-	-	-	-	8.411	13.168	-	-	-	-	18.052	2.533	2.586	-	-	6.232	3.926	-	-	-
		Kolowell head stream	-	-	-	-	-	-	-	-	-	7.589	2.284	-	-	2.827	4.230	5.025	3.396	2.004	4.006	1.969	-	-	-
		Etebele River	-	-	-	-	-	-	2.832	3.941	4.311	2.901	4.523	6.979	5.358	5.327	5.399	4.865	3.732	3.448	2.206	-	-	-	-
2	Delta	Ijala Creek	-	-	-	-	-	-	-	-	-	1.255	0.951	-	1.547	0.976	0.938	0.876	-	-	-	-	-	-	-
		Ugheli Creek	-	-	0.003	0.002	0.006	0.003	0.004	0.002	0.656	0.001	0.014	0.001	0.010	0.002	11.731	0.010	0.225	0.002	0.001	0.004	-	-	-
		Ubeji creek	-	-	-	-	-	-	-	-	-	-	-	-	-	-	-	1.687	-	-	-	-	-	-	-
		Ifie-Kporo river	-	-	-	-	-	-	0.184	0.009	-	1.999	0.131	0.009	1.297	63.154	63.154	1.421	0.501	0.006	0.114	0.002	0.0310	0.0170	0.011
		Ekure itsekiri crack	-	-	-	-	-	-	-	-	-	0.003	0.004	5.699	0.033	3.195	59.744	0.002	0.001	0.233	0.002	0.005	-	-	-
		Olumoro Burropit	-	-	-	-	-	-	-	-	-	0.015	0.119	0.037	0.206	1.488	0.006	0.002	0.360	0.001	0.360	0.001	0.002	0.001	0.007
		Egbo stream	-	-	0.025	0.001	0.001	0.001	0.015	0.006	0.012	0.042	6.087	0.161	4.528	0.068	0.003	0.004	0.228	0.001	0.002	0.002	0.012	0.100	0.12
		Ughewhe stream	-	-	-	-	-	-	-	-	-	0.118	0.005	7.927	0.049	4.735	0.016	0.011	0.001	0.156	0.002	0.001	0.002	0.014	0.045
3	Rivers	Orash river	-	-	-	1.940	8.959	12.599	9.369	23.718	2.985	5.198	1.638	68.106	2.270	-	-	14.893	-	-	-	-	-	-	-
		Abwoma river	-	-	-	-	-	4.068	5.413	10.200	8.735	13.967	6.554	11.424	9.840	9.200	11.103	8.896	7.641	6.179	2.412	2.364	-	-	-
		Iwofa river	-	-	-	-	-	-	-	-	1.835	7.338	2.908	5.264	3.930	4.061	4.054	3.850	3.082	2.819	1.947	-	-	-	-
		Okirika river	-	-	-	-	1.884	-	-	-	4.060	1.753	1.773	2.998	2.352	3.297	4.110	5.320	6.114	2.981	2.797	-	2.443	-	-
		Woji river	-	-	-	-	-	5.585	4.166	4.093	3.098	1.817	2.481	112.822	2.093	-	-	271.033	-	-	-	-	2.127	-	-

Table 2. Concentration (Mg/L) Of Straight Chain Aliphatic Hydrocarbon In Some Selected Surface Water Of Niger Delta (Nduka and Orisakwe, 2011).

			Poly aromatic hydrocarbon (PAH) degraders		Straight chain	Both PAH's and straight chain degraders		
S/no	State	River/stream/creeks	Achromobacter	Aspergillus	Proteus	Bacillus	Pseudomoners	Penicillum
1	Bayelsa	Akiplai well head	-	$2.0x 10^{1}$	$3 0X10^{2}$	$1.8x10^{3}$	$9.0x10^{2}$	-
		Nembe creek	-	$1.0x 10^{1}$	-	$8.3X10^{3}$	$1.4X10^{3}$	-
		Otensigha wellhead stream	-	$4.0 x10^{1}$	$7.0X10^{1}$	$1.43x10^{4}$	$2.3x10^{2}$	$2.0x10^{1}$
		Kolowellhead stream	-	1.6x102	$3.0x10^{2}$	$8.0x10^{2}$	$5.3 x 10^{2}$	-
		Etebele river	-	$7.0x10^{1}$	$6.0X10^{2}$	$1.36 X 10^{4}$	$1.5 X 10^{2}$	-
2	Delta	Ijala creek	-	-	-	$6.0x10^{1}$	$3.0x10^{1}$	-
		Jeddo river	$1.0 x10^{1}$	-	-	$1.0 x10^{2}$	-	1.0x10
		Ekpan river	-	-	-	$2.0 x 10^{2}$	-	-
		Ughelli river	$1.0 x 10^{0}$	$2.0 x 10^{1}$	-	$3.3 x 10^{2}$	$4.0 x 10^{1}$	$2.0x10^{1}$
		Ubeji creek	-	-	-	$1.5 x10^{2}$	$2.0 x 10^{1}$	-
		Ifie kporo river	$2.0 x 10^{1}$	-	-	$2.0 x 10^{2}$	$1.3 x 10^{1}$	-
		Ekurede itsekiri river	-	-	-	$2.0 x 10^{2}$	$8.0 x10^{1}$	-
		Aja-Etan	$5.0 x 10^{1}$	-	$3.0x 10^{1}$	3.0x10	$2.0x10^{1}$	-
		Olomoro burrow pit	-	-	-	$5.0 x10^{1}$	$3.0x10^{1}$	-
		Egbo stream	-	$2.0 x10^{1}$	-	$4.0x10^{1}$	$1.0x10^{1}$	-
		Ughewhe stream	-	-	-	$2.0X10^{1}$	-	$1.X10^{1}$
3	Rivers	Anieze river	-	-	-	$1.4x10^{2}$	$1.9x10_{2}$	-
		Orash river	$2.0x10^{2}$	$3.,x10^{2}$	$3.0x10^{2}$	$4.8x10^{3}$	1.910^{2}	-
		Abuloma river	-	-	-	$1.7x10^{2}$	$1.0x10^{2}$	-
		Iwofe river	-	-	$2.0x10^{1}$	$6.0x10^{1}$	$1.2x10^{2}$	-
		Okirika river	-	$3.0x10^{1}$	-	$5.0x10^{1}$	$2.0x10^{1}$	-
		Ozubuko Amawell stream	-	$1.4x10^{1}$	-	$3.0x10^{1}$	-	-
		Ozubuko borehole stream	-	-	-	$1.0x10^{2}$	$3.0x10^{1}$	-
		Tombia river	-	$4.0x10^{1}$	-	$1.6x10^{2}$	$1.1x10^{2}$	-
		Ozuba river	-	-	-	$2.0x10^{1}$	$3.2x10_{2}$	-
		Woji river	-	$4.0x10^{1}$	-	$4.0x10^{1}$	$8.0x10^{1}$	-

Table 3. Amount (Mg/L) Of Bacterial Count In Selected Water Samples Of Niger Delta (Nduka and Orisakwe, 2011)Impact of Crude Oil Spill on the water organism.

6.1 Planktons

Planktons are the basis of marine food chain and it includes eggs and larvae of fish and other animals. It is very sensitive to the toxicity of oil and lots of planktonic organism will be killed, particularly if the concentration of oil in the water is increased by the wave action or chemical dispersants, therefore it is possible in a worst case scenario for the effect on the plankton to lead to some longer term effect (Adeyemi, 2004).

6.2 Aqua culture

Fish in fish farms are trapped by the nets and cannot escape the oil while it is quite likely that the fish may not die since they are tolerant to oil; they will be likely tainted. Tainting is where the fish absorbs oil from the water and its flesh when eaten tastes off, if the tainting is severe which is most likely to happen if a lot of oil has been dispersed into the water, the fish may be indelible leading to economic loss (SPDC Nigeria 1997, Adeyemi, 2004).

6.3 Fish

Adult fish have very sensitive sense organs so they taste the oil and quickly leave the area of the spill, unless they are trapped there in some way. It is very rare to have large fish kills due to marine oil spills. Juvenile fish however are more vulnerable since they often live in shallow water near shore areas known as "nursery areas" until they become adult. If a toxic oil impact a shallow bay and the oil is dispersed down into the water, the increased concentrations could have severe effects on juveniles and will result in long term impacts on adult fish stocks. This been evidenced in the Niger-delta region of Nigeria. Polyaromatic hydrocarbon (a carcinogen), been lipophylic tend to accumulate more in fish than in the sediment and least in the water samples of the region. Fish therefore is the best biomarker for the levels of PAHs contamination in marine samples (Anyakora et al, 2005). Nitrate pollution which has been reported in the area (Nduka and Orisakwe, 2011), at increased concentrations are harmful to aquatic animals. Marine invertebrates and fish exposed to nitrate may be smaller in size, have reduced maturity rate and lower reproductive success. In extreme high exposure levels, aquatic invertebrates and fish may die (Ohio State University, 2008). Early life stages of aquatic animals are more sensitive to nitrate than juvenile and adult animals. Early studies by Gbadebo et al, 2009, showed that petroleum oil, whether crude oil or spent oil is very toxic to fish (Clarias garipinus fingerlings), but spent oil produced more toxic effects on the clarias fingerlings at every concentration than that of crude oil. Their findings is supported by the fact that the water-soluble components of crude oil are toxic and could affect the survival and metabolism of aquatic animals like fish (Cote, 1976).

6.4 Sea birds

Sea birds are very easily affected by oil on the surface water because they spend long period sitting on the water. The oil is soaked up by their feathers and cannot fly, feed or preen themselves. They ingest oil which then damages their gut, die of starvation, cold, poisoning or shock. The sensitivity of species of birds to presence of oil vary. The impact on local population of birds can be serious and result in reduced population for many years (SPDC, Nigeria, 1997). In their work, Fry et al, 1986, showed that when wedge-tailed shearwaters birds were exposed to weathered Santa Barbara crude oil, it resulted in a greatly reduced number of eggs laid and complete hatching failure of 60 pair of exposed birds. Oral doses of oil in gelatin capsules reduced laying and breeding success but to a lesser extent than external exposure, also survival of chicks of dosed birds were reduced. It has also been observed that exposure of birds to crude oil before breeding delayed sexual maturation or onset of lay of eggs in captive mallards (Holmes et al, 1978, Coon and Dieter, 1981). Some crude oils are embryotoxic and if ingested during the period of egg formation may affect the development of progeny (Gorsline and Holmes, 1982b).

6.5 Dispersant problem

To stem the effect of oil spill, different grade of chemicals (dispersants) were developed to dispose the oil and subsequently hasten its sinking. The use of dispersants to deal with oil spills should be seen as an aid to the process of natural dispersion. Dispersants reduce surface tension of oil/water interface thereby breaking the oil slick into fine droplets that are then dispersed into the water column, for each dispersant, there will be a different range of effectiveness. The range of effectiveness is related to the concentration of the emulsifiers in the formulation, the effectiveness of the emulsifiers and the type of solvent used.

One of the disadvantages of dispersants is that they introduce a second source of pollution into the marine environment and indeed most of the opposition to their use stems from this fact. If a dispersant works effectively, then it is removing the oil from the water surface and distributing it throughout the body of the water. The upper section (normally top 3 metres) of the water column will contain fairly high concentrations of oil and this may reach toxic levels (SPDC, Nigeria 1997).

7. Chemical pollution associated with oil spill

Apart from environmental devastation associated with oil spill, the chemical constituents of the spilled oil induce toxic effect to the plant, animals, man and aquatic organism. Polyaromatic hydrocarbon (PAHs) are a class of compounds composed of two or more aromatic rings. They are a component of crude and refined petroleum products. Petroleum production, import and export of petroleum products also contribute a lot to the extent of PAH contamination especially in the marine samples (Baek et al 1991, Lorber et al 1994, Nwachukwu 2000, Nwachukwu et al 2001). They have been reported in water samples, fish species and soil sediments of the Niger Delta region of Nigeria which has had extensive petroleum production activities over the past few decades (Anyakora et al 2005, Anyakora and Coker 2006, Anyakora et al 2005). PAHs are classified as environmentally hazardous organic compounds due to their known or suspected carcinogenicity and are included in the European community (EC) and United States Environmental Protection Agency (USEPA) priority pollutant list (Nieva-Cano et al, 2001). These, with metals, volatile organic compounds (VOCs), and others have been previously discussed under composition of crude oil.

8. Air pollution resulting from crude oil refining and gas flaring

Crude oil exploration in the Niger delta region Nigeria has resulted in gas flaring, fire and gas explosions and bush burning, results in air borne particulate matter, also called suspended particulate matter (SPM) can be found in ambient air in the form of dust, smoke or other aerosols. SPM can occur as a secondary aerosol resulting from atmospheric transformation of gaseous pollutants emitted from combustion sources such as power plants and automobiles or natural sources such as forests. Particles can also result from condensation of volatile elements and species in the atmosphere and form very small particles or absorbed on the surface of already formed, finely divided particles. The combustion of crude oil associated gas (flaring) releases several gases into the atmosphere. Carbon monoxide (CO) is emitted into the atmosphere mainly as a product of the gas, it is accompanied by release of carbon (iv) oxide(CO_2) (a green house gas) (Ideriah and Stanley, 2008). Average value of 69.25 μgm^{-3} of CO_z level have been reported in Niger-delta. Release of CO_2 has potential to adversely affect the health and well-being of nearby organisms (Nwaichi and Uzazobona 2011). Hydrogen sulphide is also present in air as a result of gas flaring, although Nigerian crude oil is reputed to have few milligram per litre of sulphur. H_2S can also result from microbial decay of organic matter and sulphate ion reduction

$$SO_4^{2-} + 2(CH_20) + 2H^+ \text{-----------}\rightarrow H_2S + H_2O + 2CO \quad (2)$$

Sea water as found in the Niger delta region has high sulphate ion, bacterially induced formation of H_2S causes pollution problems in coastal regions as in our study area and is a major source of atmospheric sulphur (Manahan, 1979). Oxides of nitrogen ($N0_x$) and ammonia can also result from crude oil associated gas flaring and refining operations Ammonia in polluted atmosphere reacts readily with acidic materials. Such as sulphuric acid aerosol droplets to form ammonium salts (Ideriah and Stanley, 2008).

$$NH_3 + H_2SO_4 \text{------------}\rightarrow NH_4\,HSO_4 \quad (3)$$

Oxides of sulphur (SO_x) and that of nitrogen (NO_x) primarily affect the respiratory system and studies on laboratory animals and humans show that these pollutants irritate the lining of the lungs and cause respiratory stress (Coffin and Stokinger, 1976). Organisms are exposed to air pollutants enroute three pathways: (1) Inhalation of gases or small particles, (2) Ingestion of particles suspended in food or water and (3) absorption of gases through the skin. Organisms response to pollutants varies and depends on the type of pollutants involved, exposure time and volume of pollutants taken up by the animal. Organism's age, sex, health and reproductive condition also play a role in its response (Maniero, 1996). The most outstanding effect of air pollution resulting from flaring of crude oil associated gas and petroleum refining activities in the Niger-delta region of Nigeria is acid rain. Over 2.5 billion Fe^3 of crude-associated gas is flared in Nigeria daily with an estimated yearly financial loss of $2.5billion. Acid rain pre-cursor gases – NO_2 and SO_2 are products of high temperature reactions and gas flaring in the study, area makes this possible, hence we have reported acid rain in our previous studies (Nduka et al 2008, Nduka and Orisakwe 2010). Acid precipitation can wear away the waxy protective coatings of leaves, damage them and reduce photosynthesis. Important cations such as K^+, Ca^{2+}, mg^{2+} and Na^+, which are very important to the welfare of green plants and the aquatic ecosystem, are leached out and become unavailable to plants, also toxic cations such as Pb^{2+}, Cd^{2+}, Hg^{2+} are demobilized into the aquatic ecosystem where they bioconcentrate in lugworms, barnacles, algae and other planktonic and benthic organisms enroute to food webs. Sulphates and nitrates, which form in the atmosphere from sulphur dioxide (SO_2) and nitrogen oxide ($N0_2$) emissions, contribute to visibility impairment. The ubiquitous soot (black carbon) or Charcoal

(Goldberg, 1985) which normally give rise to black rain in Niger-delta region of Nigeria is one of the major components of air pollution worldwide.

9. Public health impact resulting from crude oil exploration and exploitation

Apart from acidic precipitation and its negative impact, acid rain precursor gases ($N0_2$ and SO_2) are part of six (6) common outdoor pollutants (ALA, 1996). Noise and thermal pollution from numerous gas flare points and extensive fire disasters that characterize the study area have led to extinction of some exotic species that are hunters delight. Inhalation of fine particulates that are fallouts from acid precursor gases have been linked to illness and premature death from heart and lung disorders such as asthma and bronchitis (USEPA, 2001). Nitrogen dioxide (NO_2) poses a health threat itself as well as playing a major role in the formation of the photochemical pollutant ozone. Previous studies has shown that animals exposed to NO_2 have diminished resistance to both bacterial and viral infection (Gardner, 1984), while children exposed to high indoor levels of NO_2 may become more susceptible to critical infections of the lower respiratory tract, bronchial tubes and lungs, and may develop bronchitis and chest cough with phlegm (Neas et al, 1991). Sulphur dioxide (SO_2) is a temporary irritant, though research have shown that increased levels of SO_2 in conjunction with particulate matter may trigger small, but measurable, temporary deficits in lung function (Dockery et al, 1986). Epidemiological studies have found that the impact of SO_2 is inseparable from that of particulate matter, but effects of the two classes of pollutants have been differentiated by an analysis exemplified by studies in Uta valley (Pope et al, 1992). The major health impact of SO_2 is on population groups susceptible to pollutant effect due to pre-existing conditions, such as asthma. Measurable atmospheric levels of SO_2 and NO_2 exist in the Niger-delta region of Nigeria which gave rise to acidic precipitation we have reported in the region (Nduka and Orisakwe, 2010). In the area, serious health problems such as skin cancers and lesions may be linked to acid rain. Stomach ulcers could also occur, as consumption of acidic water can alter the pH of the stomach and leach the mucous membrane of the intestinal walls, this is more so as Nigerians depends heavily on rain water for drinking, cooking, laundry and other domestic uses.

Degradation processes such as photolysis, hydrolysis, oxidation and biodegradation are all involved in the chemical transformation of a compound upon its entry into the environment. The value of C_{10} to $C_{3}o$ straight chain hydrocarbon and the concentrations of chloride ion in the water samples of the Niger Delta region of Nigeria, shows that the formation of polychlorinated-n-alkanes (PCAs) of the general formular $C_nH_{2n}+_2$-zcl_2 is possible (Nduka and Orisakwe, 2011). Polychlorinated-n-alkanes (PCAs) are normally manufactured in the presence of ultra violet (UV) light (Tony et al. 1998). The surface water of the region where crude oil spill occurs are wide open to direct ultra violet (UV) light from the sun, its formation is by simple substitution of hydrogen atoms by chlorine atoms, an example of chain reaction involving free radicals (Morrison and Boyo, 1983), also for the fact that the temperatures of water are above ambient values even in rainy season (Nduka and Orisakwe, 2011) shows that their formation is possible. Those of great interest are the C_{10} - C_{13} PCAs, which have the greatest potential for environmental release (Environmental Canada 1993a) and the highest toxicity of PCA products (Serrone et al, 1987, Wills et al 1994, Mukherjee, 1990). Taken together, the carcinogenic effect of Polyaromatic hydrocarbons (PAHs) and the fact that study has shown that the C_{10} - C_{13} PCAs inhibit intercellular communication in rat liver epithelial cells, a phenomenon that suggests these chemical may be acting as tumor promoters (Kato, 1996), could accelerate environmental and public health hazards such as malignant lymphomas (Omoti and Halim 2005, Omoti, 2006) and soft tissues sarcomas (Seleye-fubara et al, 2005) already reported in the Niger Delta region of Nigeria. In addition to these above, high nitrate levels in water samples from the region can also result in incidences of some cancers and lesions (Gulis et al, 2002), spontaneous abortion and ectopic pregnancy also reported in the area (Gbaforo and Igbafe, 2002). Pollution keratoconjuctivitis have been reported among children in oil-producing areas of Niger Delta, Nigeria. It has adverse consequences due to accumulation of differed categories of pollutants from drilling, production, refining of crude oil and production of petrochemicals, mainly black carbon. Persistent itching, foreign body sensation and specified areas of conjunctiva/limbal discoloration were used as markers for Pollution Kerato Conjunctivitis (PKC) (Asonye and Bellow, 2004). In their work investigation into the pharmacological basis for some of the folkloric uses of bonny light crude oil in Nigeria, Orisakwe et al, 2000, discovered that Nigerian bonny light crude oil caused complete inhibition of histamine-induced smooth muscle contraction while producing only a partial inhibition of the acetylcholine-induced contraction. It had no effects on the acetylcholine-induced skeletal muscle contraction, but proved good to analgesic effect that is comparable to aspirin. Because of complex composition of crude oil, it has multiple potential type of toxic effect which may include long term petroleum pollution on individual organism such as impaired reproduction (Feuston et al, 1997), reduced growth (Eisler, 1987), tumours and lesions (Malins and Ostrander, 1994), blood disorders (Yamato et al 1996) and morphological abnormalities (Kennish, 1997). Crude oil showed changes in the hypothalamo-pituitary-thyroid adrenal axis when male rats were exposed to it, and the effect is believed to be due to stress (Vyskocil et al, 1988). The Nigeria nation has an outrageous twelve million infertile people though not restricted to oil producing areas alone (Giwa-Osagie, 2003, whose infertility is believed to be due to infection (Cates et al, 1985), although some infection do persist after treatment. There are elevated consistence oligospermia or azoospermia in Nigeria than most other causes of infertility and less resources for its management (Osegbe and Amaku, 1985). Therefore it has been established that the Nigerian bonny light crude oil is a testicular toxicant and its use as a folklore medicine (which is very ubiquitous) in Nigeria may cause infertility (Orisakwe et al, 2004). They have also reported that the kidney cells of an adult albino rats were damaged by bonny light crude; crude oil caused a destruction of the renal reserve capacity (Orisakwe et al 2004).

10. Conclusion

Commercial exploration and exploitation of huge crude oil deposits and gas reserves in the Niger–Delta region of Nigeria has resulted in the alteration of the regions environment in certain negative manner. Vegetation is cleared to make way for seismic lines, roads are built, drilling mud and oil may reach surface water. The effect is the millions of barrels of crude oil and its lower fraction spills that results from various operations, damaging valuable commodities in the environment, plants and animals, harbours, beaches, marinas are devastated making them unfit for use. Due to its complicated composition, petroleum hydrocarbon has the potential to eliciting various toxic effects, which can cause acute lethal toxicity, sub lethal chronic toxicity or both depending on the exposure, dosage and type of organism exposed. Toxicity is not restricted to the immediate surrounding of the spill due to dispersion, dilution and deposition of oil into the water column and onto shores and sediments through various mechanisms, all organisms within the influence may be exposed to adverse effect associated with the oil. We can rightly conclude that crude oil exploitation in the Niger–Delta region of Nigeria is a major environmental and public health concern.

11. References

Achi SS and Shide EG (2004), Analysis of Trace Metals by Wet ashing and spectrophotometric Techniques of crude oil Samples. *J. chem. Soc. Nigeria*, 29 (11):11-14

Abam TKS, Okagbue PI (1997). The cone penetrometer and soil characterization in the deltas; *Journal of Mining and Geology*; 33(1): 15-24

Adeyemi O.T (2004), Oil exploration and environmental degradation. The Nigerian experience. *Environ. Inform.* Arch. 2: 387 - 393

Albers, P.H. (1995). Petroleum and individual polycyclic aromatic hydrocarbons in: Handbook of Ecotoxicology. Hoffman D.J, Rattner BA, Burton Jr. J.A, Cairns Jr J, eds. Lewis Publishers, Boca Raton, FL, 330 - 355

Amadi UMP, Amadi, PA (1990). Salt water migration in the coastal aguifers in the Nigeria; *Journal of Mining and Geology*; 26(1): 35

Amadi P.A (1986). Characteristics of some natural waters from the Port-Harcourt area of Rivers State, Nigeria. Unpublished MCG Thesis, University of Ibadan, 98

Amaize, E (2007), Crises from the Creeks. Saturday vanguard Newspapers (www.vangardngr.com). March 17. pp 11-13

American Lung Association (ALA.1996). Health Effects of outdoor Air Pollution. 13, 18-20

Anyakora C, Ogbeche A, Palmer P, Coker H (2005). Determination of Polynuclear aromatic hydrocarbons in marine samples of siokolo fishing settlement. *J. of chromatography A*; 1073: 323 - 330

Anyakora C, Ogbeche A, Palmer P, Coker H, Ukpo G, Ogah C. (2005). GC/MS analysis of polynuclear aromatic hydrocarbons in sediments samples from Niger-Delta region. *Chemosphere*; 60: 990 - 997

Anyakora C, Coker H (2006). Determination of polynuclear aromatic hydrocarbons (PAHs) in selected water bodies in the Niger Delta. *A.J. Biotechnologyl* 5(21): 2024 - 2031

Anyakora CA, Ogbeche KA palmer P, Coker H, Ukpo G, Ogah C (2004). A screen for Benzo (a) pyrene a carcinogen, in the water samples from the Niger Delta region. Nigeria. *J. Hosp. Med.*, 14: 288-293

Asonye C.C and Bello E.R (2004). The blight of pollution kerato conjunctivitis among children in oil – Producing industrial areas of Delta State, Nigeria. Ecotoxicology and Environmental safety. 59:244 - 248

Awosika LF (1995). Impacts of global climate change and sea level rise on coastal resources and Energy development in Nigeria. In: Umolu, J.C (ed). Global climate change: impact on Energy Development. DAMTECH Nigeria Limited Nigeria

Atlas RM (1991). Microbial hydrocarbon degradation: Bioremediation of oil spills. Biotechnology. 52:149-156

Atlas RM, and Bartha R (1973). Fate and effects of pollution in marine environment. Residues Review; 49: 49 – 85

Bowman J.A and Wills J.B (1967). Some application of the nitrous Exide–acetylene flame in chemicals analytical by atomic absorption spectrometry, *Analytical chemistry*, 39 (11); 1210 - 1216

Baek SO, Field R.A , Gold Stone RA, Kirk. PW, Lester JN, Perry R, (1991). A review of atmospheric polycyclic hydrocarbons, sources, fate and behavior. *Water, Air, Soil Pollut*; 60: 279

Barwise I.G. (1990). Role of nickel and vanadium in petroleum classification. *J. American Chem.* Soc. 47 - 65

BICNEWS (2005). Bioremediation: nature's way to a cleaner Environment. Malaysian Biotechnology information centre (MABIC). 9: 1-6 (www.bic.org.my)

Blokker PC (1964). Spreading and Evaporation of Products on water, 4th International Harbour conference, Antwerp

Bob L. (1990). Waste and pollution Basi/Blackwell Ltd. 108 Cowley Road. Oxford, OX41JF, England. 40 - 51

Burwood R, and Speers G.C. (1974) Photo-oxidation as a factor in environmental dispersal of crude oil. *Estuarine coastal Mar Sci*; 2: 117 - 135

Cates W, Farley TMM. Rowe P.J (1985). Worldwide patterns o infertility, is Africa different? Lancet; 2:596 - 598

Clark R.B (1992). Marine pollution. Clarendon Press, Oxford

Coffin DL and Stokinger H.E (1976). Air Pollution, 3rd ed. Vol. 11, A.C. Stern ed., Academic Press, New York. 231-360

Coon N.C and Dieter MP (1981), Responses of adult mallard ducks to ingested south Louisana crude oil. *Environ. Res*; 224: 309 -314

Cote, R.P (1976). The effects of Petroleum industry liquid wastes on aquatic life with special emphasis on the Canadian environment. National Research council Canada, Canada

Delaune R.D, Gambrell R.P, Pardue J.H, Patrick Jr. JH (1990). Fate of Petroleum Hydrocarbons and toxic organics in Louisana coastal environments, Estauries 13: 17 - 21

Dockery, D.W et al (1986). Change in pulmonary function in children association with air pollution episode. J. Air pollut. Control Assoc. 32:937-942

Eisler R. (1987). Polycyclic aromatic hydrocarbon hazard to fish, wild life and invertebrates: a synoptic review, vol. 85. Washington, Dc: fish and wild life service, US Department of interior, 1-11

Ekundayo EO, Obuekwe CO (2001). Effects of an oil spill on soil physico-chemical properties of a spill site in a typic udipsamment of the Niger Delta basin of Nigeria. *Environmental Monitoring and Assessment*; 60(2): 235-249

Ellrich J, Hirner A, Stark H (1985). Distribution of trace elements in crude oil from southern Germany, *J.Chem. Geol.* 48; 313 - 323

Environment Canada (1993a). Priority substances program. CEPA assessment report, chlorinated Parafins, commercial chemicals Branch, Hull, Quebec

Fay TAC (1969) The Spread of oil slick on a calm sea.Oil on the sea (D.P. Hoult, Ed), Plenum. 53 - 63

Feuston MH, Hamilton CE, Schreiner CA, Mackerer CR (1997).Developmental toxicity of dermally applied crude oil in rats. *J. Toxicol. Environment Health*; 52; 79 - 93.

Fry DM. Swenson J, Addiego L.A, Grau CR, Kang A (1986). Reduced Reproduction of Wedge-tailed shearwaters Exposed to Weathered Santa Barbara Crude oil. *Arch Environ. Contam. Toxicol.* 15: 453 - 463

Gardner, D.E (1984). Oxidant- induced enhanced Sensitivity to infection in animal model and their extrapolation to man. *J. Toxicol. Environ. Health* 13,423-439

Galt J.A, Lehr W.J, Payton D.L (1991). Fate and transport of the Exxon Valdez oil spill. *Environ Sci Technol;* 25: 202 - 205

Garner HF, Fellow OBE, Green SJ, Harper FD, Pegg RE (1953), J. Inst. Petroleum, 39: 279 - 293

Gharoro EP, Igbafe AA (2002). Ectopic Pregnancy revisited in Benin City, Nigeria: analysis of 152 cases. *Acta Obstet Gynecol Sand;* 81 (12): 1139-1143

Gbadebo, A.M, Taiwo, A.M and Ola OB (2009). Effects of crude oil and spent oil on clarias garipinus, a typical marine fish. *American journal of Environmental Sciences;* 5(6): 752 - 758.

Giwa-Osagie O.O (2003). Nigeria has twelve million infertile persons *Pharmanews;* 25(7): 48 - 49

Goldberg E.D. (1985). Black Carbon in the Environment. Properties and Distribution. Wiley, New York

Gorsline J and Holmes WN (1981). Effects of petroleum and adreno-cortical activity and on hepatic naphthalene-metabolizing activity in mallard ducks. *Arch. Environ contam Toxicol;* 10: 765 - 777

Gulis G, Gomolyova M, Cerhan JR (2002). An ecologic study of nitrate in municipal drinking water and cancer incidence in Trnava District, Slovakia. *Environ Res;* 88: 182 -187

Gundlack E.R, Boehm P.D, Marchland M, Atlas R.M, Ward D.M, Wolfe D.A. (1983). The fate of Amoco cardiz oil. *Sciences.* 221: 122 - 129

Hamlat, MS, Djeffal S and Kadi H (2001) Assessment of radiation Exposures from naturally occurring radioactive material in the oil and gas industry. *Applied Radiation and Isotopes;* (1):141-146

Handson W.E. (1994) Chemical Technology of Petroleum, McGraw-Hill Books company-London. 141 - 162

Hettinger (Jr) WR, Keith CD, Ging JL and Teter W (1955), Ind. Eng, Chem. 47 719-730

Holmes W.N, Cavanaugh K.P, Cronshaw J. (1978). The effects of ingested petroleum on oviposition and some aspects of reproduction in experimental colonies of Mallard ducks (Anas platyrhynchos). *J. Reprod. Fert;* 54; 335 - 347

Huang CP and Elliot H.A. (1977). The stability of emulsified crude oils as affected by suspended particles in: fates and effects of petroleum and hydrocarbons in marine organisms and Ecosystem. Wolfe D.A, ed. Pergamon Press, New York. 413 - 420

Ibru, G (2001), Hotel and tourism development Potentials in Delta State, A paper presented at the 2nd anniversary of the administration of James Onanefe Ibori, Executive governor of Delta State at the conference hall; Hotel excel, NNPC Road, Effurun, Nigeria, June and 2001

Ideriah TJK and Stanley H.O (2008). Air quality around some cement industries in Port Harcourt, Nigeria. Scientia Africana 7(2): 27 -34

Jeffrey, P.G (1973). Large scale Experiments on spreading of Oil at sea and its Disappearance by natural factors. In conference on prevention and control of oil spills. API and EPA, Washington, 469

Jibiri N.N, Emelue HU (2008). Soil radionuclide concentrations and radiological assessment in and around a refining and petrochemical company in Warri, Niger Delta, Nigeria. J. Radiol. Prot. 28(3): 361 - 368

Karchmer JH and Gunn EL (1952), Determination of Trace metals in petroleum fractions, Analytical chemistry, 24(11); 1733 - 1734

Kawchan J.A (1955), Quantitative spectrographic determination of vanadium in petroleum products by logarithmic rector method, Analytical Chemistry, 27 (12): 1873-1874.

Kennish MJ (1997). Practical handbook of estuarine and marine pollution Boca Raton, FL: CRC Press: 83 – 95

Lambert-Aikhionbare DO, Bush PR, Ibe AC (1990). Integrated geological and geochemical interpretation of source rock studies in the Niger Delta. *Journal of Mining and Geology*; 26(1): 97-106

Lewis FH and Sani M (1981), from Hydrocarbons to Petrochemicals, 1st ed, Gulf Publishing, Saudi Arabia, 38-47

Lorber M, Cleverly D, Sehuam J, Philips L, Schweer G, Leighton T (1994). *Environ Sci. Technoil*. 156:39

Mackay D. and McAuliffe C.D (1989). Fate of hydrocarbons discharged at sea. *Oil chem. Pollut*; 5: 1 – 20

Malins DC and Ostrander GK (1994). Aquatic toxicology. Molecular, biochemical, and Cellular perspectives. Chelsea, MI: Lewis Publishers

Manahan SE (1979). Environmental Chemistry, 3rd edition; Willard Grant Press, Boston, Massachusetts 331-398

Maniero T.G. (1996). The effect of Air Pollutions on Wild life and implication in class I Areas. National Park Service Air Resources, Denver, Colorado. 1-9

Milner O.I (1963), Analysis of Petroleum for Trace Metals, International series of monographs on analytical chemistry Pergamon press, oxford, England 14 35 – 81

Ming ET and Both LL (1956), Petroleum Refiner, 35 192-194

Morrison RT, Boyd RN (1983). Organic chemistry, 4th ed. Allyn and Bacon. Newton, MA

Mukherjee AB (1990). The use of chlorinated paraffins and their possible effects in the environment. National board of waters and the environment, Helsink, Finland. Series A66. National Board of waters and the environment, Helsinki, Finland

National Atlas (1987). Atlas of the Federal Republic of Nigeria. Lagos Government press, 136.

National Research Council (1985). Oil in the sea: imputs, fates, and effects. National Academy Press, Washington, D.C, 601

Nduka JKC, Orisakwe O.E. (2007). Heavy metal levels and physic-chemical quality of potable water supply in Warri Nigeria. Nigeria. *Annali di chemica*; 97 (9): 867 – 874

Nduka J.K and Orisakwe O.E (2011).Assessment of pollution profile of selected surface water in the Niger Delta region of Nigeria. Lambert Academic publishers, Germany

Nduka J.K, and Orisakwe O.E (2010) Precipitation chemistry and occurrence of acid rain over the oil-producing Niger Delta Region of Nigeria. *The scientific world journal* 10: 528 -534

Nduka JKC, Orisakwe O.E, Ezenweke L.O. (2010). Nitrate and Nitrite levels of potable water supply in Warri Nigeria: A public Health concern. *J. Environ Health*. 72 (6): 28 – 31

Nduka JKC, Orisakwe OE, Ezenweke LO, Ezenwa TE, Chendo MN, Ezeabasili N.G (2008), Acid Rain Phenomenon in Niger Delta Region of Nigeria: Economic, Biodiversity and Public Health Concern. *The Scientific World Journal*; 8: 811 – 818

Neas L.M, Dockery, D.W, et al (1991). Association of indoor nitrogen dioxide with respiratory symptoms and pulmonary function in children. *American J. Epidemiology*; 134: 204-209

Neff J.M. (1987). Biological effects of oil in the marine environment. *Chem Eng Progr*; 83: 27 33

Neilson AH (1994). Organic chemicals in the Aquatic Envrionment: Distribution, persistence and Toxicity. Lewis, Boca Raton, FL

Neumann H.J, Lahme B, Severein D (1981). Geology of Petroleum: Composition and properties of petroleum. Vol 5, Halstead Press, New York, 131

Nieva - Cano M.J, Rubio-Barroso S, Santos-Delgado MJ (2001). Determination of PAH in food samples by HPLC with flourimetric detection following sonication extraction without sample clean-up. *The Analyst*. 126; 1326 - 1331

Nitrates in surface water. Ohio state University Extension, Department of Horticulture and crop Science 2021 Coffey Road, Columbus Ohio 43210-1044. AGF-204-95. http://ohioline.osu.edu/agf-fact/0204.html (assessed March 26, 2008)

Nwaichi E.O. and Uzazobona M.A. (2011). Estimation of the CO_2 level due to gas flaring in the Niger-Delta. *Research journal of Environmental Sciences* 5(6): 565 - 572.

Nwachukwu SCU (2000): *J. Environ. Biol.* 21, 241

Nwachukwu SCU, James P. Gurney TR (2001). *J. Environ. Biol.* 22:29.

Nwandinigwe C.A, Nworgu O.N (1999). Metal contamination in some Nigeria well heads Crudes. Comparative Analysis. *J. Chem. Soc. Nigeria*; 24:118 - 121

Okoro D, Ikolo A.O (2005). Compositional patterns and sources of polynulear Aromatic hydrocarbon in water and sediment samples of Ogunu creek of the Warri River. Proc. 28th Annual Int. Conf. Chem.. Soc. Nigeria. 2(1): 168-171

Olajire AA, Altenburger R, Kuster E, Brack W (2005). Chemical and ecotoxixological Assessment of polycyclic aromatic hydrocarbon-contaminated sediments of the Niger Delta, Southern Nigeria *Science Total Environment.* 340 (1-3): 123-136

Omati CE, Halim NK (2005). Adult lymphomas in Edo State, Niger Delta region of Nigeria-Uinicopathological profile of 205 cases. *Clin lab Haematol*; 27(5): 302-306

Omati CE (2006). Socio-demographic factors of adult malignant lymphomans in Benin City, Nigeria, *Niger Postgrad Med. J*; 13(3): 256-260

Orisakwe O.E, Akumaka DD, Njan A.A and O.J Afome (2004). Testicular toxicity of Nigerian bonny light crude oil in male albino rats. *Reproductive Toxicology*; 18: 439-442

Orisakwe O.E, Njan A.A, Afonne O.J, Akumka DD, Orish VN and Udemezue OO (2004). Investigation into the Nephrotoxicity of Nigerian Bonny light crude oil in Albino rats. *Int. J. Environ. Res. Public Health:* 1 (2); 106 - 110

Osagbe D.N and Amaku E.O. (1985). The causes of male infertility in 504 consecutive Nigerian patients. *Int. Urol Nephrol;* 17:349

Oteze GE (1983). Ground water levels and Ground movement. In S.A Ola (ed). Tropical soils of Nigeria in Engineering practice, Rotterdam: A.A. Balkema publishers, 172-195

Payne JR and Philips CR (1985). Petroleum spills in the marine Environment: the chemistry and formation of water-in-oil emulsions and Tar Balls. Lewis Publishers, Inc. Chelsea, MI

Pope. C A, Schwartz J, and Ransom, M.R (1992). Daily Mortality and PM_{10} Pollution in Utah Valley. *Arch. Environ. Health.* 47, 211-217

Reijers TJA, Petters SW, Nwajide CS (1996); The Niger Delta Basin. In Selected chapters on Geology, Sedimentary Geology and Stratigraphy in Nigeria. SPDC, Nigeria. 103-117

Rosenberg E (1993). Microorganisms to combat pollution. Kluwer Academic publishers; Boston, USA

Seleye-fubara D, Nwosu S.O, Yellowe BE (2005) Soft Tissue sarcomas in the Niger Delta Region of Nigeria (a referral hospital study). *Niger. J. med.* 14(2): 188 - 194.

Serrone DM, Birtley RDN, Weigand W, Millisiher R (1987). Toxicology of chlorinated paraffins. *Food Chem. Toxicol 25*: 553-562

Shell Petroleum Development Company (SPDC), Nigeria (1997). Oil/chemical spill contingency plan and procedures Bulletin

Speight .J. (1999). The Chemistry and Technology of Petroleum. Marcel Dekker.

Teal JM (1993). A local oil spill revisited. Oceanus: 36: 65 – 70

Tiratsoo .E.N (1973). Oil fields of the word, Scientific press Ltd, England, 1- 68

Tony GT, Fisk AT, Westmore JB, Muir DCG (1998). Environmental chemistry and Toxicology of polychlorinated n-Alkanes. *Rev Environ Contam Toxicol;* 158: 53-128

USDHHS and ATSDR, (1995), U.S. Department of Health and Human Services and Agency for Toxic substances and diseases Registry (ATSDR), Toxicological profile for polycyclic Aromatic Hydrocarbons. August, 1995. ATSDR; Atlanta

U.S. Environmental Protection, Agency, clean Air markets Division, Acid Rain Programmers -2001 progress Report. EPA-430-R-02-009

Vyskocil A, Tusi M, and Obstril J. (1988) A subchronic inhalation study with unleaded petrol in rats. *J. Appl Toxicol:* 8: 239 – 242

Williams FB and Robert FD (1967), Petroleum Processing Hand book, McGraw –Hill, New York, 11 -19

Wills B, Crookes MJ, Diment J, Dobson SD (1994). Environmental hazard assessment: chlorinated paraffins. Toxic substances Division, Dept. of the Environment Garston, UK

Wilson VS, and LaBlanc GA (1999/2000). Petroleum Pollution. *Rev Toxicol.* 3: 77 – 112.

Yamato O, Goto I, and Maede Y (1996). Hemolytic anemia in wild seaducks caused by marine oil pollution. *J. Wildlife Dis;* 32: 381 -384

Young LY, Cerninglia CE (1995) Microbial Transformation and Degradation of Toxic Organic Chemicals, Wiley, New York

Magnetic Susceptibility of Petroleum Reservoir Crude Oils in Petroleum Engineering

Oleksandr P. Ivakhnenko
Kazakh-British Technical University,
Department of Petroleum Engineering
Kazakhstan

1. Introduction

Magnetic methods are prominent in the area of petroleum engineering and geoscience. They have clear advantages for use in the petroleum industry, including high resolution and rapidity of measurements, non-destructive analysis and cost-effectiveness (Ivakhnenko, 1999). However, there is very limited available data about the magnetic susceptibility of the crude oils. The most resent and complete studies were by Ivakhnenko (2006), Ivakhnenko & Potter (2004) and Ergin & Yarulin (1979). These studies showed that the mass magnetic susceptibility of the crude oils varied from -0.942 to -1.042 (10^{-8} m^3/kg). Analysis of the crude oil components, which are plotted in Figure 1, showed that the most diamagnetic compounds are the alkanes, and the least diamagnetic hydrocarbon compounds are aromatic hydrocarbons such as benzol with its homologues, naphtheno-aromatic and polycyclic aromatic hydrocarbons. Cyclopentane and cyclohexane hydrocarbons populate intermediate positions between the most and the least diamagnetic hydrocarbon compounds. Alkanes with cyclopentanes and cyclohexanes ranged in value from about -1.00 to -1.13 (10^{-8} m^3 kg^{-1}). Aromatic hydrocarbons have a magnetic susceptibility range of -0.85 to -0.97 (10^{-8} m^3 kg^{-1}). In contrast, the oxygen and nitrogen compounds were significantly less diamagnetic. In general, the sulphur crude oil components exhibit a mass susceptibility range similar to that of the benzol homologues.

Also, the ferrimagnetic nature of naturally altered hydrocarbons has been observed by a few authors. The results of a study by McCabe et al (1987) show an association of secondary magnetite with biodegraded liquid crude oil at two localities from the Thornton Quarry (Illinois) and the Cynthia Quarry (Mississippi). The final product of microbial biodegradation, solid bitumen, has been found to be strongly magnetic. Other research also confirms the magnetic nature of gilsonite and hydrocarbon fluid inclusions within crystals of calcite in speleothems (Elmore et al., 1987), and biodegraded hydrocarbons in Venezuela (Aldana et al., 1996). The spatial connection of the hydrocarbon fluids in reservoirs with ferrimagnetic magnetite concentration has been occasionally reported by Gold (1990, 1991).

In this Chapter we detail a systematic study of the mass magnetic susceptibilities of natural crude reservoir oil. These included crude oils from various oil provinces worldwide.

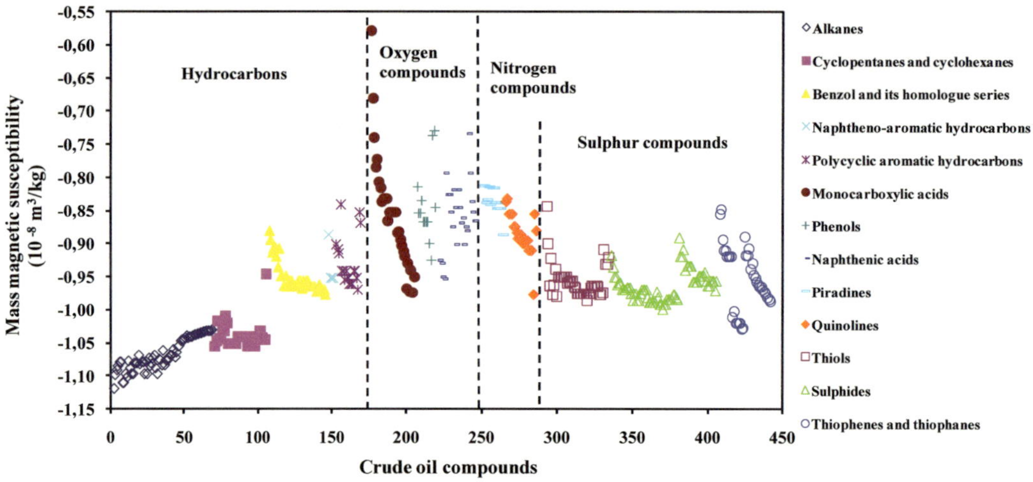

Fig. 1. Mass magnetic susceptibility of crude oil compounds, based on the data of Ergin and Yarulin (1979).

2. Magnetic susceptibility

All substances, solids and fluids, are affected by the application of a magnetic field. The direction of magnetic force lines changes in the presence of different types of magnetic substances. Diamagnetic substances cause a reduction in the density of magnetic force lines making the applied field weaker. In contrast paramagnetic and ferrimagnetic substances cause an increase in the density of magnetic force lines making the applied field stronger. The analogue of such a change of magnetic field inside a substance, named magnetic induction B, is described as follows in the SI system:

$$B = \mu_0(H+M) \quad (1),$$

where H is the magnetic field strength, μ_0 is the permeability of free space, and M is the intensity of magnetisation of a substance, related to the unit of volume. The μ_0 is equal to 4π 10^{-7} Henry/m. It is convenient to have the parameters in equation (1) independent of magnetic field strength. Thus dividing equation (1) by H we have:

$$\mu = B/H = \mu_0+\mu_0 M/H = \mu_0+\mu_0\chi_v \quad (2)$$

In this equation μ is the relative permeability, which is proportional to the dimensionless coefficient the volume magnetic susceptibility (χ_v). The volume magnetic susceptibility is a measure of how magnetisable a substance can become in the presence of a magnetic field (Equation 3).

$$\chi_v = M/H \text{ [dimensionless]} \quad (3)$$

Magnetic susceptibility is one of the most informative fundamental magnetic parameters (Ivakhnenko, 1999). Besides the volume susceptibility, there exists specific or mass magnetic susceptibility, χ_m,

$$\chi_m = Mm/H = \chi_v/\rho \quad [10^{-8}\ m^3\ kg^{-1}] \tag{4}$$

measured in $m^3\ kg^{-1}$. Mass magnetic susceptibility is defined as the ratio of the mass magnetisation (M_m) to the magnetic field (H) or as the volume magnetic susceptibility (χ_v) divided by the density (ρ) of the substance (Equation 4).

3. Experimental measurements and procedures

Hydrocarbon reservoirs contain a variety of naturally occurring fluids (heavy and light oils, gas condensates, formation waters). Drilling and injecting operations will mean that the formation will also contain drilling mud and other injected fluids. The results in this Chapter consider only the main types of natural reservoir fluids in liquid form: crude oils and formation waters. A suite of representative samples of fresh crude oil were collected mainly from different oil fields from world oil provinces such as the Middle East, North America, Europe, the Far East and Russia. The samples of crude were chosen with a range of distinctive physical and chemical differences. The fluids were kept in their sealed containers until a few days before the measurements when they were poured into glass sealed tubes.

We also studied formation waters and a sample of sea water, which was pumped through the injected wells into the reservoirs, and a sample of distilled water for comparison. All the fluid samples were supplied free of solid (sand, clay, parts of metals, and carbonate sediment) or other fluid contamination.

For magnetic susceptibility measurements were used very sensitive measuring equipment comprising a Sherwood magnetic susceptibility balance (MSB). The MSB or Evans magnetic balance is designed as a reverse traditional Gouy magnetic balance. The Evans method uses the same configuration as the Gouy method except that instead of measuring the force with which a magnet exerts on the sample, the equal and opposite force which the sample exerts on a moving permanent magnet is measured. The magnets are at the end of a beam, and when the sample is placed in one of the magnets, the beam is not in equilibrium. A current is applied to the other magnet until the beam is back in equilibrium, and by measuring the current it is possible to measure the magnetic susceptibility.

The calibration of the MSB was made using distilled water, produced in the presence of air. A value of -0.9043 ($10^{-8}\ m^3\ kg^{-1}$) for the mass magnetic susceptibility of water at 20 °C (Selwood, 1956) was used for the calibration. Repeat calibration measurements were regularly made throughout the measurement period. The values of χ_m for the studied fluids were determined at room temperature (normally about 18-20 °C), and corrected for the displaced air in the measuring tube. The volume magnetic susceptibility of the air displaced in the measuring tube has been taken as 0.364 (10^{-6} SI) due to 20.9% paramagnetic oxygen contribution at 20 °C temperature and atmospheric pressure.

4. Mass magnetic susceptibility of crude oils

The analysis of the mass magnetic susceptibilities of the natural reservoir fluids shows that the mass magnetic susceptibilities of all the studied fluids are diamagnetic. In Figure 2 shown the measurements of the reservoir fluids from world-wide locations made on the Sherwood Scientific MSB balance. It shows that the mass magnetic susceptibility of crude oils is changing from a value of -0.9634 (Russia) to -1.0401 (the Far East) (10^{-8} m^3 kg^{-1}). The mass magnetic susceptibility of this range of crude oils differs by no more than 7.37% in their magnetic susceptibility values. The formation waters from the North Sea oil province exhibit a distinctly different range from -0.8729 to -0.8862 (10^{-8} m^3 kg^{-1}), differing by only 1.5%.

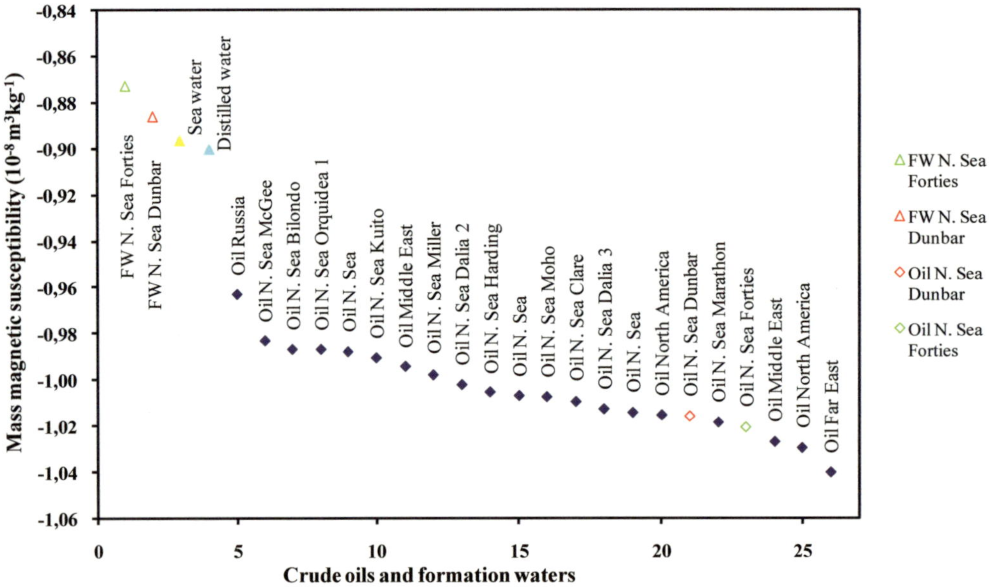

Fig. 2. Mass magnetic susceptibility of crude oils determined using a Sherwood MSB balance. The measurement errors are of the order of ±0.0035 (10^{-8} m^3 kg^{-1}), close to the size of the symbols.

As we can see from Figure 2 the formation waters are distinctly different from the crude oils (around 12%) in terms of their average mass magnetic susceptibilities. This difference is exemplified by the Dunbar and Forties results, where there are clear differences between the mass magnetic susceptibility values for crude oil and formation water from the same oilfield. The Forties formation water differs (less diamagnetic) from the Forties crude oil by around 14%, while the Dunbar formation water differs (again less diamagnetic) from the Dunbar crude oil by around 13%. This demonstrates that there is a real difference between the mass magnetic susceptibility of the crude oils and the formation waters, which may have been less clear-cut had we only measured crude oil from one site and compared it with formation water from another site.

Differences in the mass susceptibility values for the different water samples may be related to the solutes they contain. The studied formation waters contain significant amounts of sodium chloride (NaCl), calcium chloride hexahydrate ($CaCl_2 \cdot 6H_2O$), magnesium chloride monohydrate ($MgCl_2 \cdot H_2O$), potassium chloride (KCl), barium chloride ($BaCl_2$) and strontium chloride ($SrCl_2$). The injected sea water sample also contains sodium sulphate

(Na_2SO_4) and is characterised by the absence of the barium and strontium chlorides. The higher concentration of solutes in Forties formation water causes it to have a less negative susceptibility (-0.8729, 10^{-8} m^3 kg^{-1}) than Dunbar formation water (-0.8862, 10^{-8} m^3 kg^{-1}). This observation is also consistent with the lower value of the mass magnetic susceptibility of sea water, which contains the lowest quantity of sodium chlorite and calcium chloride hexahydrate among the studied waters. This is also consistent with the fact that the mass susceptibility of all formation and sea water solutes (Figure 3) was found to be less negative than susceptibility of distilled water at 20 °C. The mass susceptibility of the distilled water at 20 °C is taken as a reference point of water based types of reservoir fluid. The presence of NaCl in the formation waters tends to decrease the water diamagnetism. Pure sodium chloride has a susceptibility value of -0.6451 (10^{-8} m^3 kg^{-1}).

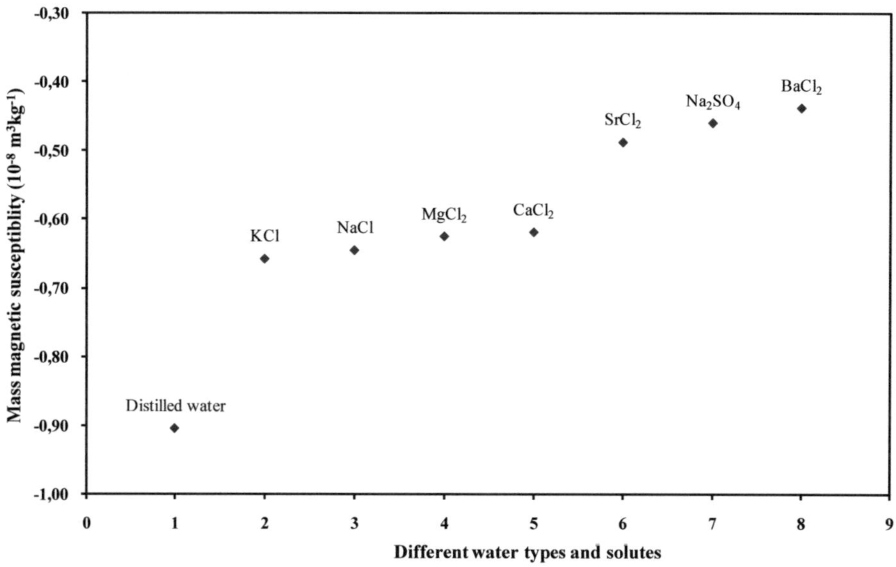

Fig. 3. Mass magnetic susceptibility of distilled water and formation water solutes.

It seems clear from Figures 2 that there are variations between the different crude oil samples, and these are related to their physical and chemical properties as described below.

4.1 Mass magnetic susceptibility and physical properties of reservoir crude oils

Mass magnetic susceptibility exhibits a good correlation with the density of the natural reservoir fluids. Figure 4 shows a plot of mass magnetic susceptibility versus density for the crude oils, formation waters, and other water samples. There is a trend of higher density corresponding to less negative mass magnetic susceptibilities, with a clear difference between the oils and the formation waters. The r^2 correlation coefficient of the density and mass susceptibility of the crude oils from my study (North Sea, North America, Russia, the Middle and Far East) shown in Figure 4 is 0.78.

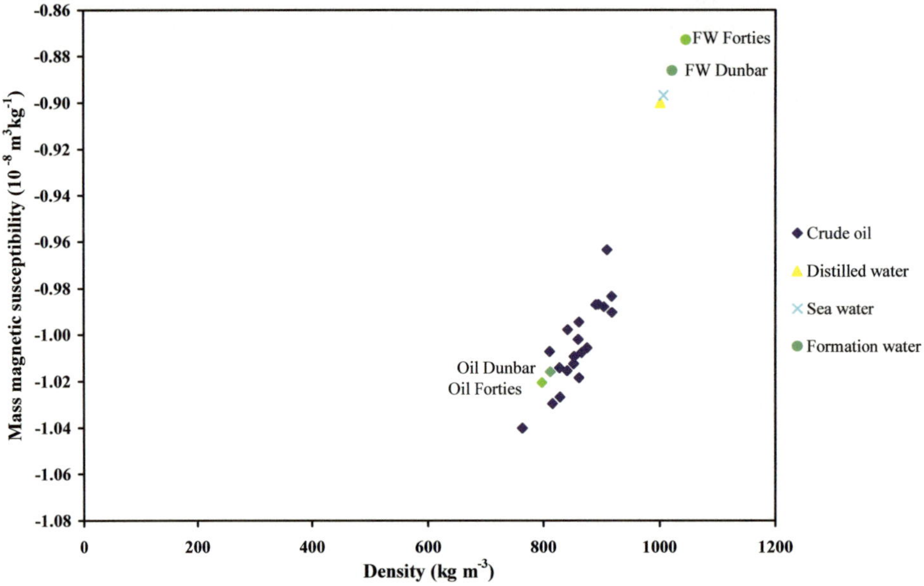

Fig. 4. Mass magnetic susceptibility versus density of crude oils and formation waters from this study. The mass susceptibility was determined using Sherwood MSB balance.

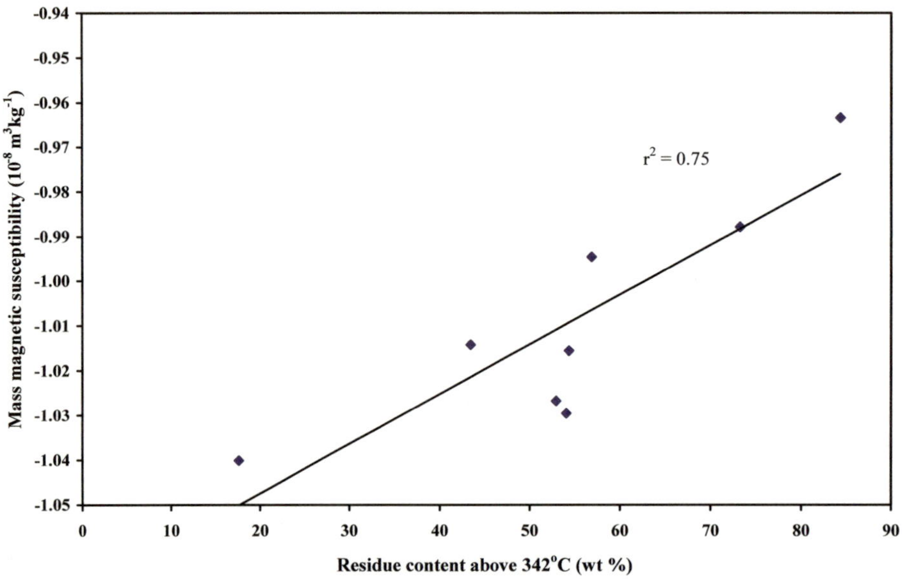

Fig. 5. Mass magnetic susceptibility versus residue content above 342 °C of crude oil samples. The measurement errors are of the order of ±0.0035 (10^{-8} m^3 kg^{-1}), close to the size of the symbols.

The mass magnetic susceptibility is strongly correlated with the content of the crude residue fraction. Figure 5 shows the residue content above 342 °C versus mass magnetic susceptibility for the crude oils for which we had compositional data. The residue is what remains after fractional distillation of the lighter hydrocarbon components. It is evident that the higher the residue content the higher is the mass magnetic susceptibility. The correlation coefficient r^2=0.75. The density of crude oils is primarily dependent on the residue content in the crude oils (Figure 6). The r^2 correlation coefficient between residue content above 342 °C and the density of crude oils is very high at 0.91. Therefore, the samples with higher residue content are also the samples with higher density, so the trend given in Figure 5 is consistent with the mass susceptibility versus density results in Figure 4.

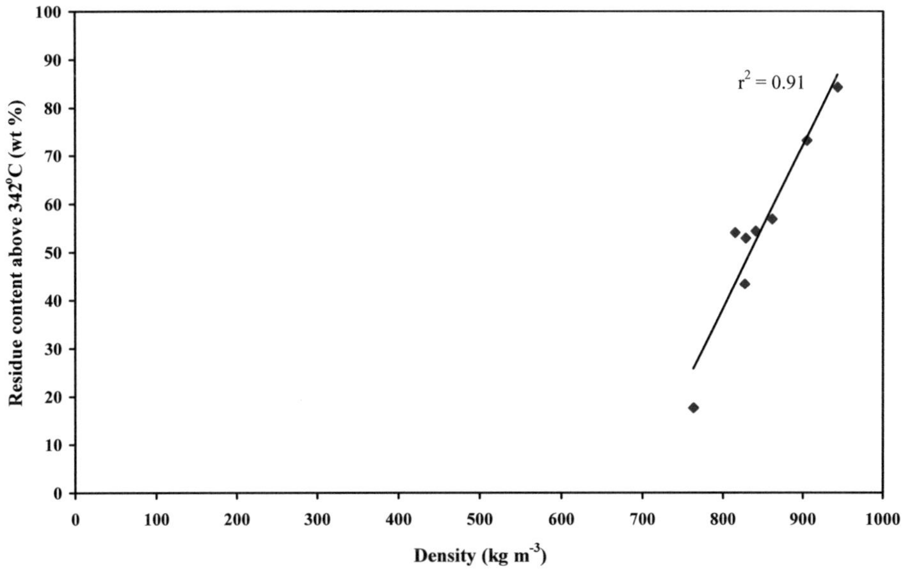

Fig. 6. Density versus relative residue content above 342 °C of crude oil.

The mass magnetic susceptibility versus stock tank oil gravity is given in Figure 7 for the crude oil samples. Stock tank oil is oil as it exists at atmospheric conditions in a stock tank (it tends to lack much of the dissolved gas present at reservoir temperatures and pressures). The gravity is expressed as API degrees as follows:

$$°API = [141.5 / S_o] - 131.5 \quad (5),$$

where S_o is the stock tank oil specific gravity, or relative density, to water at 298 K, and API is an acronym for American Petroleum Institute. For a value of 10° API, S_o is 1.0, the specific gravity of water. Figure 7 shows that there is a distinct trend of decreasing mass magnetic susceptibility with increasing gravity. The linear regression coefficient r^2 is 0.72. This trend is consistent with the expected trend on the basis of the susceptibility versus density results.

The values of the API gravity form the common basis of the oil classification system (Table 1). In general I found that the mass susceptibility and API gravity are related via the

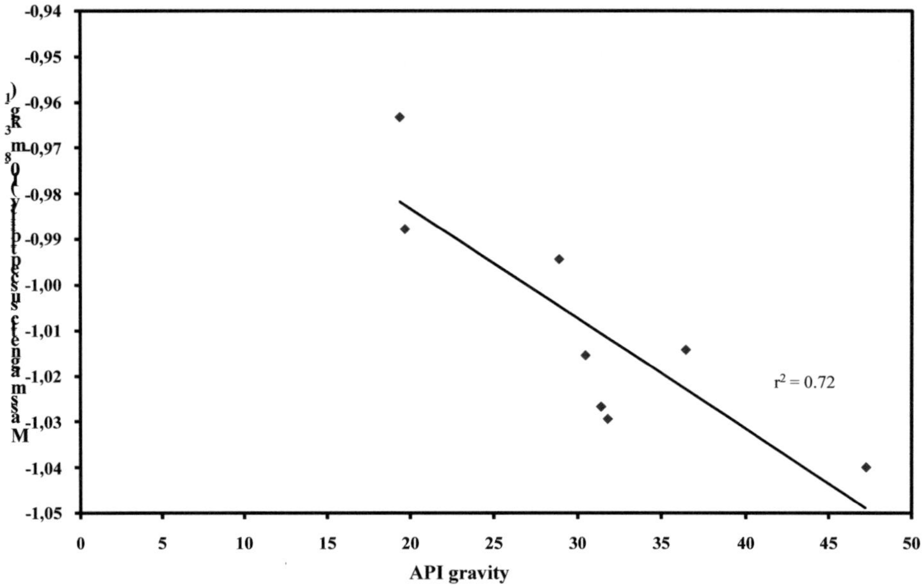

Fig. 7. Mass magnetic susceptibility versus stock tank oil gravity of crude oil samples.

empirical equation (1) below. Therefore, it appears possible use mass magnetic susceptibility for oil type classification.

$$\chi_m \cdot 10^{-8} = -0.0024 \text{ API} - 0.94 \tag{6}$$

Table 2 shows in general that tar with API gravity less than 10° is characterized by mass susceptibility values higher than -0.9592 (10^{-8} m^3 kg^{-1}). Heavy oils correspond to susceptibility in the range -0.9592 to -0.9952 (10^{-8} m^3 kg^{-1}). "Black" low shrinkage and "volatile" high shrinkage oils are characterized by mass susceptibility ranges of -1.0072 to -1.0312 and -1.0312 to -1.0552 (10^{-8} m^3 kg^{-1}) respectively.

Reservoir fluid	Surface appearance	API gravity	Typical composition (mole %)					
			C_1	C_2	C_3	C_4	C_5	C_6
Tar	Black substance	<10°	-	-	-	-	-	90+
Heavy oil	Black viscous liquid	10°-25°	20	3	2	2	12	71
"Black" low shrinkage oil	Dark brown to black viscous liquid	30°-40°	49	2.8	1.9	1.6	1.2	43.5
"Volatile" high shrinkage oil	Brown liquid - various yellow, red or green hues	40°-50°	64	7.5	4.7	4.1	3	16.7

Table 1. API gravity ranges of various oil types and their typical mole composition.

Reservoir fluid	API gravity	Mass magnetic susceptibility (10^{-8} m^3 kg^{-1})
Tar	<10°	>-0.9592
Heavy oil	10°-25°	-0.9592 - -0.9952
"Black" low shrinkage oil	30°-40°	-1.0072 - -1.0312
"Volatile" high shrinkage oil	40°-50°	-1.0312 - -1.0552

Table 2. Mass magnetic susceptibility ranges of oil types.

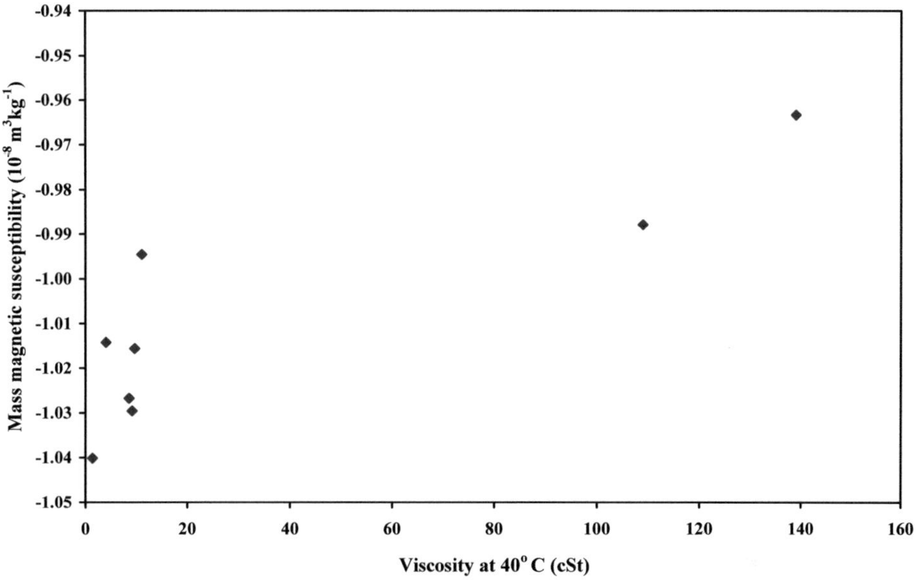

Fig. 8. Mass magnetic susceptibility versus viscosity at 40 °C of crude oil samples.

Figure 8 shows results for the mass magnetic susceptibility versus the viscosity at a temperature of 40 °C for crude oils. This also showed a trend of increasing mass susceptibility with increasing viscosity. I have omitted the linear regression line, where $r^2=0.73$, since it is fairly meaningless given that there appear to be two clusters, and the correlation may be non-linear. The broad trend we observed might be expected since the samples with higher viscosity are also the ones with higher density, which gave higher (less negative) values of magnetic susceptibility.

4.2 Mass magnetic susceptibility and concentration of sulphur and other chemical compounds

Mineral compounds in crude oils consist of complex organometallic compounds, salt solutions of organic acids and also colloidal mineral substances. At present more than sixty different chemical elements have been found in crude oils, such as Ni, V, Ca, Fe, Si, Ln, Cu, Al, Mg, Na, Sn, Ti, Sr, Pb, Co, Ag, Mn, Cd, As, Hg, etc. Vanadium and nickel exhibit relatively high concentrations of 10^{-3} to 10^{-6}% in oil samples around the world. These two elements usually are concentrated in the asphalt-resin fraction of crude oil.

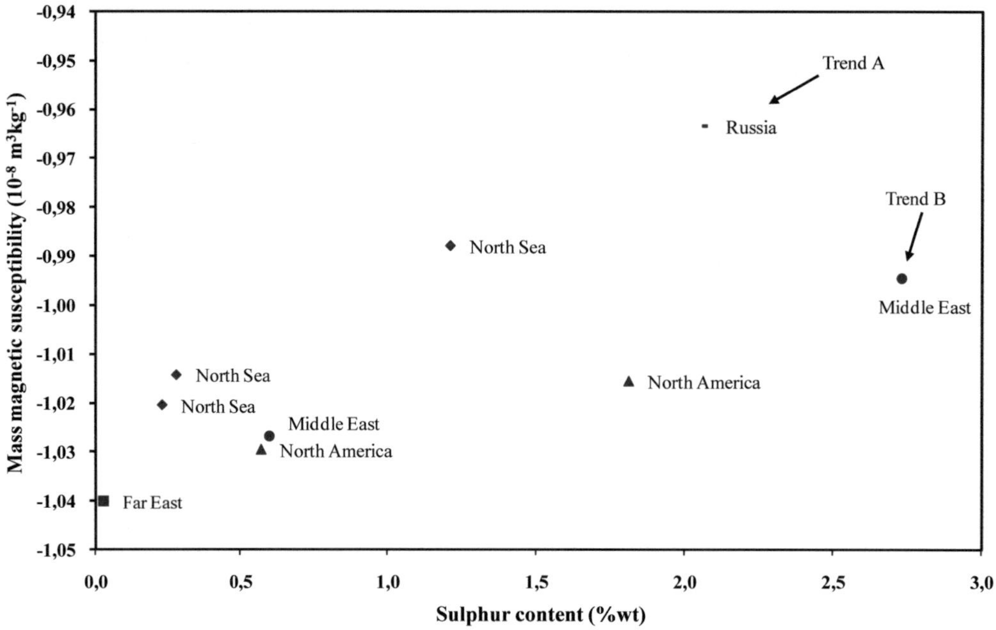

Fig. 9. Mass magnetic susceptibility versus sulphur content of crude oil samples.

It might therefore be expected that the mass magnetic susceptibility of crude oils will reflect their chemical composition. Figure 9 shows the mass magnetic susceptibility versus sulphur content for the crude oils for which I had compositional data. In general, a higher sulphur content corresponds to a higher (less negative) mass susceptibility. There is a suggestion of possibly two trends: one (trend A) including the Russian and North Sea samples, and the other (trend B) containing the North American and Middle East samples. The Russian and uppermost North Sea sample have higher residue concentrations and higher densities than the North American and Middle East samples. Trend A gave a correlation coefficient $r^2=0.95$, and for trend B $r^2=0.97$. Higher sulphur content also generally corresponds to higher residue content and density within each of the two trending groups.

The carbon residue content and mass magnetic susceptibility show a good correlation (Figure 10) with a linear regression coefficient $r^2=0.72$.

The relationship between metal content (usually in the form of organometallic compounds) in crude oils and mass magnetic susceptibility appears to yield certain correlations. The higher the metal content the higher (less negative) the susceptibility. For instance, mass magnetic susceptibility and vanadium content exhibit a linear coefficient of correlation $r^2=0.78$ (Figure 11). Lead content and mass magnetic susceptibility are correlated with $r^2=0.75$ (Figure 12). The relationships of mass magnetic susceptibility with nickel and cadmium is characterised by similar linear regressions. Nickel displays a slightly higher $r^2=0.69$ (Figure 13) than cadmium, where $r^2=0.63$ (Figure 14). The correspondence of mass susceptibility and iron content are consistent with the trend of other metals. However, Figure 15 shows that the linear correlation coefficient is very low with $r^2=0.37$.

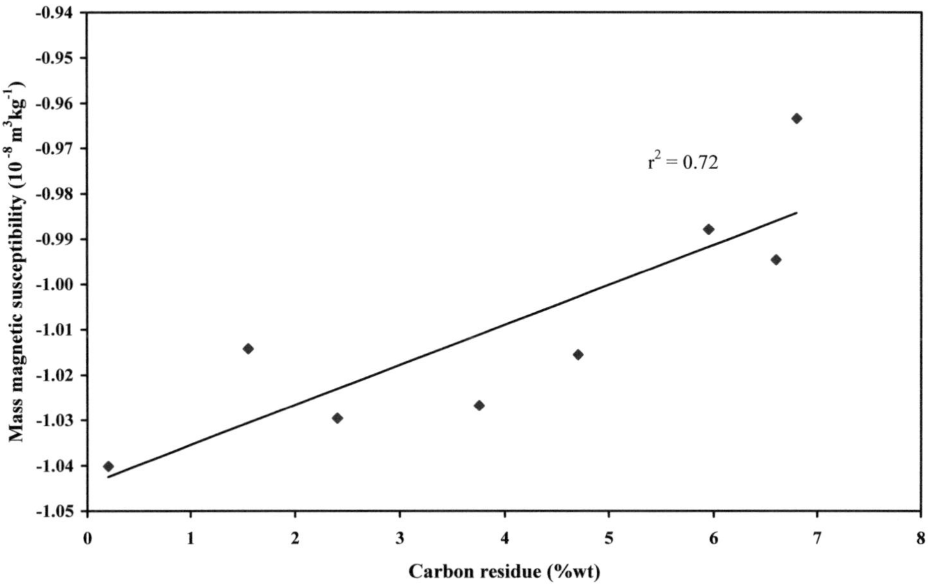

Fig. 10. Mass magnetic susceptibility versus carbon residue content of crude oil samples.

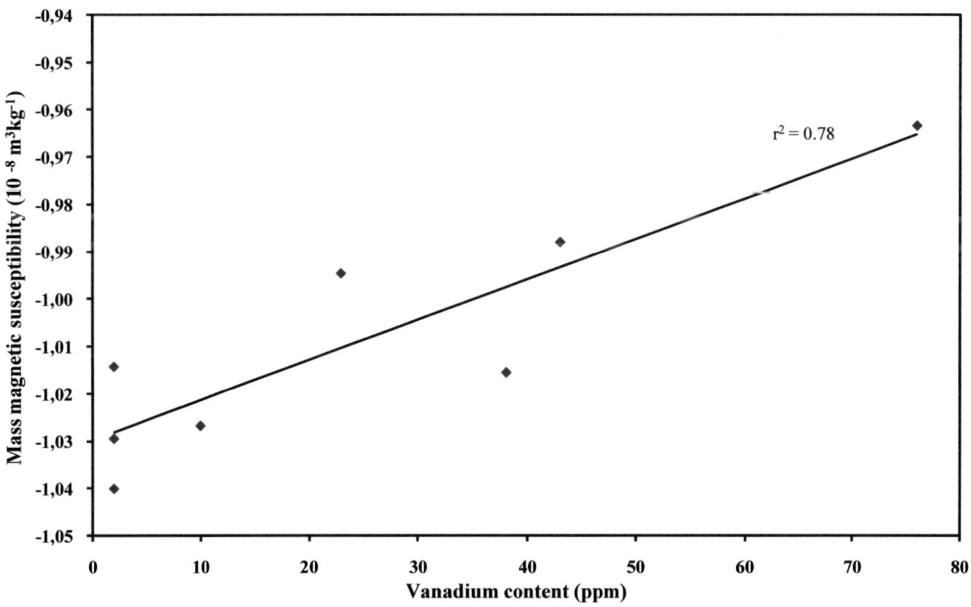

Fig. 11. Mass magnetic susceptibility versus vanadium content of crude oils.

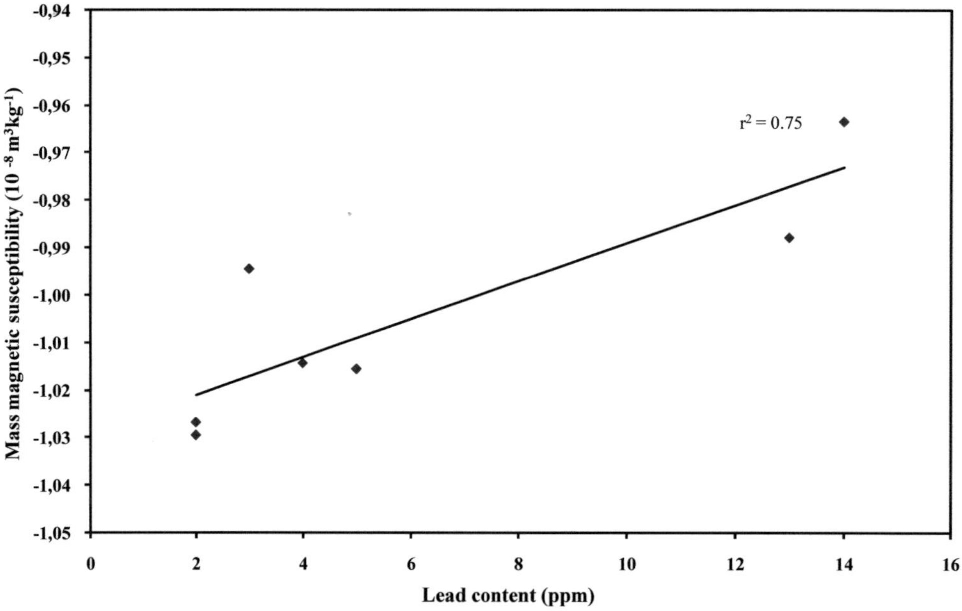

Fig. 12. Mass magnetic susceptibility versus lead content s of crude oils.

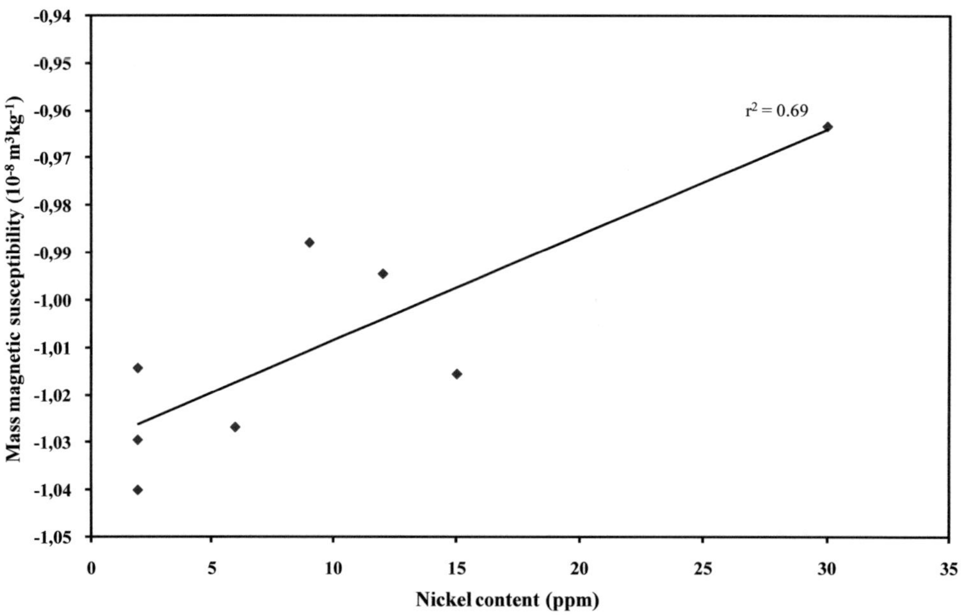

Fig. 13. Mass magnetic susceptibility versus nickel content of crude oils.

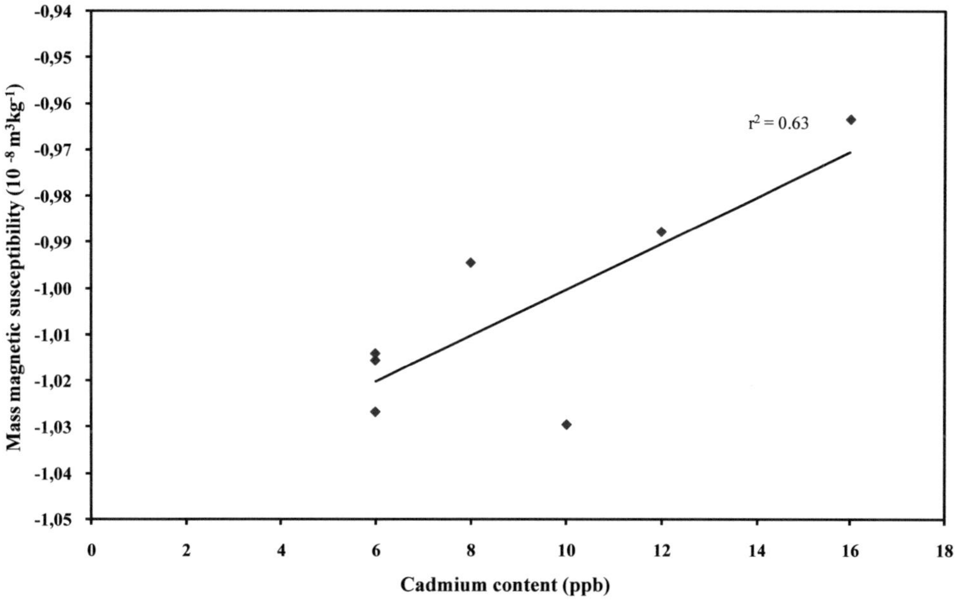

Fig. 14. Mass magnetic susceptibility versus cadmium content of crude oils.

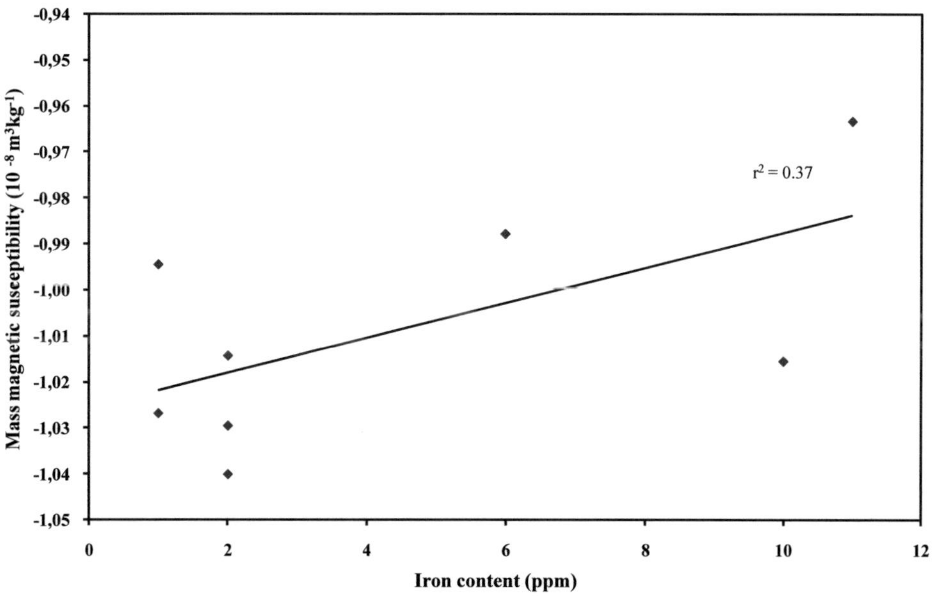

Fig. 15. Mass magnetic susceptibility versus iron content of crude oils.

The results show that in each case there appears to be a trend of higher mass magnetic susceptibility with increasing metal content. This trend might ordinarily be expected. However, the results should be treated with some caution as it was noticed that samples with higher metal content also had higher density, which also corresponds to higher mass susceptibility. The relative roles of the metal content versus the intrinsic fluid density are presently unclear. It seems that crude oil samples with higher density have higher residue content and that these contain greater amounts of organometallic compounds (Ivakhnenko & Potter, 2004). If the metal content was due to elemental metal, then trace amounts would have a significant effect on the susceptibility. For instance, just 10 ppm by weight of ferromagnetic elemental iron would increase the mass susceptibility of the sample by about 0.7 (10^{-8} m^3 kg^{-1}). In reality the metals are likely to be components in organometallic compounds, which would have substantially lower intrinsic values of magnetic susceptibility, and without knowing the exact composition of these compounds their precise influence on the magnetic susceptibility of the crude oils remains uncertain.

4.3 Regional characteristics of the mass magnetic susceptibility of crude oils

Any magnetic differences between crude oils of different regions may reflect specific features of the geological and geochemical history of the oil provinces, and might provide some support to the suggestion by Ergin and Yarulin (1979) that crude oils from different provinces might be distinguished on the basis of their magnetic susceptibility.

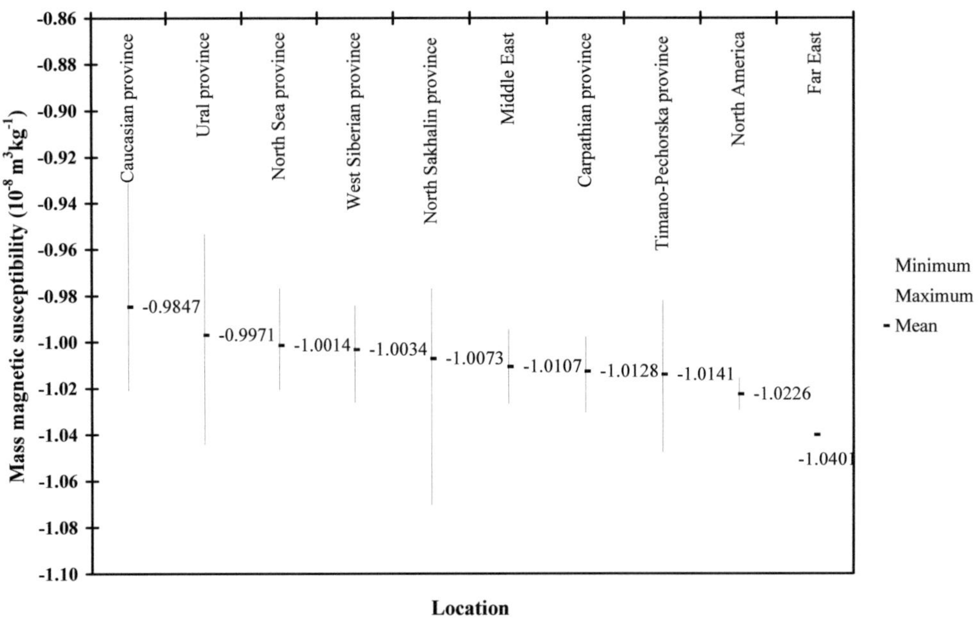

Fig. 16. Average, minimum and maximum values of mass magnetic susceptibility in relation to specific oil provinces.

Our analysis of the mass magnetic susceptibility data for crude oils from different oil provinces (the North Sea, Caucasian, Ural, Timano-Pechorska, West Siberian, Middle East, North America, Far East, and Carpathian provinces) seems to show some differences between the various oil provinces (Figure 16). The crude oils of the Caucasian province demonstrate the highest average value of mass susceptibility (-0.9847, 10^{-8} m^3 kg^{-1}), whereas the mass susceptibility of oils from North America and the Far East appear to be the most diamagnetic with -1.0226 and -1.0401 (10^{-8} m^3 kg^{-1}) respectively. However, some of the oil provinces are represented by few samples. Clearly more samples need to be studid in order to confirm any broad consistent differences between the various oil provinces.

4.4 Comparison between the mass magnetic susceptibility of reservoir fluids and petroleum reservoir minerals

It may be useful in terms of rock-fluid interactions to compare the mass magnetic susceptibilities of crude oils and formation waters in relation to some typical petroleum reservoir minerals, such as the diamagnetic matrix minerals and the paramagnetic clays. The comparisons are shown in Figures 17 and 18. The reservoir mineral data comes from previously published experimental results. The average values of the mass magnetic susceptibilities of the reservoir fluids are comparable to the diamagnetic susceptibilities of the main matrix minerals (quartz, calcite; Figure 17). However, they are distinctly different from the higher positive values of the paramagnetic permeability controlling clays such as illite (Figure 18). The magnetic properties of reservoir rocks and minerals are described in more details by Ivakhnenko (2006). The experimental values for the minerals were taken from (Borradaile et al., 1990; Hunt et al., 1995; Thompson & Oldfield, 1986).

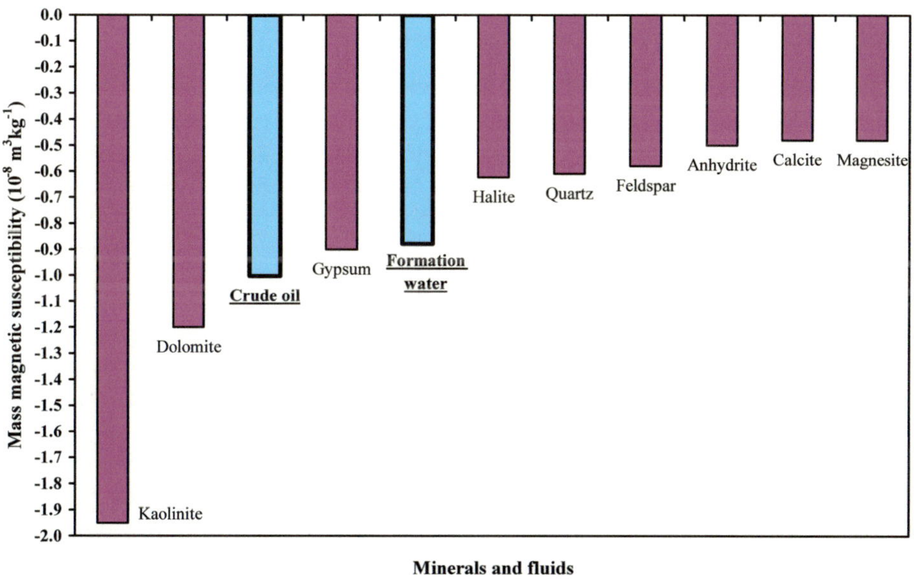

Fig. 17. Mass magnetic susceptibility of typical reservoir diamagnetic minerals in relation to average crude oil and formation water values.

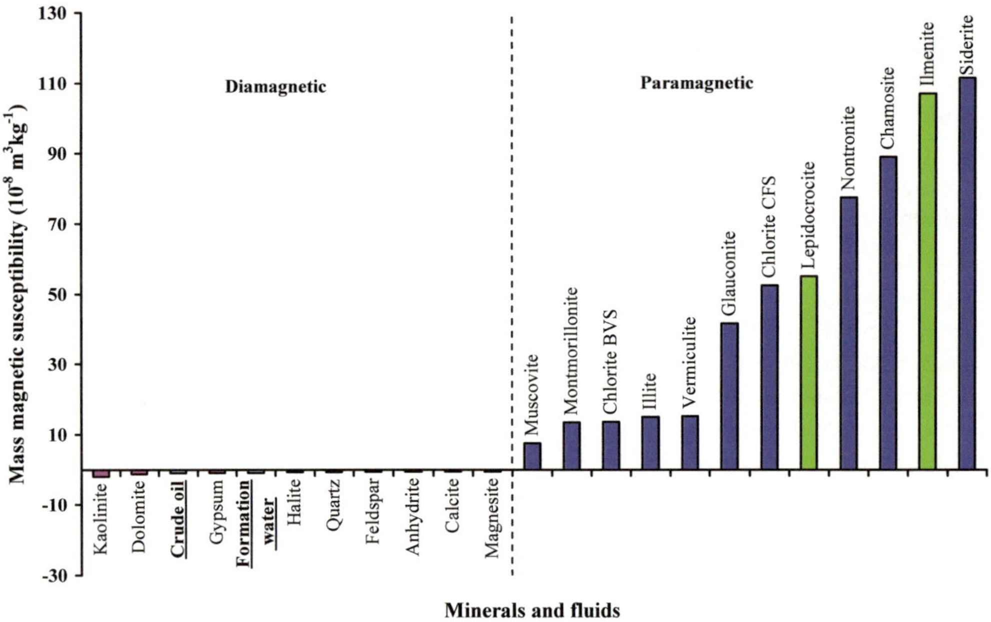

Fig. 18. A comparison of the mass magnetic susceptibility of typical reservoir diamagnetic and paramagnetic minerals in relation to average crude oil and formation water values.

5. Potential applications for petroleum reservoirs

Magnetic susceptibility measurements might find a use in passive sensors in reservoirs for distinguishing between formation waters and crude oils (for example, in helping to determine the onset of water breakthrough). Such sensors would provide an environmentally friendly alternative to radioactive tracers. Although viscosity meters might also distinguish between formation waters and crude oils, the magnetic sensors have a further advantage in that they could also rapidly detect small concentrations of ferrimagnetic minerals or migrating fines, such as paramagnetic clays.

Since mass magnetic susceptibilities of the natural reservoir fluids are more negative than the majority of the diamagnetic matrix reservoir minerals such as quartz, feldspar and calcite, and at the same time the values are significantly less diamagnetic than the clay kaolinite, it appears that magnetic properties may possibly play some role in rock-fluid interactions. The relative magnetic forces between quartz and formation water and between quartz and crude oil, in the Earth's field, might be a factor in determining the wettability (water wet or oil wet) of the reservoir rock. For reservoir rocks containing significant amounts of paramagnetic clays, such as illite, the relative magnetic roles of formation water and crude oil could be reversed (compared to the quartz case) according to Figures 17 and 18, possibly influencing the changes in wettability that one often observes between clean sandstone (quartz rich with little clay) and muddy sandstones containing paramagnetic clays. Recent work has shown links between nuclear magnetic resonance (NMR) and wettability (Guan et al, 2002) and so a link between magnetic susceptibility and wettability may also be a possibility.

The magnetic susceptibility of reservoir fluids appears to be important parameter for rapidly characterising the following:

- Types of reservoir fluids: formation water and crude oil.
- Quantitatively distinguishing different crude oils. Further work may show that this is better than other methods, such as refractive index or fluorescence.
- Major physical properties.
- Fractional and compositional constitution of crude oils, and content of chemical elements, such as metal content.
- The solute (or anion and cation) composition in the water based types of fluids (formation waters, injected waters).
- Detecting the presence of fines and contaminants.
- Detecting changes in fluids in-situ in the presence of the background reservoir rock matrix signal.

6. Conclusions

The following conclusions can be drawn from this Chapter:

- There were distinct differences between the mass magnetic susceptibilities (χ_m) of crude oils and formation waters. All the samples studied were diamagnetic, but the values for the crude oils were more negative.
- The values of χ_m for the crude oils and formation waters correlated with their densities. The values for crude oil also correlated with other physical properties, namely residue content (and group hydrocarbon content), stock tank oil gravity, and viscosity. The results suggest that the magnetic measurements could potentially be used to rapidly characterise the physical differences between various petroleum reservoir fluids.
- The values of χ_m for the crude oils also showed correlations with trace amounts of chemical components, namely the contents of sulphur, vanadium, lead, cadmium, nickel and iron. The results, however, should be treated with some caution, since the samples with higher contents of these elements also generally have higher density, which also correlates with the mass susceptibility. It appears that crude oils with higher density have higher residue content, and also contain higher concentrations of the above components. The relative contributions of intrinsic fluid density and these trace components to the total magnetic susceptibility signal is presently unclear. The values of χ_m for the formation waters were related to their solute composition.
- There is suggestion from the results that crude oils from different world oil provinces might be distinguished on the basis of their magnetic susceptibility. However, there are significant ranges and overlaps between the results for some provinces, and more samples need to be measured before consistent differences can be confirmed.
- The mass magnetic susceptibilities of the reservoir fluids are much closer in value to the main diamagnetic matrix minerals (for example quartz and calcite) than the paramagnetic permeability controlling clay minerals.

7. Acknowledgment

Author is grateful to D. Potter for useful discussions.

8. References

Aldana, M., Costanzo-Alvarez, V., Vitiello, D., Diaz, M., & Silva, P. (1996). Causal relationship between biodegradation of hydrocarbons and magnetic susceptibility in samples from La Victoria oil field (Venezuela). *AGU Fall Meetings*, San Francisco. Supplement Vol. 77 (46), F163.

Borradaile, G. J., MacKenzie, A., & Jensen E. (1990). Silicate versus trace mineral susceptibility in metamorphic rocks: *Journal of Geophysical Research - Solid Earth*, 95, pp. 8447-8451.

Elmore, R. D., Engel, M. H., Crawford, L., Nick, K., Imbus, S., & Sofer, Z. (1987). Evidence a relationship between hydrocarbons and authigenic magnetite. *Nature*, 325, pp. 428-430.

Ergin, Y. V., & Yarulin, K. S. (1979). *Magnetic properties of oils*. Nauka., Moscow, USSR, pp. 1-200.

Gold, T. (1990). Discussion of asphaltene-like material in Siljan ring well suggests mineralized altered drilling fluid. *JPT*. SPE Publication 20322, 269.

Gold, T. (1991). Sweden's Siljan ring well evaluated. *Oil and Gas Journal*, Vol.89, No.2, pp. 76-78.

Guan, H., Brougham, D., Sorbie, K. S., & Packer, K. J. (2002). Wettability effects in a sandstone reservoir and outcrop cores from NMR relaxation time distributions. *Journal of Petroleum Science and Engineering*, Vol. 34, pp. 33-52.

Hunt, C. P., Moskowitz, B. M., & Banerjee, S.K. (1995). Magnetic properties of rocks and minerals, In: *Rock Physics and Phase Relations: a Handbook of Physical Constants*, T.J. Ahrens, (Ed.): American Geophysical Union reference, Shelf 3, pp. 189 – 204.

Ivakhnenko, O. P. (1999). Mineralogical aspects of magnetic-mineralogical model. *Journal of Geology*, 16, Kiev, pp. 24-32.

Ivakhnenko, O. P., 2006. Magnetic analysis of petroleum reservoir fluids, matrix mineral assemblages and fluid-rock interactions. Thesis, Institute of Petroleum Engineering, Edinburgh, UK, 210 pp.

Ivakhnenko, O. P. & Potter, D. K. (2004). Magnetic susceptibility of petroleum reservoir fluids. *Physics and Chemistry of the Earth*, Vol.29, No.13-14, pp. 899-907.

McCabe, C., Sassen, R., & Saffer, B. (1987). Occurrence of secondary magnetite within biodegraded oil. *Geology*, Vol.15, No.1, pp. 7-10.

Selwood, P.W. (1956). *Magnetochemistry*, Second edition. Interscience Publishers, New York, pp. 1-435.

Thompson, R. & Oldfield, F. (1986). *Environmental Magnetism*. Allen & Unwin, London, pp. 1-277.

5

Environmental Bases on the Exploitation of Crude Oil in Mexico

Dinora Vázquez-Luna
Colegio de Postgraduados
México

1. Introduction

Oil is one of the most important energy resources for the global economy. Regarding this, Mexico has an economic dependency of the oil industry that has being going on for decades. Based on the current oil production, the proven and possible reserves might last for ten years; still, if we consider probable and possible oil reserves, Mexico could count with an oil production for more than 30 years. Under this consideration, Mexico requires state of the art technologies for exploring, exploiting and processing crude oil.

Nowadays the technological and industrial development has to be environmentally responsible in accordance with the global needs. That is why the creation of new technologies and resources exploitation must be based on a responsible energetic development. Regarding this matter, being environmental friendly is a main goal for the society (Rodríguez *et al.*, 2009), and gets to our attention that Mexico -until the seventies- did not apply any environmental criteria while exercising its oil activities (Ortínez *et al.*, 2003). In consequence, oil production, leading, transportation, storage and processing had a negative impact on soils (Trujillo *et al.*, 1995; Rivera-Cruz & Trujillo-Narcía, 2004), waters (Adams *et al.*, 1999), and ecosystems at the southeast of the country (Santos *et al.*, 2011).

In Mexico there are environmental regulations about hydrocarbons pollution, although these bypass criteria about the hydrocarbons' chronic effects on the ecosystems. However, in accordance with the current development conditions, it is necessary to introduce the *environmental basis for the oil exploitation*, whose main goal is to analyze the effects of the traditional oil related activities. The aim is to lay down the foundations for the creation of new technologies that contribute to a responsible and affordable energetic development for the country.

2. Environmental effects of the oil exploitation in Mexico

The world's economic sustenance, as based on the oil industry, has originated serious environmental issues (Hall *et al.*, 2003). In Mexico, the oil industry has worn down the Southeastern natural resources, thus altering properties of soils (Rivera-Cruz & Trujillo-Narcía, 2004), sub-soils (Iturbe *et al.*, 2007) and water (Ortiz *et al.*, 2005), as a consequence of problems related oil extraction, processing and transportation (George *et al.*, 2011).

Most soils affected by hydrocarbons are located in tropical zones with high rain precipitation, augmenting the pollutants' dispersion through mangroves or zones with deficient drainage (Gutiérrez & Zavala, 2002; Rivera-Cruz & Trujillo-Narcía, 2004; García-López *et al.*, 2006; Vega *et al.*, 2009). Additionally, this situation gets worse due to the age of the oil facilities, their lack of maintenance, as well as clandestine oil valves that have caused chronic spills of already weathered oil (Rivera-Cruz *et al.*, 2005) which contains compounds of high molecular weight, endangering the resource's sustainability. The affected states with the highest number of environmental emergencies occurred in Mexico are Veracruz, Campeche and Tabasco, representing 78.7% of the events related to PEMEX activities mainly because of its deteriorated pipelines, clandestine oil valves, corrosion and mechanical impacts (PEMEX, 2003; Olivera-Villaseñor & Rodríguez-Castellanos, 2005).

The water bodies and the coastal zone are also affected by wastes derived from oil exploring, offshore production, sea and submarine transportation, shipment and storage operations, accidents during operations such as submarine oil pipes cracks, tankers accidents, spill outs and explosions at oil rigs (García-Cuéllar *et al.*, 2004; Mei & Yin, 2009).

During the seventies, Mexico developed oil exploitation technology and intensified its crude oil production and transportation via underwater pipelines to cargo floating buoys to storage ports located at Tabasco, as well as the oil's transformation and refining at Coatzacoalcos, Veracruz, and Salina Cruz. As a consequence, this created industrial networks all over the country that raised pollution issues at coastal zones, thus impacting the ecosystems of the Gulf of Mexico and the Pacific southeast coast (Carbajal & Chavira, 1985; Botello, 1996; González-Lozano *et al.*, 2006; Salazar-Coria *et al.*, 2007). Nevertheless, conscience was not made until the IXTOC-I accident regarding the potential risks of the industry's activities (Jernelöv, 2010). Currently the main oil and gas production zone is located at the Gulf of Mexico, in the Campeche maritime zone where there are severely affected ecosystems after more than three decades of oil exploitation (García-Cuéllar *et al.*, 2004). The last Gulf of Mexico disaster was estimated three times that of the Valdez spill (Trevors & Saier, 2010).

On the other hand, in Mexico was installed the first and largest oil refinery (until 2004) in Latin America, dating from 1908; the environmental deterioration is evident after more than one hundred years of oil exploitation. The effects are observed at the lower Coatzacoalcos River (Toledo, 1995; González-Mille *et al.*, 2010) as it has suffered the impacts of the oil refining and transportation processes since the swamp areas surrounding the oil refinery are used as waste traps. In addition, accidents related to carelessness during the load and cleaning of the tankers, as well as the discharge of the cooling water from the Minatitlán refinery into the river have created a complex mixture of hazardous materials that pollute the lower Coatzacoalcos River area directly, affecting both the fishing resources and the inhabitants (Toledo, 1983; Rosales-Hoz. & Carranza-Edwards, 1998; Cruz-Orea *et al*, 2004; Ruelas-Inzunza *et al*, 2009; Ruelas-Inzunza *et al.*, 2011).

Concurrently, we must add the petrochemical activity to the oil exploring and refining processes (Rao *et al.*, 2007a). With the creation of petrochemical complexes since 1960, these activities increased the impact of the area conditioning where the industrial zones were located (Ortiz *et al.*, 2005). Additionally, huge amounts of materials were dredged in order to build the artificial dock of Pajaritos. The industrial plants operation, the numerous transportation networks (petrochemical ducts and pipelines) and the linking earth systems built at the lower areas caused other activities with an environmental impact for the zone (Toledo, 1995; Adams *et al.*, 2008). Another environmental effect of the intensity of the oil exploitation has been the loss of swamps, mangroves, and other elements of the water coastal systems that must be attended given its importance for the environmental services (Gutiérrez & Zavala, 2002; Bahena-Manjarrez *et al.*, 2002).

3. Crude oil toxicity

Toxicity is the ability of a chemical substance to damage and alter certain functions of the biological systems (Rivero *et al.*, 2001). There are two toxicity criteria for the natural systems. *Acute toxicity* is produced by large, short-termed, accidental polluting agents' discharges (Roth & Baltz, 2009), although it may have long term effects. On the other hand, *chronic toxicity* is expressed by effects noticed on the long term due to relatively small amounts of a toxic compound found on air, water or soil (Scarlett *et al.*, 2007).

Crude oil is constituted by a complex hydrocarbon mixture and a wide range of n-alkanes (C_6-C_{60}), alkenes, aromatic hydrocarbons, as well as polar fractions formed by asphaltene and resins (Salanitro *et al.*, 2000). From these, hydrocarbons with of the highest molecular weight are the most persistent within the environment (Rivera-Cruz & Trujillo-Narcía, 2004; Kostecki *et al.*, 2005). The oil toxicity within the ecosystems depends on the physical and chemical characteristics of former's components (Vega *et al.*, 2009), discharges time and types (Romaniuk *et al.*, 2007), weathered degree (Rivera-Cruz *et al.*, 2005), biological (Rivera-Cruz *et al.*, 2002) and environmental (King *et al.*, 2006) factors. In this sense, oil can affect the natural systems differentially.

In Mexico, the main oil pollution sources are the oil-well pits and the deficiencies in their maintenance, the discharges of the processing facilities, petrochemical plants and oil ducts cracks since most of the facilities are sixty years ago (Botello, 1990). There is a systemic and synergic interaction of the effects of the hydrocarbon polluted soils; in this sense, all of the ecosystems' components are altered, which affects the soils' properties, the present microorganisms, and even the plants' growth and reproduction, endangering the ecosystems' sustainability (Palma-López *et al.*, 2007).

3.1 Ground ecosystem: soil, microorganisms and plants

The oil affectations are due mainly to oil spills; the negative effects of crude oil depend on the spill type (Kolesnikov *et al.*, 2010), the zone's ecological characteristics, the amount and type of spilled oil, as well as the time over the soil (Hernández-Acosta *et al.*, 2004) and weathered degree (Rivera-Cruz *et al.*, 2005). In this sense, the pollution levels vary in accordance with the hydrocarbons' source, the age of the oil facilities and their deterioration (Adams *et al.*, 1999). Hydrocarbon polluted soils also experience physical, chemical and biological processes (Li *et al.*, 1997; Martínez & López, 2001; Rivera-Cruz *et al.*, 2002; Rivera-Cruz, 2004), altering the sustainability and productivity of the systems (FAO *et al.*, 1980).

Nevertheless, there are properties inherent to soils that favor the fixation and toxicity of the pollutants (Charman & Murphy, 2007). For instance, soils with a clay texture take a long time to recover from an oil spill, while the thick texture soils recover in short time

(Hernández-Acosta *et al.*, 2006). Still, this last texture can favor mobility towards the phreatic surfaces by the infiltration of the pollutant (Iturbe *et al.*, 2007), thus widening the range of the hydrocarbon's toxicity (Srogi *et al.*, 2007) towards the water tables (Fig. 1). Therefore, the aquatic organisms and the trophic chain can be severely affected, in consequence, the inhabitants consuming these products.

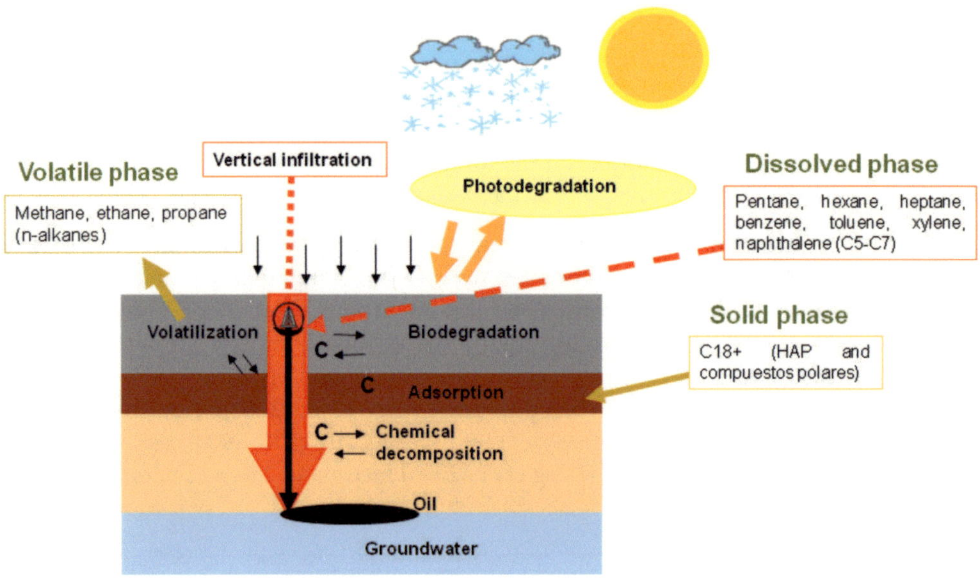

Fig. 1.Vertical infiltration process of crude oil (Eweis *et al.*, 1998).

There are synergic effects due to the crude oil soil pollution, since the oil blocks the gas interchange with the atmosphere -given its anoxic properties- (Leitgi *et al.*, 2008) and the change of the physical and chemical properties of the soils (Martínez & López, 2001) severely diminish the microbial communities benefic to the soil (Labud *et al.*, 2007). These microbiological variables are indeed a good indicator of the impact of a pollutant on the soil (Eibes *et al.*, 2006).

The toxicity mechanisms caused by the oil on soils is not limited to the microorganisms, since it also includes plants that suffer from hydric stress (Chaîneau *et al.*, 1997) due to the lack of water and nutrients. Concurrently, the lipid structures within the cells of the plants may be affected if the former are not quickly metabolized. In this sense, the oil has diverse effects over the plants since it inhibits the germination, growth and the biomass accumulation, reflecting these effects on a smaller plants production and, with time, in detriment of the natural resources sustainability.

The variables that have a determining effect over the plants affected by the soil hydrocarbons pollution are the soil ecology, the rhizosphere, emergence and germination, aerial and radical growth, biomass accumulation and salts present on the soil (Fig. 2).

Fig. 2. Variables that have a determining effect over the plants affected by hydrocarbon polluted soils are: a) soil ecology, b) rhizosphere, c) emergence and germination, d) aerial and radical growth, e) biomass accumulation and f) salts presence.

a. **Soil ecology**: Oil also has an effect over the soil's biological composition (Tang *et al.*, 2011), since toxic concentrations of oil on the soil inhibit the development of different species of nematodes, protozoa, rotifers, algae, fungi, bacteria and actinomycetes (Chaîneau *et al.*, 2003; Ilarionov *et al.*, 2003). Likewise, it induces the **loss of biodiversity** of microbial communities, which are of significant relevance in the biogeochemical cycles of the ecosystem affecting, as a consequence, its productivity (Rhodes & Hendricks, 1990) and the nutrients availability.
b. **Rhizosphere**: It is the soil area surrounding the plant's root containing, 10 to 100 times more exudates than a soil lacking plants (Rao *et al.*, 2007b). The root exudates (sugars, alcohols, and enzymes) provide enough carbon and energy for the rhizospheric microorganisms (Olguín *et al.*, 2007; Muñoz *et al.*, 2010) and it is precisely on this region where the intense and complex interactions between the roots systems, the microorganisms and the environment occur, with an increase of the total microbial activity (De la Garza *et al.*, 2008). Oil blocks the gases interchange between the soil and the atmosphere, causing death or diminishment of the bacteria and nematodes (Tynybaeva *et al.*, 2008). Nevertheless, Freedman (1989) and Germida *et al.* (2002) report some microorganisms can increase its population in the presence of hydrocarbons thanks to their capabilities for surviving under such conditions.
c. **Emergence and germination**: Oil forms a hydrophobic layer diminishing the hygroscopic water retention (Quiñones *et al.*, 2003). This reduces the plants' water retention capability, directly affecting the seeds emergence and germination (Vázquez-Luna *et al.*, 2010a). Other effect is reflected due to the volatile oil fractions that penetrate and damage the seeds embryo (Banks & Schultz, 2005), diminishing its viability (Chaîneau *et al.*, 1997) and affecting the ecosystem balance (Labud *et al.*, 2007).
d. **Aerial and radical growth**. Recent researches find out that high hydrocarbons concentrations damage the plants growth and development since the pollutants diminish the radicle elongation and the vegetative growth (Vázquez-Luna *et al.*, 2010a). García (2005) found a growth and dry weight reduction in rice seedlings after a 25 day exposure to 90,000 $mg \cdot kg^{-1}$ of weathered oil. This effect could be attributed to the oil since it forms a hydrophobic layer, limiting the root's water and nutriments absorption.

e. **Biomass accumulation**. When the plant grows and its needs increase, the lack of absorbed water diminishes the cellular swelling, reduces or inhibits the nutriments incorporation processes and affects the vegetative growth (Inckot *et al.*, 2011), and the later grains or fruits harvest. On this regard, researches by Rivera-Cruz & Trujillo Narcía (2004) found that the exposure to oil hydrocarbons concentrations of 2791, 9025 and 79,457 mg·kg^{-1} on the soil inhibited the vegetative growth and reduced the plants biomass in the seedlings of Echinochloa polystachya, Brachiaria mutica and Cyperus spp grasses. The biomass reduction is due to the damage caused in the root system (Langer *et al.*, 2010), making more difficult the plants' growth and, consequently, reducing their biomass (Zavala-Cruz *et al.*, 2005) (Fig. 3). Additionally, there are visual reports of chlorosis and reddish hues characteristic of phenolic compounds on the leaves, due to the roots' stress and damages (Harvey *et al.*, 2001; Peña-Castro *et al.*, 2006) as shown on Figure 4.
f. **Salts presence**. Regarding salts presence, Adams *et al.* (2008) state that, when an oil spill occurs, high salinity is commonly associated to the production water or to the formation of the oil well. Soluble salts of calcium carbonates, nitrates and sulfates increase in presence of hydrocarbons (Ke *et al.*, 2011), as shown on Figure 5.

a)

b)

Fig. 3. Differences in grasses growth and biomass accumulation in a) polluted zone with 12,276 mg·kg^{-1} of TPH and b) zone with 82 mg·kg^{-1}.

Fig. 4. Oil toxicity visible effects due to: a) chlorosis and reddish hues in cotyledons (*Crotalia incana)* and b) damages in *Leucaena leucocephala* roots exposed to 80,000 mg·kg^{-1} of HTP.

Fig. 5. Salts presence in polluted soils (with 12,276 mg·kg^{-1} of HTP) as a collateral effect of the oil industry pollution.

3.2 Marine ecosystems

The oil toxicity in the marine environment is very complex due to the great diversity of factors intervening during an environmental risk event. When oil is spilled or introduced to the marine ecosystem (Mercer & Trevors, 2011), it becomes weathered (Fig. 6), during which several processes take place, such as the **evaporation** of the volatile compounds, **dispersion** by means of waves, winds and turbulences (Chang *et al.*, 2011), **emulsification**, which constitutes the main cause of the persistence of light and medium crude oils on the sea surface, **dilution** depending of the crude oil type, temperature, turbulence and dispersion, **sedimentation** or sinking of the particles by adhesion to the sediments or organic matter, and **biodegradation** (Botello, 1995; Nikolopoulou & Kalogerakis, 2010; Prince, 2010). In Mexico, the Tonalá River in Veracruz and the Laguna de Términos in Campeche have shown the highest levels of dissolved hydrocarbons in the Gulf of Mexico (Botello *et al.*, 1996).

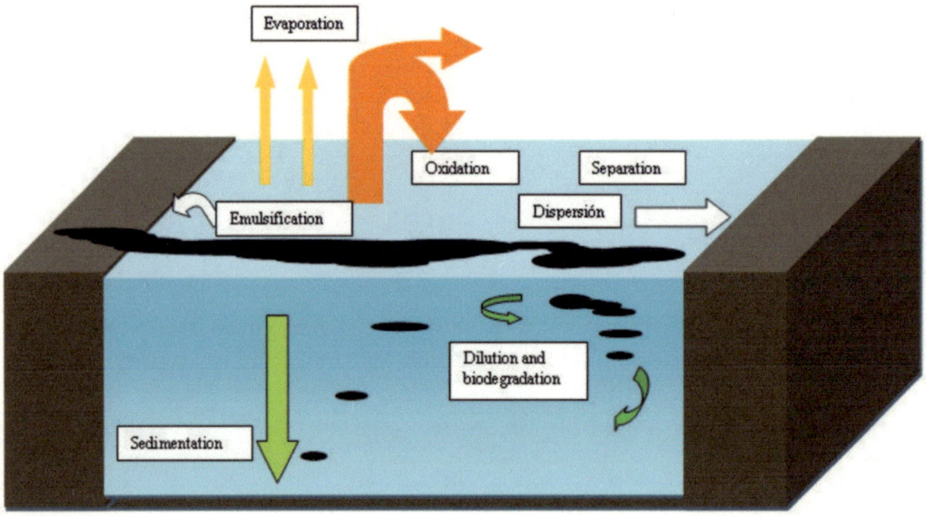

Fig. 6. Weathering process of the oil spilled in the sea, taken from Botello (1995).

Regarding the hydrocarbons assessment in the lower Coatzacoalcos River basin, the classical works of Botello & Páez (1986) are outstanding since they found the highest polycyclic aromatic hydrocarbons (PAH) levels at the zones of fixed discharges or of intense oil activity (García-Ruelas *et al.*, 2004). Other studies determined the presence of PAH on 19 organic species such as fish, crustaceans and mollusks (sea bass, native sea bream, scallop and prawns) and the presence of benzo(a)pyrene and benzo(ghi)perylene, which are the most hazardous given its carcinogenic potential (Sharma *et al.*, 2002). On the other hand, oil refinement and processing also contribute to broadening the range of toxicity caused to the fishing industry.

The marine oil pollution effects are harder to estimate (Trevors & Saier, 2010), since there is no information previous to the beginning of the oil-industry activities over the effects on the ecosystems and its components; as a consequence, it is not possible to measure the magnitude of the oil industry in the marine seaway. Additionally, it has not been possible to determine the chronic effects of pollution over the ecosystems since the data available is very precise in a given time and determined concentrations (García-Cuéllar *et al.*, 2004). Currently, there are studies proving the existence of considerable disturbances in the environment (Scarlett *et al.*, 2007; Denoyelle *et al.*, 2012); however, it is important to highlight that pollutants bioaccumulate in water organisms destined to human consumption (Webb, 2011).

3.3 Human beings

Mexico's oil industry history has been characterized by conflicts, power clashes, interests differences, competence between diverse companies, oilers and political leaders discrepancies, jealous preservation of the acquired rights by foreign interests, international consumer interest to have access a strategic product to a low cost, struggle for applying the 1917 Constitution over a given industry and the pressure from the workers to obtain a higher participation of the profit sharing (Brown, 2005).

The social conflicts attributed to the oil-zones development has been studied by different researchers; among them, Bustamante & Jarrín (2005) point that the presence of oil-related activities does not improve the population life standard nor destroys it, although it must be said that such study did not consider the analysis of violence indicators nor the environment toxicity. On this regard, Avellaneda (2004) expresses that there cannot be an "environmental conflict" study which excludes the social dimension and vice versa, since the critics toward the oil-industry activities shall be discussed from several social, political, economic and cultural angles (Avellaneda, 2005). About this, a recent study by Vázquez *et al.* (2010b) found negative effects over the equitable development in areas near oil zones and considers the population health as a primordial entity for development. Regarding this, Elliot (1994) says that when the degree of residues generation exceeds the atmosphere, oceans, vegetation and soils natural capacity, these are absorbed affecting human health and the ecological systems.

A study made at the Ecuador's Amazonia found that women living near oil wells zones (up to 5 km) presented symptoms such as tiredness, nasal and throat irritation, headache, eye irritation, earache, diarrhea and gastritis; these were associated with the proximity to oil wells and stations (San Sebastián *et al.*, 2001). The main effects found in other studies regarding the acute exposure to oil after an oil spill at sea have been headache, throat and eyes irritation, tiredness, in addition to disorders such as anxiety and depression (Lyons *et al.*, 1999). However, a repeated or long (chronic) exposure to low concentrations of oil volatile compounds can produce nausea, drowsiness, and headache (Kaplan *et al.*, 1993). In accordance with Sánchez (2003), certain cases of pediatric intoxications are due to short chain hydrocarbons acting as asphyxiating agents given their high volatility and low viscosity, replacing the alveolar gas and producing hypoxia. When going through the alveolar membrane, they create symptoms such as the diminishment of the conscience threshold progressing towards convulsions, epileptic status or coma; additionally they induce the apparition of arrhythmias. On their side, the long chain hydrocarbons have a lower toxic power and large amounts are needed for them to produce central depression. The symptoms related range from symmetric sensorial dysfunction in the distal zones of the extremities, weakness of fingers and toes, loss of deep sensitive reflexes, to central nervous system depression, drowsiness and motor incoordination.

The hydrocarbons pollution might have teratogenic, mutagenic and carcinogen effects on human health (Neff, 2004). The oil hydrocarbons toxicology has been reviewed by the Agency for Toxic Substances and Diseases Registry (ATSDR) in the USA. According to this environmental agency, a major part of the total repercussion over the human health of all the chemical products on the soil is due to specific compounds called *worrying chemical substances* with a significant toxicological power. The worrying chemical products for gasoline, kerosene, and overall fuel polluted areas are benzene, toluene, ethilbenzene and xylene (BTEX) depending on the spill and the oilfield nature, which may contain heavy metals such as nickel and vanadium (López *et al.*, 2008). Additional to the polycyclic (or polynuclear) aromatic hydrocarbons such as benzo(a)pyrene (EA, 2003), most of these hydrocarbons of high-molecular weight are present in soils with weathered oil or previous spills (Rivera-Cruz, 2004; Rivera-Cruz & Trujillo-Narcía, 2004). About this, the International Agency for Research on Cancer (IARC), dependent of the World Health Organization (WHO), assessed benzo(a)anthracene, benzo(a)pyrene, and dibenz(a,h)anthracene as probable human carcinogens, and benzo(b)fluoranthene, benzo(k)fluoranthene, indene, pyrene, and naphthalene as possible human carcinogens (Kirkeleit *et al.*, 2008).

Some compounds act as xenoestrogens, having the ability of stimulating the mammary gland tissue (De Celis *et al.*, 2006); others may work as endocrine disruptors having effects on the reproduction (Chichizola, 2003). On this regard, cancer is the third cause of death on people aged 1 to 19 years; while only 5 to 10% of the malign tumors have been actually related to genetic causes; the rest might be influenced by a wide range of environmental factors (Anderson, 2001).

Because of this, it is important to regulate the oil refinement processes since they can pour into the atmosphere a large number of chemical compounds such as naphthalene, considered a dangerous airborne compound in accordance with the US Environmental Protection Agency (USEPA) since it can cause eye, skin and respiratory tract irritation. If inhaled for long periods it can damage the kidneys and the liver, in addition to skin allergies and dermatitis (Baars, 2002), cataracts, retina damage and also can attack the central nervous system. In high concentrations it can destroy red blood cells, causing hemolytic anemia (USEPA, 2003); as well, it is considered as a possible human carcinogen (Carmichael *et al.*, 1991; ATSDR, 2004).

The main health risks are due to contact, inhalation and ingestion, and may increase depending on age, gender and exposure degree (Chen & Liao, 2006). Some surveys show that among the 13 fractions of TPH, the aliphactics EC8-16 and aromatics EC10-21 are the main contributors to human health risks along all of the exposure routes (Park & Park, 2010).

4. The challenge: responsible energy development

The Responsible Energy Development in Mexico is an immediate need in view of the holistic analysis of the effects of the oil industry over the natural resources (Patín, 2004). For a proposal to work, it is required the joint efforts of the scientific, technological, industrial, political, regulatory, legal and social sectors. In this sense, the environmental problems must be attacked from different angles, involving all of its elements (Fig. 7).

Mexico needs a holistic political view of the oil hydrocarbons pollution including the social, ecological and economical aspects, and –most of all– to create the conditions required for the Responsible Energy Development with an integral regulatory, economical and legislative frame.

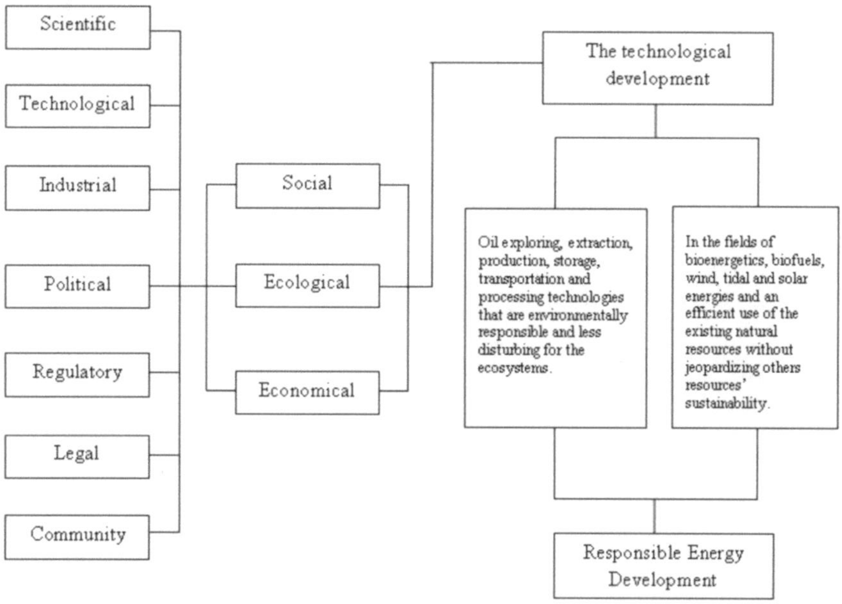

Fig. 7. Holistic approach in order to achieve a Responsible Energy Development.

On one hand, the oil spills and accidents that may occur due to maintenance deficiencies can be prevented with the proper maintenance and technological design on the side of the industry. On the other hand, citizens must avoid the clandestine oil valves and inform the authorities about fails; this information has been added into the environmental Mexican regulations. Nevertheless, there is still the need to increase the ecosystems protection due to the chronic effects of the polluting agents via strict legislation and regulations regarding the industrial wastes poured into the atmosphere, water bodies and soils, incorporating the periodic evaluation to the fishing resources for human consumption.

The economy is a subsystem within the whole development system and it is necessary that part of the oil surplus is used to finance research as well as for the technological development of adequate oil exploring, extraction, production, storage, transportation and processing technologies that are environmentally responsible and less disturbing for the ecosystems.

Finally, it is urgent to invest in the creation, research and development of new alternative energy sources for them to be gradually introduced within the country's energetic development. Such investments would enable the exploitation of profitable and sustainable energy sources, as well as the technological development in the fields of bioenergetics, biofuels, wind, tidal and solar energies and an efficient use of the existing natural resources without jeopardizing others resources' sustainability.

5. Conclusion

The environmental bases for the oil exploitation are the mainstays in order to analyze the effects of the traditional exploitation of oil in Mexico. The technological and scientific advances must be built over these bases, towards the development of new and better technologies for the resources exploitation. Consequently, this must cover the knowledge of the effects of pollution and thus promote the energetic sustainability without compromising human health, the environmental balance and the national economy.

Oil is fundamental for the economy of many countries. Therefore, it is not convenient to propose a radical change in the energetic system; rather, the proposal should consider the scientific and technological development based on two actions. The first one must focus on the exploring, extraction, processing, storage and transportation of crude oil with less risk and minimal disturbances of the natural resources. The second must bear in mind the gradual and integrated incorporation of the sustainable energies, allowing the efficient exploitation of resources without endangering others.

6. Acknowledgment

The author wants to especially thank Dr. Macario Vázquez Rivera and Irma Luna Capetillo for their valuable support and the Technical Council of COLPOS (Dr. Pilar Alberti Manzanares, Dr. Joel Zavala Cruz, Dr. Elizabeth Hernández Acosta, Dr. Miguel Escalona Maurice and Dr. Ruth De Celis Carrillo) for their scientific consultancy. Finally, the author wants to thank the "Fideicomiso Institucional 2010" of the Colegio de Postgraduados for financing the publication of the present article.

7. References

Adams, R. H.; Domínguez, V. I. & García, L. (1999). Potencial de la biorremediación de suelo y agua impactados por petróleo en el trópico mexicano. *Terra,* 17 (2): 159-174

Adams, R. H.; Zavala-Cruz, J. & Morales-García, F. (2008). Concentración residual de hidrocarburos en el suelo del trópico II: Afectación a la fertilidad y su recuperación. *Interciencia,* 33: 483-489

Anderson, R.N. (2001). Deaths: leading causes for 1999. *Natl Vital Stat Rep,* 49(11):1-87

ATSDR (Agency for Toxic Substances and Disease Registry). (2004). Public health statement for naphthalene, 1-methylnaphthalene,2-ethylnaphthalene. Available online at http://www.atsdr.cdc.gov/ToxProfiles/phs9018.html

Avellaneda, C. A. (2004). Petróleo, ambiente y conflicto en Colombia. In: *Guerra, sociedad y medio ambiente.* Martha Cárdenas Ed. Foro Nacional Ambiental. Bogotá, Colombia. 545 p.

Avellaneda, C. A. (2005). Petróleo, seguridad ambiental y explotación petrolera marina en Colombia. *Revista de Ciencias Sociales,* 21: 11-17

Baars, B-J. (2002). The wreckage of the oil tanker 'Erika'—human health risk assessment of beach cleaning, sunbathing and swimming. *Toxicology Letters,* 128(1-3): 55-68

Bahena-Manjarrez, J.; Rosales-Hoz L. & Carranza-Edwards, A. (2002). Spatial and temporal variation of heavy metals in a tropical estuary. *Environmental Geology,* 42:575–582

Banks, M.K. & Schultz, K.E. (2005). Comparison of plants for germination toxicity tests in petroleum-contaminated soils. *Water, Air, & Soil Pollution,* 167(1-4): 211-219

Brown, J. (2005). Los archivos del petróleo y la revolución mexicana. América Latina en la Historia Económica. *Revista de Fuentes e Investigación,* 23: 49-60

Botello, V. A. (1990). *Impacto ambiental de los hidrocarburos organoclorados y microorganismos patógenos específicos en lagunas costeras del Golfo de México.* Informe final 1989-1990,

OEA-CONACYT. Inst. Cienc. Mar y Limnol., Univ. Nal. Autónoma de México. 69 p.

Botello, V. A. (1995). Fuentes, transformación y caracterización geoquímica del petróleo en el ambiente marino, p. 211-223. In: *Golfo de México. Contaminación e impacto ambiental: diagnóstico y tendencias.* Botello, V. A., J. L. Rojas G., J. A. Benítez y D. Zárate-Lomelí (Eds). EPOMEX. Serie Científica 5. Universidad Autónoma de Campeche. México 666 p.

Botello, V. A. (1996). Características, composición y propiedades fisicoquímicas del petróleo. In: *Golfo de México. Contaminación e impacto ambiental: diagnóstico y tendencias.* Botello, V. A., J. L. Rojas G., J. A. Benítez y D. Zárate-Lomelí (eds). EPOMEX. Serie Científica 5. Universidad Autónoma de Campeche. México 666 p.

Botello, V. A. & Páez, F. (1986). *La Contaminación: el problema crucial.* Vol. I. Centro de Ecodesarrollo. México.140p.

Bustamante, T. & Jarrín, M. C. (2005) Impactos sociales de la actividad petrolera en Ecuador: un análisis de los indicadores. *Revista de Ciencias Sociales,* 21: 19-34

Carbajal, P. J. L. & Chavira, M. D. (1985). La contaminación de los sistemas lagunar-estuarios de las costas mexicanas. *Elementos,* 2(10): 58-64

Carmichael, P. L.; NíShé, M. & Phillips, D. H. (1991) DNA adducts in human and mouse skin maintained in short-term culture and treated with petrol and diesel engine lubricating oils. *Cancer Letters,* 57(3): 229-235

Chaîneau, H. C.; Morel, J. L. & Oudot, J. (1997) Phytotoxicity and uptake of fuel oil hydrocarbons. *J. Environ. Qual.* 26: 1478-1483

Chaîneau, C. H.; Yepremian, C.; Vidalie, J. F.; Ducreux, J. & Ballerini, D. (2003). Bioremediation of a Crude Oil-Polluted Soil: Biodegradation, Leaching and Toxicity Assessments. *Water, Air, and Soil Pollution,* 144: 419–440

Chang, Y-L.; Oey, L-Y.; Xu, F-H.; Lu, H-F. & Fujisaki, A. (2011). 2010 oil spill: trajectory projections based on ensemble drifter analyses. *Ocean Dyn* 61(6):829–839

Charman, P. E. V. & Murphy, B. W. (2007). Soils their properties and management. Oxford University Press. Third edition. Hong Kong. 461 p.

Chen, S-C. & Liao, C-M. (2006). Health risk assessment on human exposed to environmental polycyclic aromatic hydrocarbons pollution sources. *Science of the Total Environment,* 366: 112-123

Chichizola, C. (2003). Disruptores Endocrinos. Efectos en la Reproducción. *Revista Argentina de Endocrinología y Metabolismo,* 40(3): 172-188

Cruz-Orea, A.; Tomás, S.A.; Guerrero-Zuñiga, A. & Rodríguez-Dorantes, A. (2004). Detection of an aromatic compound at the roots of Cyperus Hermaphroditus by photoacoustic techniques. *International Journal of Thermophysics,* 25(2): 603-610

Denoyelle, M.; Geslin, E.; Jorissen, F. J.; Cazes, L. & Galgani, F. (2012). Innovative use of foraminifera in ecotoxicology: A marine chronic bioassay for testing potential toxicity of drilling muds. *Ecological Indicators,* 12: 17-25

De Celis, R.; Morgan, G.; Bravo A. & Feria A. (2006). Cáncer de mama y exposición a hidrocarburos aromáticos. *e-Gnosis,* (4): 1-8

De la Garza, E. R.; Ortiz, Y. P.; Macias, B. A.; García, C. & Coll, D. (2008). Actividad biótica del suelo y la contaminación por hidrocarburos. *Revista Latinoamericana de Recursos Naturales,* 4: 49-54.

EA (Environment Agency). (2003) Principles for Evaluating the Human Health Risks from Petroleum Hydrocarbons in Soils: A Consultation Paper. 43p.

Eibes, G.; Cajthaml, T.; Moreira, M. T.; Feijoo, G. & Lema, J. M. (2006). Enzymatic degradation of anthracene, dibenzothiophene and pyrene by manganese peroxidase in media containing acetone. *Chemosphere,* 64, 408–414

Elliot, A. J. (1994). *An introduction to sustainable development.* Zed Books. Londres. pp. 34-68

Eweis, J. B.; Ergas, S. J.; Chang, D. P. Y. & Schroeder, E. D. (1998). Bioremediation principles. Series in Water Resources and Environmental Engineering. McGraw Hill. New York.

FAO, PNUMA & UNESCO. (1980). *Metodología provisional para la evaluación de la degradación de los suelos*. Roma. 86 p.

Freedman, B. (1989). Environmental Ecology: the impacts of pollution and other stresses on ecosystem structure and function. Academic Press, Inc. San Diego California. USA. 424 p.

García, R. D. E. (2005). Estudio de la toxicidad de los petróleos nuevo e intemperizado en el cultivo de arroz (*Oryza sativa* L.). Tesis de Ingeniería Ambiental. División Académica de Ciencias Biológicas. Universidad Juárez Autónoma de Tabasco. Villaermosa , Tabasco. 87 p.

García-Cuellar, J. A.; Arreguín-Sánchez, F.; Vázquez, S. H. & Lluch-Cota, D. B. (2004). Impacto ecológico de la industria petrolera en la sonda de Campeche, México, tras tres décadas de actividad: una revisión. *Interciencia,* 29: 311–319

García-López, E.; Zavala-Cruz, J. & Palma-López, D. J. (2006). Caracterización de las comunidades vegetales en un área afectada por derrames de hidrocarburos. *Terra Latinoamericana,* 24: 17-26

García-Ruelas, C., Botello, A.V., Ponce-Vélez, G. & Díaz-González, G. (2004). Polycyclic aromatic hydrocarbons in coastal sediments from the subtropical Mexican Pacific. *Marine Pollution Bulletin,* 49: (5-6) 514-519

George, S.J.; Sherbone, J.; Hinz, C. & Tibbett, M. (2011). Terrestrial exposure of oilfield flowline additives diminish soil structural stability and remediative microbial function. *Environmental Pollution,* 159: (10) 2740-2749

Germida, J. J.; Frick, C. M. & Farrell, R. E. (2002). Phytoremediation of oil-contaminated soils. In: *Developments in Soil Science,* Volume 28B, A. Violante, P.M. Huang, J.-M. Bollag and L. Gianfreda (Eds). Elsevier Science B.V. 169-186

González-Lozano, M. C.; Méndez-Rodríguez, L. C.; López-Veneroni, D. G. & Vázquez-Botello, A. (2006). Evaluación de la contaminación en sedimentos del área portuaria y zona costera de Salina Cruz, Oaxaca, México. *Interciencia,* 31(9): 647-656

González-Mille, D. J.; Ilizaliturri-Hernández, C. A.; Espinosa-Reyes, G.; Costilla-Salazar, R.; Díaz-Barriga, F.; Ize-Lema, I. & Mejía-Saavedra, J. (2010). Exposure to persistent organic pollutants (POPs) and DNA damage as an indicator of environmental stress in fish of different feeding habits of Coatzacoalcos, Veracruz, Mexico. *Ecotoxicology,* 19:1238-1248

Gutiérrez, M. C. C. & Zavala, C. J. (2002). Rasgos hidromórficos de suelos tropicales contaminados con hidrocarburos. *Terra Latinoamericana,* 20: 101-111

Hall, C.; Tharakan, P.; Hallock, J.; Cleveland, C. & Jefferson, M. (2003). Hydrocarbons and the evolution of human culture. *Nature,* 426: 318-322

Harvey, P. J.; Campanella, B. F.; Castro, P. M. L.; Harms, H.; Lichtfouse, E.; Schaeffner, A. R.; Smrcek, S. & Werck-Reichhart, D. (2001). Phytoremediation of polyaromatic hydrocarbons, anilines and phenols. *Environmental Science Pollution Res.,* 9: 29-47

Hernández-Acosta, E.; Rubiños-Panta, J. E. & Albarado-López, J. (2004). *Restauración de los suelos contaminados con hidrocarburos: Conceptos básicos*. Colegio de Postgraduados. Montecillo, Estado de México. México. 148 p.
Hernández-Acosta, E.; Gutiérrez-Castorena, M. C.; Rubiños-Plata, J. E. & Alvarado-López, J. (2006). Caracterización del suelo y plantas de un sitio contaminado con hidrocarburos. *Terra Latinoamericana,* 24: 463-470
Ilarionov, S. A.; Nazarov, A. V. & Kalachnikova, I. G. (2003). The Role of Micromycetes in the Phytotoxicity of Crude Oil-Polluted Soils. *Russian Journal of Ecology,* 34(5):303–308
Iturbe, R.; Flores, C.; Castro, A. & Torres, L. G. (2007). Sub-soil contamination due to oil spills in six oil-pipeline pumping stations in northern Mexico. *Chemosphere,* 68: 893-906.
Inckot., R. C.; Santos, G. de O.; De Souza, L. A. & Bona, C. (2011) Germination and development of Mimosa pilulifera in petroleum-contaminated soil and bioremediated soil. *Flora - Morphology, Distribution, Functional Ecology of Plants* 206: (3) 261-266
Jernelöv, A. (2010). The Threats from Oil Spills: Now, Then, and in the Future. *Ambio,* 39:353-366
Kaplan, M. B.; Brandt-Rauf, P.; Axley, J. W.; Shen, T. T. & Sewell, G. H. (1993). Residential release of number 2 fuel oil: a contributor to indoor air pollution. *Am J Public Health,* 83(1): 84-88
Ke, L.; Zhang, C.; Guo, C.; Hui L. G. & Fung Y. T. N. (2011). Effects of environmental stresses on the responses of mangrove plants to spent lubricating oil. *Marine Pollution Bulletin,* 63: (15-12) 385-395
Kirkeleit, J.; Riise, T.; Bråtveit, M. & Moen, B. E. (2008). Increased risk of acute myelogenous leukemia and multiple myeloma in a historical cohort of upstream petroleum workers exposed to crude oil. *Cancer Causes Control,* 19:13–23
Kolesnikov, S. I.; Gaivoronskii, V. G.; Rotina, E. N.; Kazeev K. Sh. & Val'kov, V. F. (2010). Assessment of soil tolerance toward contamination with black oil in the south of Russia on the basis of soil biological indices: A model experiment. *Eurasian Soil Science,* 43(8): 929-934
Kostecki, P.; Morrison, R. & Dragun, J. (2005). Hydrocarbons. In: *Encyclopedia of Soils in the Environment,* Daniel Hillel (Ed.). Oxford. 217- 226 pp.
King, R. F.; Royle, A.; Putwain, P. D. & Dickinson, N. M. (2006). Changing contaminant mobility in a dredged canal sediment during a three-year phytoremediation trial. *Environmental Pollution,* 143: 318-326
Labud, V.; Garcia, C. & Hernández, T. (2007). Effect of hydrocarbon pollution on the microbial properties of a sandy and a clay soil. *Chemosphere,* 66: 1863-1871
Langer, I.; Syafruddin, S.; Steinkellner, S.; Puschenreiter, M. & Wenzel, W. W. (2010). Plant growth and root morphology of Phaseolus vulgaris L. grown in a split-root system is affected by heterogeneity of crude oil pollution and mycorrhizal colonization. *Plant Soil,* 332:339-355
Leitgib, L.; Gruiz, K.; Fenyvesi, E.; Balogh, G. & Murányi, A. (2008). Development of an innovative soil remediation: "Cyclodextrin-enhanced combined technology". *Science of the Total Environmental,* 392: 12-21
Li, X.; Feng, Y, & Sawatsky N. (1997). Importance of soil-water relations in assessing the endpoint of bioremediated soils. *Plant Soil,* (192): 219-226

López, E.; Schuhmacher, M. & Domingo, J. L. (2008). Human health risks of petroleum-contaminated groundwater. *Env Sci Pollut Res,* 15 (3):278-288

Lyons, R. A.; Temple, M. F.; Evans, D., Fone, D. L. & Palmer, S. R. (1999). Acute health effects of the sea empress oil spill. *Journal Epidemiol Community Health,* 53:306-310

Martínez, E. M. & López, F. S. (2001). Efecto de hidrocarburos en las propiedades físicas y químicas de suelo arcilloso. *Terra* (19): 9-17

Muñoz, C. L. N.; Nevárez, M. G. V.; Ballinas, C. M. L. & Peralta, P. M. R. (2010). Fitorremediación como una alternativa para el tratamiento de suelos contaminados. *Revista Internacional de ciencia y Tecnología Biomédica,* 3: 1-8

Mei, H. & Yin, Y. (2009). Studies on marine oil spills and their ecological damage. *J. Ocean Univ. China* (*Oceanic and Coastal Sea Research*), 8(3): 312-316

Mercer, K. & Trevors, J. T. (2011). Remediation of oil spills in temperate and tropical coastal marine environments. *Environmentalist,* 31:338–347

Neff, J. M. (2004). Bioaccumulation in marine organisms. Effect of contaminants from oil web produced water. Elsevier, Netherlands. pp. 241-313

Nikolopoulou, M. & Kalogerakis, N. (2010). Biostimulation strategies for enhanced bioremediation of marine oil spills including chronic pollution. In: *Handbook of Hydrocarbon and Lipid Microbiology,* K. N. Timmis (Eds.), Springer-Verlag Berlin Heidelberg. 2522-2529

Olguín, E. J.; Hernández, M. E. & Sánchez-Galván, G. (2007). Contaminación de manglares por hidrocarburos y estrategias de biorremediación, fitorremediación y restauración. *Rev. Int. Contam. Ambient.,* 23: 139-154

Olivera-Villaseñor, R. E. & Rodríguez-Castellanos, A. (2005). Estudio del riesgo en ductos de transporte de gasolinas y diesel en México. *Científica,* 9: (4) 159-165

Ortiz, P. M. A.; Siebe, C. & Kram, S. (2005). Diferenciación ecogeográfica de Tabasco. Cap. 14. Pp. 305-322. In: *Biodiversidad del estado de Tabasco.* Bueno J., Álvarez F. y Santiago S. (Eds.) Instituto de Biología, UNAM-CONABIO. México, D. F.

Ortínez, B. O.; Ize, L. I. & Gavilán, G. A. (2003). La restauración de los suelos contaminados con hidrocarburos en México. Instituto Nacional de Ecología. *Gaceta ecológica,* 69: 83-92

Quiñones, A. E. E.; Ferrera-Cerrato, R.; Gavi, R. F.; Fernández, L. L.; Rodríguez, V. R. & Alarcón, A. (2003) Emergencia y crecimiento de maíz en un suelo contaminado con petróleo crudo. *Agrociencia,* 37: 585-594

Palma-López, D. J.; Cisneros, D. J.; Moreno, C. E. & Rincón-Ramírez, J. A. (2007). *Suelos de Tabasco: su uso y manejo sustentable. Instituto del Trópico Húmedo.* Colegio de Postgraduados, Fundación Produce Tabasco A. C. Villahermosa, Tabasco. México. 195 p.

Park, I-S. & Park, J-W. (2010). A novel total petroleum hydrocarbon fractionation strategy for human health risk assessment for petroleum hydrocarbon-contaminated site management. *Journal of Hazardous Materials,* 179(1-3): 1128-1135

Patín, S. A. (2004). Assessment of anthropogenic impact on marine ecosystems and biological resources in the process of oil and gas field development in the Shelf area. *Water Resource,* 31(4): 413-422

PEMEX (2003). *Anuario estadístico. Exploración y producción.* PEMEX. México. 64 pp.

Peña-Castro, J. M.; Barrera-Figueroa, B. E.; Fernández-Linares, L.; Ruiz-Medrano, R. & Xoconostle-Cázares, B. (2006). Isolation and identification of up-regulated genes in bermudagrass roots (*Cynodon dactylon* L.) grown under petroleum hydrocarbon stress. *Plant Science,* 170: 724–731

Prince, R. C. (2010). Bioremediation of marine oil spills. *Handbook of Hydrocarbon and Lipid Microbiology*, 24: 2617-2630

Rao, P. S.; Ansari, M. F.; Gavane, A. G.; Pandit, V. I.; Nema, P. & Devotta, S. (2007a). Seasonal variation of toxic benzene emissions in petroleum refinery. *Environ Monit Assess*, 128:323-328

Rao, N. C. V.; Afzal, M.; Malallah, G.; Kurian, M. & Gulshan, S. (2007b). Hydrocarbon uptake by roots of Vicia faba (Fabaceae). *Environ Monit Assess*, 132:439-443

Rhodes, A. N. & Hendricks, C.W. (1990). A continuos-flow method for measuring effects of chemical on soil nitrification. *Toxicity Assess*, 5: 77-89

Rivera-Cruz, M. del C.; Ferrera-Cerrato, R.; Volke-Haller, V.; Fernández-Linares, L. & Rodríguez-Vázquez, R. (2002). Poblaciones microbianas en perfiles de suelos afectados por hidrocarburos del petróleo en el estado de Tabasco, México. *Agrociencia*, 36: 149-160

Rivera-Cruz, M. del C. (2004). Clasificación de suelos tropicales influenciados por derrames de petróleo en Tabasco. *Tecnociencia Universitaria*, 7: 6-25

Rivera-Cruz, M. del C. & Trujillo-Narcía, A. (2004). Estudio de toxicidad vegetal en suelos contaminados con petróleos nuevo e intemperizado. *Interciencia*, (29): 369-376

Rivera-Cruz, M. C.; Trujillo-Narcía, A.; Miranda, C. M. A. & Maldonado, C. E. (2005). Evaluación toxicológica de suelos contaminados con petróleos nuevo e intemperizado mediante ensayos con leguminosas. *Interciencia*, 30: 326-331

Rivero, O.; Rizo, P.; Ponciano, G. & Oláiz, G. (2001). *Daños a la salud por plaguicidas. Introducción a la toxicología ambiental*. Editorial. Manual Moderno. México. 488 p.

Rodríguez, S. M.; Pereira, M. E.; Ferreira, D. A.; Silva, E.; Hursthouse, A. S. & Duarte, A. C. (2009). A review of regulatory decisions for environmental protection: Part I Challenges in the implementation of national soil policies. *Environment International*, 35: 202-213.

Romaniuk, R.; Brandt, J. F.; Ríos, P. R. & Giuffré, L. (2007). Atenuación natural y remediación inducida en suelos contaminados con hidrocarburos. *Ci. Suelo*, 25: 139-149

Rosales-Hoz, L. & Carranza-Edwards, A. (1998). Heavy Metals in sediments from Coatzacoalcos river, Mexico. *Bull. Environ. Contam. Toxicol*, 60: 553-561

Roth, F. A-M. &, Baltz, M. D. (2009). Short-Term Effects of an Oil Spill on Marsh-Edge Fishes and Decapod Crustaceans. *Estuaries and Coasts*, 32:565-572

Ruelas-Inzunza, J.; Páez-Osuna, F.; Zamora-Arellano, N.; Amezcua-Martínez, F. & Bojórquez-Leyva, H. (2009). Mercury in biota and surficial sediments from Coatzacoalcos estuary, Gulf of Mexico: Distribution and seasonal variation. *Water Air Soil Pollut*, 197:165-174

Ruelas-Inzunza, J.; Páez-Osuna, F.; Ruiz-Fernández, A. C. & Zamora-Arellano, N. (2011). Health risk associated to dietary intake of mercury in selected coastal areas of Mexico. *Bull Environ Contam Toxicol*, 86:180–188

Salazar-Coria, L.; Amezcua-Allieri, M. A.; Tenorio-Torres, M. & González-Macías, C. (2007). Polyaromatic Hydrocarbons (PAHs) and metal evaluation After a diesel spill in Oaxaca, Mexico. *Bull Environ Contam Toxicol*, 79:462-467

Salanitro, J. P. (2000). Bioremediation of petroleum hydrocarbons in soil. *Advances in Agronomy* 72: 53-105

Sánchez, S., C.A. (2003). Intoxicación por hidrocarburos. In: *Manual de intoxicaciones en Pediatría*. Santiago Mintegui (eds). Ediciones Ergon. Majadahonda, Madrid. Capítulo 15. pp. 151-159

San Sebastián, M.; Armstrong B. & Stephens C. (2001). La salud de mujeres que viven cerca de pozos y estaciones de petróleo en la Amazonía ecuatoriana. *Rev Panam Salud Publica,* 9(6): 375-383

Santos, H. F.; Carmo, F. L.; Paes, J. E. S.; Rosado, A. S. & Peixoto, R. S. (2011) Bioremediation of Mangroves Impacted by Petroleum. *Water Air Soil Pollut* 216:329–350

Scarlett, A.; Galloway, T. S. & Rowland, S. J. (2007). Chronic toxicity of unresolved complex mixtures (UCM) of hydrocarbons in marine sediments. *J Soils Sediments* 7(4): 200-206

Sharma, V. K.; Hicks, S. D.; Rivera, W. & Vazquez, F. G. (2002). Characterization and degradation of petroleum hydrocarbons following an oil spill into a coastal environment of South Texas, U.S.A. *Water, Air, and Soil Pollution,* 134: 111-127

Srogi, K. (2007). Monitoring of environmental exposure to polycyclic aromatic hydrocarbons: a review. *Environ Chem Lett,* 5:169–195

Tang, J.; Wang, M.; Wang, F.; Sun, Q. & Zhou, Q. (2011). Eco-toxicity of petroleum hydrocarbon contaminated soil. *Journal of Environmental Sciences,* 23(5): 845-851

Toledo, O. A. (1983). *Como destruir el Paraíso. El Desastre Ecológico del Sureste.* Centro de Ecodesarrollo-Océano, México. 149p.

Toledo, O. A. (1995). *Economía de la Biodiversidad.* PNUMA. Serie Textos Básicos. México. 273p.

Trevors, J. T. & Saier, M. H. (2010). The legacy of oil spills. *Water Air Soil Pollut* 211:1-3

Trujillo, N. A.; Zavala, C. J. & Lagunes, E. L. del C. (1995). Contaminación de suelos por metales pesados e hidrocarburos aromáticos en Tabasco. In: *Memoria VII Reunión Científica-Tecnológica Forestal y Agropecuaria,* INIFAP.Villahermosa, Tabasco, México. pp. 45-52

Tynybaeva, T. G.; Kostina, N. V.; Terekhov, A. M. & Kurakov, A. V. (2008). The microbiological activity and toxicity of oil-polluted playa solonchaks and filled grounds within the Severnye Buzachi Oil Field (Kazakhstan). *Eurasian Soil Science,* 41(10): 1115–1123

USEPA. (2003). Health effects support document for naphthalene. EPA 822-R-03-005, Office of Water, Health and Ecological Criteria Division, Washington, DC. (Available online at http://www.epa.gov/safewater/ccl/pdf/naphthalene.pdf

Vázquez-Luna, D.; Castelán-Estrada, M.; Rivera-Cruz, M. C.; Ortiz-Ceballos, A. I. & Izquierdo, R. F. (2010a). Crotalaria incana L. y Leucaena leucocephala Lam. (Leguminosae): Especies indicadoras de toxicidad por hidrocarburos del petróleo en suelo. *Revista Internacional de Contaminación Ambiental* 26: 183-191

Vázquez-Luna, D.; Manzanares, A. P.; Zavala, C. J; Hernández, A. E.; Escalona M. M. & De Celis, C. R. (2010b). Impacto de la industria petrolera sobre el desarrollo equitativo en cuatro zonas de Huimanguillo, Tabasco. *Naturaleza y Desarrollo,* 8 (2): 6-22

Vega, A. F.; Covelo, F. E.; Reigosa, J. M. & Andrade, M. L. (2009). Degradation of fuel oil in salt marsh soils affected by the Prestige oil spill. *J. Hazard. Mater,* 166: 1020-1029

Webb, J. (2011). Environmental contamination of fish and humans through deforestation and oil extraction in Andean Amazonia. Ph.D. diss., McGill University (Canada), In: *Dissertations & Theses: The Sciences and Engineering Collection* [database on-line]; available from http://www.proquest.com (publication number AAT NR72714; accessed September 29, 2011).

Zavala-Cruz, J.; Gavi-Reyes, F.; Adams-Schroeder, R. H.; Ferrera-Cerrato, R.; Palma-López, D. J.; Vaquera-Huerta, H. & Domínguez-Ezquivel, J. M. (2005). Derrames de petróleo en suelos y adaptación de pastos tropicales en el Activo Cinco Presidentes, Tabasco, México. *Terra Latinoamericana,* 23: 293-302.

6

Spreading and Retraction of Spilled Crude Oil on Sea Water

Koichi Takamura[1], Nina Loahardjo[1], Winoto Winoto[1], Jill Buckley[1], Norman R. Morrow[1], Makoto Kunieda[2], Yunfeng Liang[2] and Toshifumi Matsuoka[2]
[1]University of Wyoming
[2]Kyoto University
[1]USA
[2]Japan

1. Introduction

The spreading of a liquid as a thin film on another liquid has fascinated the pioneers of modern surface science for over two centuries, including two pioneering women scientists, Agnes Pockels and Katharine B. Blodgett (Rayleigh & Pockels, 1891; Rayleigh, 1899; Harkins & Feldman, 1922; Harkins, 1941; Langmuir, 1933; Blodgett, 1935; Transue et al., 1942; Zisman, 1941a, 1941b, & 1941c; Shewmaker et al., 1954; Derrick, 1982; Covington, 2011). Davies & Rideal (1961) noted, "As long ago as 1765 Benjamin Franklin observed that olive oil spreads over water to a thickness of 25 Å". Harkins & Feldman (1922) defined the spreading coefficient, S, as:

$$S = W_A - W_C \tag{1}$$

thus a liquid will not spread if its work of cohesion W_C is greater than the work of adhesion W_A for the interface of the liquid and another liquid or solid upon which spreading is to occur. The W_A and W_C values are related to interfacial tensions by;

$$W_A = \gamma_a + \gamma_b - \gamma_{ab} \tag{2}$$

$$W_C = 2\gamma_b \tag{3}$$

and the spreading coefficient can be defined as;

$$S = \gamma_a - \left(\gamma_b + \gamma_{ab}\right) \tag{4}$$

Here, b represents the liquid for which spreading upon a is under consideration. The validity of equation (4) was confirmed by the spreading behaviour of 89 organic liquids on water. The description by Harkins & Feldman (1922) of the effect of placing a drop of oleic acid at the centre of a lens of petroleum (refined) oil on the water surface is of particular interest. "The lens is broken up into a great number of fragments which seem to be projected with almost explosive violence toward the edges of the tray".

Effects of polar compounds on spreading of mineral oil on water were investigated in detail by Zisman (1941a, 1941b). Fatty acids and amines having over 13 carbon atoms per molecule adsorbed permanently at the oil-water interface if the pH of the water was adjusted to values greater than 10.5 or less than 3, respectively. Ionization of the adsorbed fatty acid or amine molecules caused the petroleum oil to spread as a thin disk of diameter directly proportional to the amount of fatty acids or amines in the oil. This observation provided an estimate of the cross-sectional area of adsorbed molecules at the interface. The results were consistent with the conclusions of Danielli on the effect of pH on the interfacial tension lowering for oleic acid adsorbed at the brombenzene-water interface (Danielli, 1937).

Zisman (1941a) also demonstrated that acids and amines of lower molecular mass would spread out from the leading edge of the oil lens across the water surface, so that *S* of the oil slowly rises. This edge diffusion was tracked visually by spreading a small amount of hydrophobic talc particles on the water surface. A schematic diagram of an apparatus used for measuring spreading rates and edge diffusion can be found in Davies & Rideal (1961).

The phenomenon of interfacial tension lowering was applied to a method for recovering residual oil from an oil reservoir (Squires, 1921; Atkinson, 1927). There are now numerous reports on laboratory research and field testing of caustic flooding. The subject was reviewed by Johnson (1976) and Mayer et al. (1983). The range of pH over which fatty acids cause interfacial tension change occurs is about 3 units of pH more alkaline than the range over which dissociation occurs in the bulk phase (Zisman, 1941a & 1941b; Danielli, 1937). The dissociation behaviour of the ionizable charge groups as a function of both bulk pH and electrolyte concentration was shown to be quantitatively predictable by the Ionizable Surface-Group Model (Takamura & Chow, 1985; Chow & Takamura, 1988).

Langmuir observed that low levels of calcium or magnesium ions prevented the spreading of oil droplets containing stearic acid (Langmuir, 1936). Zisman (1941a) conducted systematic studies of the effects of various polyvalent metallic ions and confirmed that rigid films were formed by the salts of the fatty acids at the oil/water interface. The amount of the metallic salt necessary to cause rigidity of the interfacial film was found to depend on the pH and the concentration of the acid in the oil. This behaviour closely resembles the formation of visco-elastic films at the crude oil/brine interfaces. The physical properties of calcium surfactants are the topics of a recent review article (Zapf et al., 2003). There have been many studies of the effect of interfacial film properties on the snap-off of oil blobs in porous media during the course of water-flooding (Chatzis et al., 1983; Laidlaw & Wardlaw, 1983; Yu & Wardlaw, 1986). The presence of visco-elastic films is a prime parameter in the stability of water-in-crude oil emulsions (Sjöblom et al., 2003).

Crude oil spreads rapidly over sea water due to the rather high values of the spreading cofficient (25-35 mN/m) (Garrett & Barger, 1970). The use of water-insoluble monomolecular films can be used to compress spilled oil into lenses of increased thickness that occupy smaller surface area, and can be retrieved mechanically (Zisman, 1943; Barger & Garrett, 1968; Garrett & Barger, 1970). In the current paper, the spreading behaviour of drops of heptane-decane-toluene ternary mixtures on the surface of water is modelled by molecular dynamics calculations. Special attention is given to the distribution of light alkane and aromatic molecules at the air/oil/water three phase line of contact. The conclusions are consistent with the results of extended detailed experimental studies of spreading and retraction of crude oil on sea water.

2. Theoretical background

The surface tension of the liquid arises from intermolecular attraction at the surface region. For liquid alkanes, this intermolecular attraction is entirely due to London dispersion forces, thus

$$\gamma_o = \gamma_o^{\ d} \tag{5}$$

where γ_o is the surface tension of the alkane oil and the superscript, d, designates the dispersion force contribution (Fowkes, 1964) which is a direct function of the refractive index of the liquid (Hunter, 1986). As seen in Figure 1, both the surface tension and refractive index of alkanes are linear functions of density.

The surface tension of water is the sum of a dispersion forces contribution, γ_w^d, and a polar (hydrogen bonding) forces contribution, γ_w^p;

$$\gamma_w = \gamma_w^{\ d} + \gamma_w^{\ p} \tag{6}$$

and γ_w^d and γ_w^p are 21.8 and 51.0 mN/m, respectively at 20°C (Fowkes, 1964). Water molecules in the oil/water interfacial region are attracted towards the interior of the water phase by water-water intractions (dispersion forces and hydrogen bonding) and towards the oil phase by oil-water interactions (dispersion forces only); likewise, oil molecules at the interfacial region are attracted to the oil phase by oil-oil dispersion forces and to the water phase by oil-water dispersion forces. The interfacial tensions were related to the geometric means of the oil-oil and water-water intermolecular interactions (Girifalco & Good, 1957);

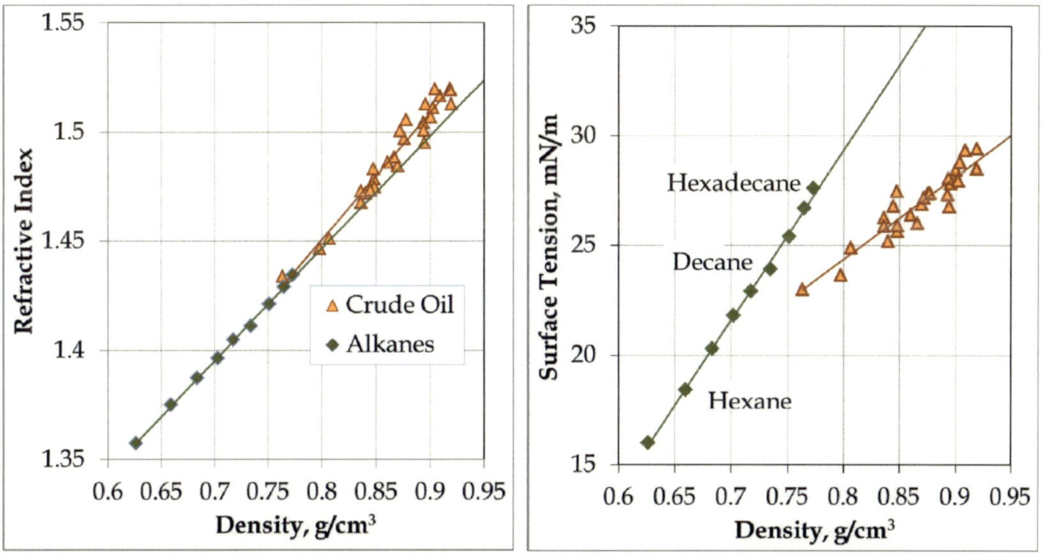

Fig. 1. Refractive index and surface tension of alkanes as a function of their density (solid diamonds). The same relationships for 23 crude oils are also shown with solid triangles. See details in the text.

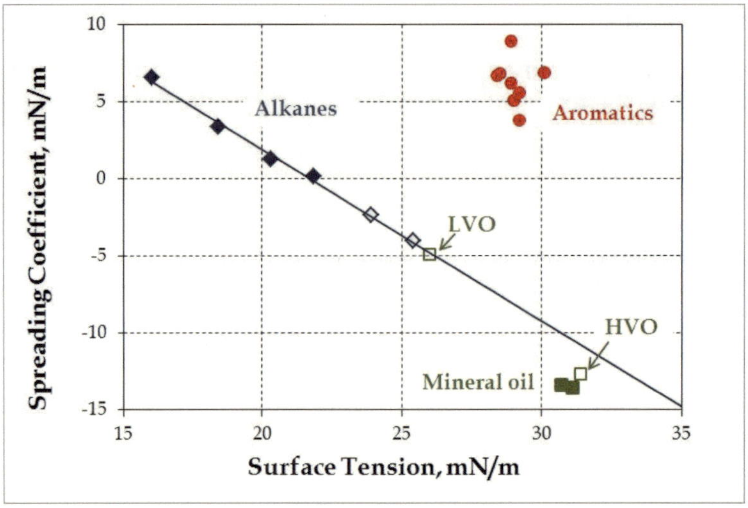

Fig. 2. Relation of spreading coefficient to surface tension of oils. Solid diamonds and circles are measured S for pure alkanes and aromatics, respectively (Pomerantz et al., 1967), other points are calculated values from γ_o and γ_{wo} data: open diamonds for alkanes (Girifalco & Good, 1957), solid squares for mineral oils referred to as "Squibb" and "Stanolax" (Harkins & Feldman, 1922), open squares for low and high viscosity mineral oils (3.9 and 173 mPa·s for LVO and HVO, respectively).

$$\gamma_{wo} = \gamma_w + \gamma_o - 2\sqrt{\gamma_w{}^d \gamma_o{}^d} \tag{7}$$

Substituting equation (6) and (7) into (4):

$$S = 2\left(\sqrt{\gamma_w{}^d \gamma_o{}^d} - \gamma_o\right) \tag{8}$$

Equation (8) shows that S for oil on water is a single-valued function of the surface tension of the oil (Fowkes, 1964) as shown in Figure 2.

The aromatic molecules are more hydrophilic than the saturated hydrocarbons (Pormerantz et al., 1967; Fowkes, 1964), resulting in lower interfacial tensions (e.g. 34.4 and 52.0 mN/m for benzene and n-heptane at 20°C, respectively). Thus, S values for the aromatics lie significantly above the straight line for the alkanes in Figure 2. Owens & Wendt (1969) have extended the equation (7) for the polar molecules:

$$\gamma_{wo} = \gamma_w + \gamma_o - 2\left(\sqrt{\gamma_w{}^d \gamma_o{}^d} + \sqrt{\gamma_w{}^p \gamma_o{}^p}\right) \tag{9}$$

Thus the spreading coefficient for the polar oil is now given as:

$$S = 2\left(\sqrt{\gamma_w{}^d \gamma_o{}^d} + \sqrt{\gamma_w{}^p \gamma_o{}^p} - \gamma_o\right) \tag{10}$$

Values of $\gamma_o{}^d$ and $\gamma_o{}^p$ for toluene are 27.8 and 1.3 mN/m, respectively (Clint & Wicks, 2001; Binks & Clint, 2002), and equation (10) predicts S=7.3 mN/m for toluene for γ_o=29.1 mN/m.

A small value of γ_o^p=1.3 mN/m for toluene results in rather substantial increase in S because of the strong polar forces of water (γ_w^p=51 mN/m).

Recent molecular dynamics simulation of the oil-water interface reveals a preferential accumulation of aromatics at the interface, due to the weak hydrogen bonding between hydrogen atoms of water and π-electrons of aromatics (Raschke & Levitt, 2004; Kunieda et al., 2010). In contrast, the low molecular mass saturates in multicomponent mixtures, e.g. mineral oil, preferentially accumulate at the surface and give low values of surface tension and the corresponding spreading coefficients fall below the line for alkanes.

Figure 1 includes the relationships for the refractive index and surface tension of the crude oil as a function of their density. Though refractive indices for the crude oil are slightly above the line for the alkanes, their surface tension values fall significantly below the line. This could be due to preferential accumulation of light end alkanes at the crude oil surface. Detailed physical properties of the crude oil can be found in Buckley & Fan (2007) and Loahardjo (2009).

This was confirmed by examining measured surface tensions of decane ($n\text{-}C_{10}H_{22}$) binary mixtures with hexadecane ($n\text{-}C_{16}H_{34}$), docosane ($n\text{-}C_{22}H_{46}$), and tetracosane ($n\text{-}C_{24}H_{50}$) (Rolo et al., 2002; Queimada et al., 2005). As seen in Figure 3, measured surface tension values of these binary mixtures at 20, 50 and 70°C, respectively, systematically fell below linear lines due to the surface excess of decane. The effect is less pronounced for lower molecular mass molecules and higher temperature. The measured surface tension of the decane/hexadecane is a linear function of the weight fraction at 60°C. The density of these binary mixtures is a linear function of the weight fraction, as expected for the ideal mixture at a wide range of temperature (Queimada et al., 2003).

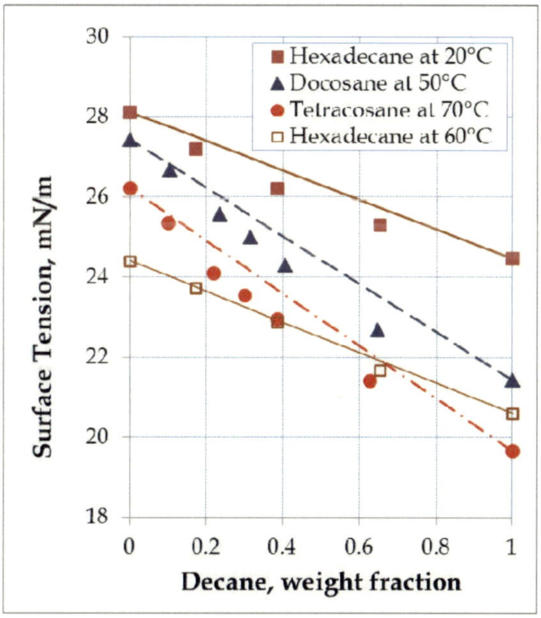

Fig. 3. Measured surface tension of decane binary mixtures with hexadecane, docosane, and tetracosane as a function of the weight fraction of decane.

3. Experimental

A non-ionic surfactant, BIO-SOFT® N91-8 (Stepan, Illinois, USA) was used as received. The surfactant is characterized as $CH_3(CH_2)_nO(CH_2CH_2O)_yH$, where n=8-10 and the average moles of ethoxylation y=8.3. The surface tension of the synthetic sea water (NaCl, KCl, $CaCl_2{\cdot}2H_2O$, and $MgCl_2{\cdot}6H_2O$ of 28.0, 0.935, 1.56 and 11.56 g/L, respectively) of pH=6-7 was measured as a function of BIO-SOFT® N91-8 concentration using the Wilhelmy plate method as adapted for the Krüss K100 tensiometer. A Krüss DVT-10 drop volume tensiometer was used to measure the interfacial tension between crude oil and sea water. For the spreading experiment, approximately 35 mL of the synthetic sea water was placed in a glass beaker of 5.5 cm inside diameter and 3.5 cm height. A drop of the crude oil (~50μL) was then placed on the water surface from a disposable glass pipette and spreading behaviour was video recorded.

4. Results and discussion

4.1 Molecular dynamics simulation of heptane-decane-toluene ternary mixtures

Spreading behaviour of a drop of the mixed solvent containing heptane, decane, and toluene of 250, 500, and 250 molecules, respectively was simulated on the surface of 18,458 water molecules. Detailed description will be given in a separate paper (Kunieda et al., 2012).

Results of the simulation are summarized in Figure 4 with a series of time lapse images, which show a spherical drop of 6.9 nm in diameter transforming to a liquid lens of maximum height of 4.1 nm after 2.97 ns of computational time. Only toluene molecules in the mixed solvent are shown in the third row of Figure 4, which clearly illustrates accumulation of toluene molecules at the water/organic solvent interface.

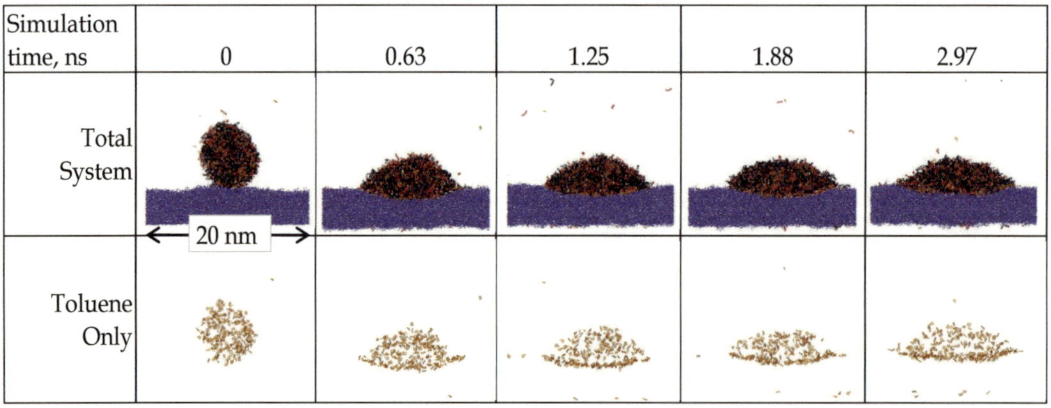

Fig. 4. Results of the molecular dynamics simulation of a mixed solvent of heptane, decane, and toluene of 0.25/0.50/0.25 mole fraction on the water surface at 25°C (the second row). Toluene molecules only in the mixed solvent are selectively shown in the third row.

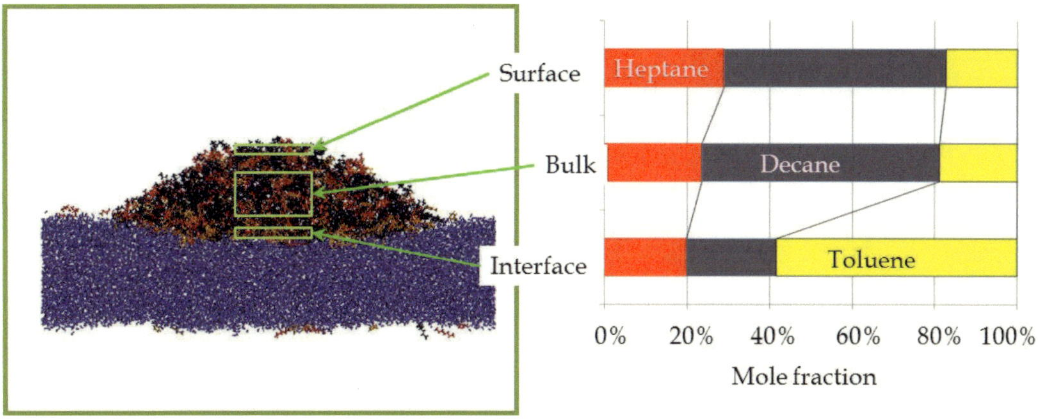

Fig. 5. Mole fraction of heptane, decane, and toluene in the bulk oil lens as well as the surface and interfacial regions after 2.97 ns computational time.

Changes in composition of the mixed solvent in the bulk, and the surface and interfacial regions are illustrated in Figure 5, which clearly demonstrates the preferential accumulation of toluene molecules at the water/oil interface; 0.25 in the bulk to slightly less than 0.6 mole fraction at the interface. Here, the interfacial region is defined as oil molecules which are at least one hydrogen atom within 0.3 nm from water molecules. The figure also confirms the surface excess of heptane against decane molecules.

4.2 Spreading behaviour of model binary and ternary mixtures on water

Density and surface tension of toluene-heptane, toluene-decane, and heptane-decane (add comma) binary mixtures were measured and reported in a previous paper (Kunieda et al., 2012). Figure 6(a) illustrates measured densities of the binary mixtures as a function of the weight fraction of either toluene or decane. The relationship for the binary mixture of alkanes (heptane and decane) follows the linear relationship against the weight fraction of decane, indicating no change in the volume upon mixing as for an ideal mixture. In comparison, those for the alkane-aromatic binary mixtures deviate systematically from the linear relationships towards lower densities, indicating the excess volume of alkanes-aromatic mixtures as reported by Qin et al. (1992). Solid diamonds in the figure represent the molecular dynamics simulation predicted density of toluene-heptane binary mixtures, which are in good agreement with measured values. This confirms that the molecular dynamics adequately simulate interactions between toluene and heptane in the bulk solution.

The measured surface tension of these binary mixtures is plotted against the weight fraction of either toluene or decane in Figure 6(b). The γ_o of a blend of alkanes, with small difference in molecular weights, follows a linear relationship against weight fraction. The same figure demonstrates pronounced reduction in surface tension from ideality of the alkane-aromatic binary mixtures. For decane/toluene, nearly half of the reduction in the surface tension of toluene occurs with addition of only 0.28 weight fraction (0.20 mole fraction) of decane. These observations are consistent with preferential accumulation of alkanes at the oil/air surface against the aromatic as shown in Figure 5 and also discussed in Kunieda et al. (2012).

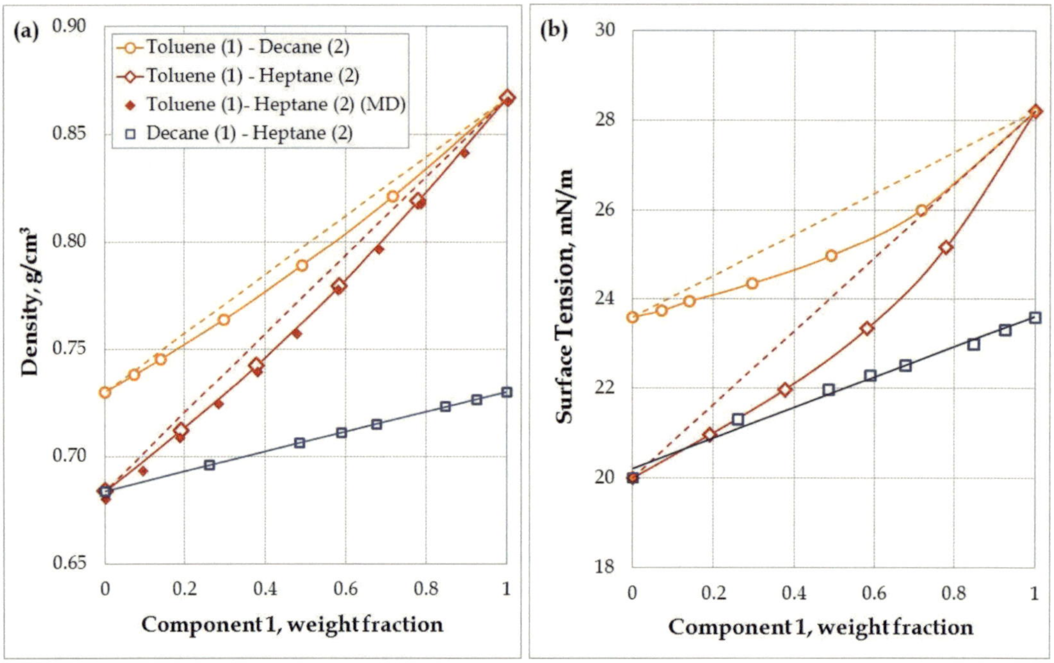

Fig. 6. Measured density (a) and surface tension (b) of toluene-heptane, toluene-decane, and decane-heptane binary mixtures as a function of the weight fraction of toluene or decane. The density of the binary mixture of alkanes is a linear function of the weight fraction, indicating an ideal mixture.

The interfacial tension of these binary mixtures was also measured against water at pH=6-7, and the spreading coefficient was calculated using equation (4). Results are summarized in Figure 7 by plotting calculated *S* against the surface tensions of the mixtures. Measured interfacial tensions of toluene, heptane, and decane are 35.2, 50.1, and 50.0 mN/m, respectively, which are about 2 mN/m lower than literature values even after removing polar components by flow through silica gel and alumina columns up to six times until the interfacial tension values were constant. Points for the heptane-decane binary mixture follow the linear relationship for a series of alkanes from the literature as discussed in Figure 2. Slightly lower values of the measured interfacial tension than the literature values of decane and heptane causes a slight deviation to larger *S* than the literature relationship.

The relationships, based on measurements for the toluene-decane and toluene-heptane binary mixtures, are highly non-linear with respect to composition, especially at above 0.4 mole fraction of toluene (see Figure 7). The *S* of pure toluene is 9.1 mN/m; addition of as much as 0.6 mole fraction of heptane has no significant effect on the spreading behaviour because *S*=9.5 mN/m for this mixture. The maximum, *S*=10.3 mN/m, is observed at 0.6-0.8 mole fraction of toluene. Thus, the binary mixture spreads more readily than either of the pure liquid components. The figure also shows addition of as much as 20% mole fraction of decane to toluene has no significant effect (*S*=9.0 mN/m). This is especially remarkable since *S*=-2.1 mN/m for the pure decane.

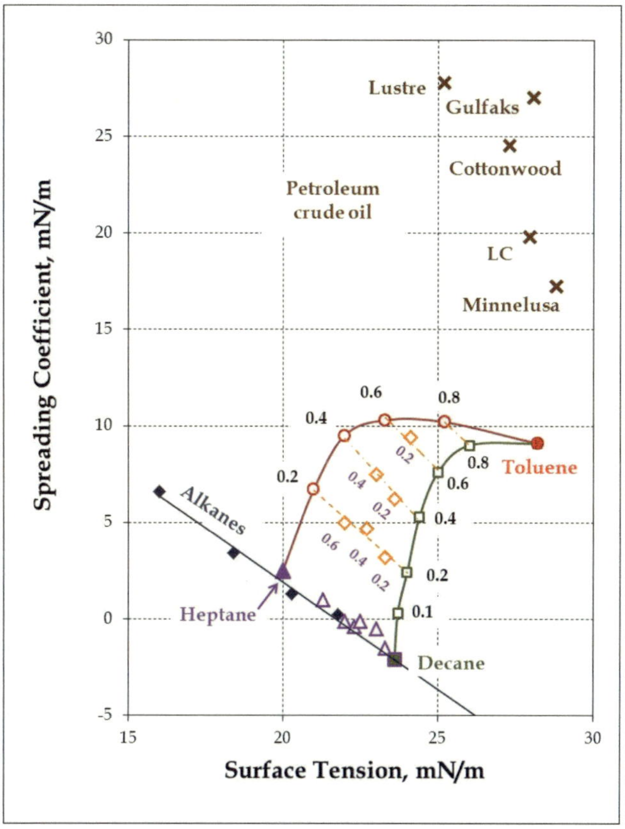

Fig. 7. Calculated spreading coefficient of toluene-heptane (circle), toluene-decane (square), and heptane-decane (triangle) binary systems as well as their ternary mixtures (open diamond). Numerical values beside data points for the binary mixtures represent the mole fraction of toluene for the binary mixtures, and mole fraction of heptane for the ternary mixtures. The same relationships for five crude oils tested for spreading behaviour on sea water are also included (cross).

Spreading coefficients for the ternary mixtures of toluene-heptane-decane are also included in the same figure. Dashed lines for the constant mole fraction of toluene are almost parallel to the linear relationship for the alkanes. The heptane-decane fractions merely affect the surface tension of the ternary mixtures and the fraction of toluene in the mixture determines the interfacial tension. S=9.4 mN/m for the ternary mixture of 0.2/0.2/0.6 mole fraction of decane/heptane/toluene, respectively, which is 0.3 mN/m higher than S for pure toluene.

These observed complexities of the spreading behaviour of the binary and ternary mixtures result from specific adsorption of the toluene and heptane at the oil/water and oil/vapor interfaces, respectively, as predicted by the molecular dynamics simulations (Figure 5).

4.3 Spreading of crude oil on sea water

Five crude oils of widely different properties were selected to examine their spreading behaviour on synthetic sea water at room temperature. Some physico-chemical properties of these crude oils at 20°C are listed in Table 1 together with the calculated spreading coefficient (S) from equation (4) using their surface and interfacial tensions and γ_w=73.5 mN/m. Additional properties of these crude oils can be found in Buckley & Fan (2007) and Loahardjo (2009). Relationships between the surface tension and S are also included in Figure 7. Values of S for the crude oil are 2-3 times larger than the aromatics, mostly because of their low interfacial tensions. This suggests polar contributions of γ_o^p=1.5-3.9 mN/m for the crude oils instead of 1.3 mN/m for toluene, suggesting the presence of weak accumulation of polar molecules in the resins and asphaltenes fractions at the oil/sea water interface at pH=6-7. In comparison, the reported value of γ_o^p for 1-undecanol, $C_{11}H_{23}OH$, is as large as 20 mN/m (Binks & Clint, 2002). Detailed discussion of the correlation between the chemical composition of the crude oil and interfacial tension can be found in (Buckley & Fan, 2007). van Oss et al. (1988) have proposed correlating the polar component to the acid-base interactions, and this could certainly be applicable to the crude oil/water interface.

Crude Oil	Aromatics %	Asphaltenes (n-C_6)%	Resins %	Density g/cm^3	Surf. Ten. mN/m	Int. Ten.[1] mN/m	S mN/m	Visc. mPa s
Lustre	18	1.0	9.0	0.840	25.2	20.5	**27.8**	5.0
Gulfaks	26	0.3	16	0.894	28.1	18.4	**27.0**	35
Cottonwood	23	2.9	17	0.893	27.3	21.6	**24.5**	26
LC	25	3.2	12	0.903	28.0	25.7	**19.8**	39
Minnelusa	20	9.1	13	0.904	28.8	27.4	**17.2**	58

1. against sea water at pH=6-7

Table 1. Selected physico-chemical properties of five crude oils used for the spreading experiments on the sea water. All measurements are at 20°C.

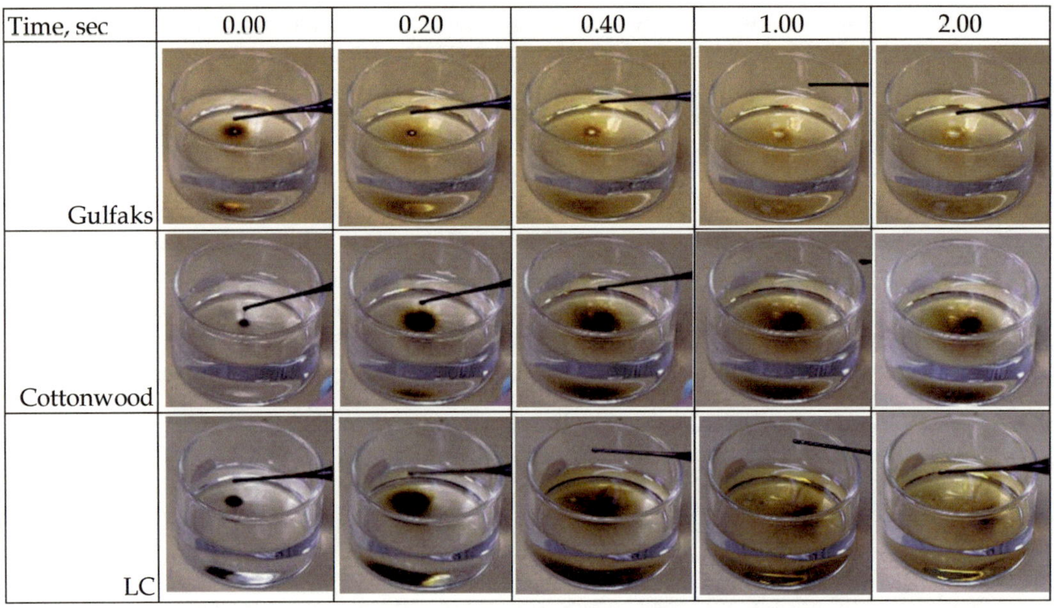

Fig. 8. Spreading behaviour of a drop (~50μL) of selected crude oils on sea water at 20°C.

Spreading behaviour of three crude oils; Gulfaks, Cottonwood, and LC are summarized in Figure 8 as a series of photographs illustrating the initial spreading of the crude oil on sea water. Calculated spreading coefficients for these crude oils are much higher than for pure alkanes and aromatics of similar γ_o, and range from 17 mN/m to 28 mN/m, as shown in Table 1. All of the tested crude oils spread rapidly on the sea water surface as a thin layer. The relationship between the thickness of the spreading oil as a function of S and density has been formulated by Langmuir (1933). Cochran & Scott (1971) developed an equation which related the spreading rate of the thin oil layer over the surface of water to the combined effect of the spreading pressure, gravitational forces, and hydrodynamic resistance which is a function of the crude oil viscosity.

4.4 Control of the spreading of crude oil on sea water

Even a very low level of surfactant concentration, especially a non-ionic, is known to effectively lower the surface tension of water. Addition of as little as 30 ppm BIO-SOFT® N91-8 lowered γ_w of the sea water by 29 mN/m from 73 to 45 mN/m, (see Figure 9(a)). Figure 9(b) represents the measured interfacial tension of 5 crude oils as a function of the concentration of the same non-ionic surfactant. The γ_{wo} of the crude oil/sea water interface is already low in the absence of this non-ionic surfactant due to adsorbed polar components as discussed above. The γ_{wo} of Minnelusa oil is reduced only 8 mN/m, from 27 to 19 mN/m, with the same level of BIO-SOFT® N91-8 concentration. Detailed discussion on the interfacial tension and the micellar structure of the surfactant molecules at the oil/water interface can be found in Hoffmann (1990).

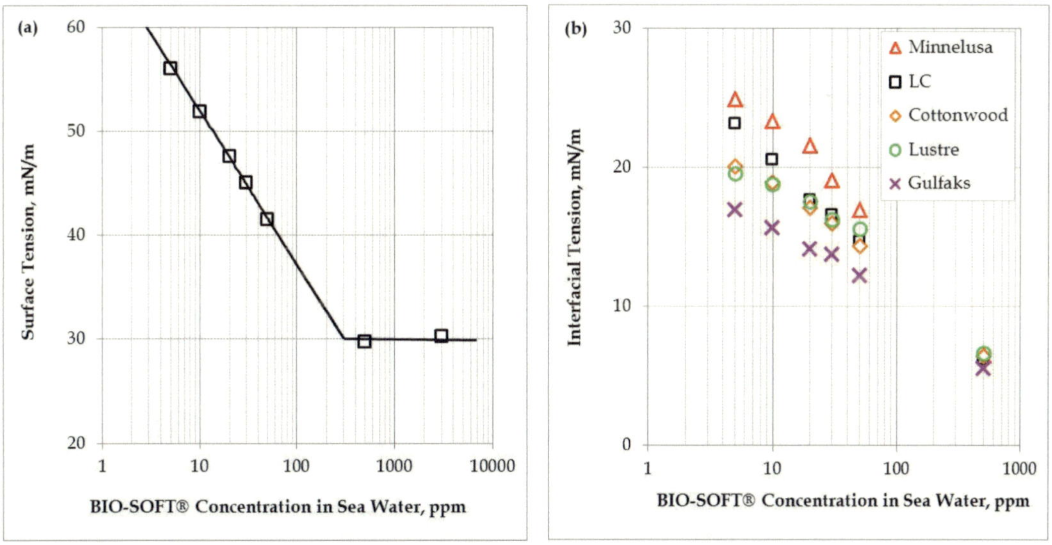

Fig. 9. The surface tension of the sea water (a) and interfacial tension of 5 crude oils (b) as a function of the non-ionic surfactant, BIO-SOFT® N91-8, concentration.

Equation (4) suggests that addition of even a small amount of non-ionic surfactant to the water would prevent spreading of crude oil ($S<0$) over the surface of sea water. The spreading coefficients for five crude oils were calculated as a function of BIO-SOFT® N91-8 concentration in Table 2. The *S* values in the table were calculated using measured γ_{wo} of each crude oil against sea water and for sea water with a given concentration of BIO-SOFT® N91-8.

Crude oil	S (with γ_{wo} of sea water)				
	Sea water with BIO-SOFT®, ppm				
	0	5	10	20	30
Lustre	28	10	6	2	-1
Gulfaks	27	10	5	1	-1
Cottonwood	25	7	3	-1	-4
LC	20	2	-2	-6	-9
Minnelusa	17	0	-4	-9	-11
	S (with γ_{wo} of sea water/BIO-SOFT®)				
Lustre	28	11	8	5	4
Gulfaks	27	11	8	5	3
Cottonwood	25	9	6	3	2
LC	20	5	3	2	0
Minnelusa	17	2	0	-3	-3

Crude oil	Initial Spreading Behavior				
	Sea water with BIO-SOFT®, ppm				
	0	5	10	20	30
Lustre	SP	INT	INT	NO	NO
Gulfaks	SP	INT	INT	NO	NO
Cottonwood	SP	INT	INT	NO	NO
LC	SP	NO	NO	NO	NO
Minnelusa	SP	INT	INT	NO	NO

Table 2. Calculated *S* and observed spreading behaviour of five crude oils on sea water. INT indicates that slow spreading of the oil droplet was still observed. *S* was calculated using the γ_{wo} of sea water in the presence (top) and absence (bottom) of BIO-SOFT® N91-8.

The spreading behaviour of Minnelusa crude oil over sea water shown in Figure 10 clearly demonstrates that as little as 5 ppm, ~1.6% of the critical micelle concentration, of the non-ionic surfactant is enough to prevent the rapid spreading of the crude oil. Observed spreading behaviour shown in Table 2 agrees better with the *S* calculated using γ_{wo} against sea water instead of sea water with a given amount of BIO-SOFT® N91-8. In the table, INT indicates that some slow spreading of the oil droplet still occurred as seen in Figure 10. This observation suggests that the surfactant molecules are not yet adsorbed at the rapidly spreading frontal perimeter of the oil/sea water interface.

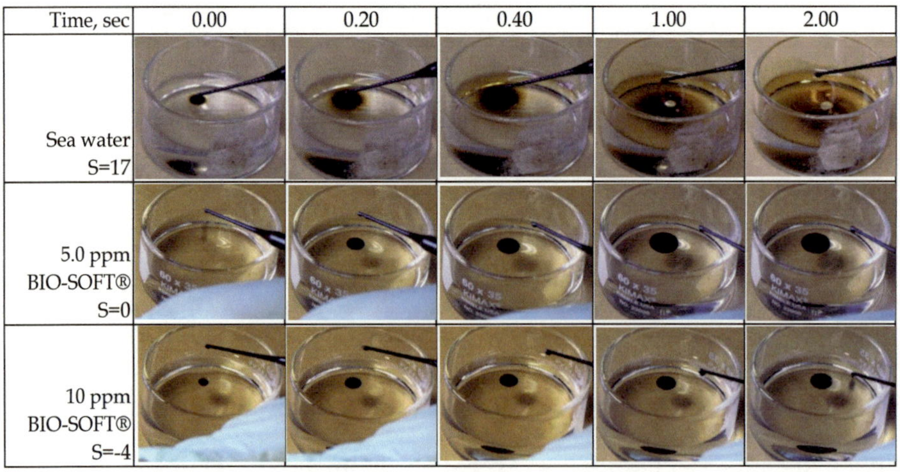

Fig. 10. Spreading behaviour of Minnelusa crude oil on sea water in the absence and the presence of 5 and 10 ppm BIO-SOFT® N91-8.

4.5 Retraction of spilled crude oil

Results shown in Figure 10 and Table 2 suggest that placement of a small amount of the aqueous non-ionic surfactant solution would effectively cause retraction of the spilled crude oil on the water surface. This is demonstrated in Figure 11, where a drop of Minnelusa crude oil (*S*=17 mN/m) quickly spread on the sea water surface (time=0 sec). The thin film of crude oil has quickly retracted to the opposite side after a drop of 500 ppm aqueous BIO-SOFT® N91-8 solution was added near the side wall of a glass beaker (top photographs). The lens of crude oil of (> 1 mm in thickness) can be skimmed mechanically from the open water surface (Cochran & Scott, 1971). The retraction of the thin film was confirmed for all of the five crude oils tested in this study.

Time, sec	0.00	0.06	0.20	0.40	1.00
Minnelusa					
Time, sec	0.00	0.02	0.06	0.15	0.40
Minnelusa					

Fig. 11. Retraction of a spilled thin film of Minnelusa crude oil by addition of a drop of 500 ppm aqueous BIO-SOFT® N91-8 solution at the side (first row) or centre (second row) of a glass beaker.

At the bottom row of photographs in Figure 11, a drop of 500 ppm aqueous BIO-SOFT® N91-8 solution was placed at the center of the thin oil disc covering the entire water surface, causing rapid retraction of the oil outward toward the perimeter of the oil disc. Build-up of a slightly thicker oil ridge can clearly be seen after 0.02 and 0.06 sec as the oil film retracts. In this demonstration, the glass wall of the beaker restricts further spreading of the oil disc, resulting in formation of a narrow ring of the oil film at 0.40 sec. This oil ring would break-up under weak mechanical disturbance (i.e. wave action) to smaller size droplets.

The outer perimeter of the oil ring is facing toward the water of high surface tension, thus a high value of *S*, whereas the inner perimeter is associated with the water of lower surface tension (and *S*<0). On an open water surface, the gradient in the capillary forces will cause rapid outward spreading of the thin oil film. This would result in break-up of the oil disc at the rapidly expanding outside perimeter of the thin oil disc as described by Zisman (1941a). This suggest the same chemical could act both as the retracting and dispersing agent based on the location of its placement; either in the oil or water phase.

4.6 Aging of thin films of crude oil on the surface of water

A drop of Lustre crude oil spread quickly (*S*=28 mN/m); a thin oil film covered the entire surface of sea water in a glass beaker within a second. The thin oil film retracted quickly (<3 sec) when a drop of the aqueous 500 ppm BIO-SOFT® N91-8 solution was added, as shown by the series of photographs in the first row of Figure 12.

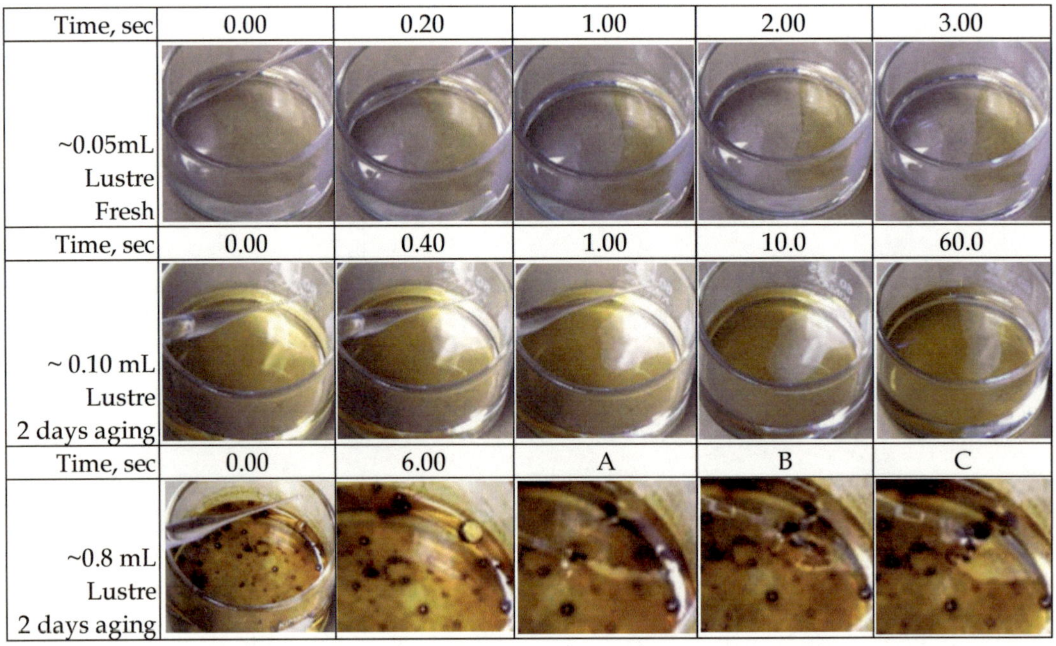

Fig. 12. Retraction of fresh and aged thin film of Lustre crude oil by addition of a drop of 500 ppm BIO-SOFT® N91-8 solution. A semi-rigid film was formed when a thicker film of the crude oil (with 0.8 mL of the oil instead of 0.05 and 0.1 mL) was aged for two days.

For the photographs shown in the second row, twice the amount (~0.10 mL) of Lustre crude oil was allowed to spread over the sea water surface and the beaker was left open in a fume hood for two days. The retraction of the aged thin film caused by a drop of 500 ppm BIO-SOFT® N91-8 solution was significantly slower than for freshly spread oil (see Figure 12).

The third row of photographs in Figure 12 were taken after, approximately 0.8 mL of Lustre crude oil was allowed to spread on the sea water surface followed by aging for two days. A thin, semi-rigid film of aromatics, rich in polar asphaltenes and resins covers the water surface. A drop of BIO-SOFT® N91-8 solution was added along the glass wall. After 6 seconds, a small round hole formed in this semi-rigid film; the rest of the film remained intact. This semi-rigid film could be broken and rolled up using the tip of a glass pipette as shown in A, B, and C of the third row of Figure 12. This observation simulates how wave action can cause formation of tar balls at sea far from the parent spill of crude oil. These tar balls are very sticky because of their high content of aromatics in the form of asphaltenes and resins.

5. Conclusion

The surface tension of alkanes is the result of the London dispersion forces, which are directly proportional to density. When the surface tensions of a series of crude oils is plotted against their densities, they also follow a close-to-linear relationship but the values fall significantly below those for alkanes of the same density due to preferential accumulation of light end alkanes at the crude oil surface.

Molecular dynamics simulation of the aromatic-water interface reveals a similar preferential accumulation of aromatics at the interface, due to the weak hydrogen bonding between the hydrogen atoms of water and the π-electrons of aromatics. This explains why aromatic-water interfacial tensions (e.g. 34.4 mN/m for benzene at 20°C) are lower than those for alkane-water (e.g. 52.0 mN/m for n-heptane).

Aromatics that are preferentially accumulated at the crude oil-water interface will promote migration of asphaltenes and polar components in the crude oil toward the interface, resulting in further reduction in the interfacial tension between crude oil and sea water to below 30 mN/m. The combination of low surface and interfacial tension values for crude oil promotes rapid spreading of crude oil on sea water when an oil spill occurs. Light end alkanes evaporate quickly due to their low vapour pressure and a thin film of aromatics, rich in asphaltenes and resins, spreads ahead of the bulk crude oil over the water surface. The spreading action results in formation of tar balls far from the parent spill of crude oil. Spreading of the crude oil can be prevented or at least greatly reduced by lowering the surface tension of the water. This can be achieved, for example by use of a very low concentration of non-ionic surfactant.

6. References

Atkinson, H. (1927). Recovery of Petroleum from Oil Bearing Sands. *U.S. Patent* No. 1,651,311

Binks, P. B. & Clint, J. H. (2002). Solid Wettability from Surface Energy Components: Relevance to Pickering Emulsions. *Langmuir*, Vol. 18, No. 4, (February 2002), pp. 1270–1273, ISSN 1520-5827

Barger, W. R. & Garrett, W. D. (1968). Modification of the Air/Sea Interface by Artificial Sea Slicks. *U.S. Naval Research Laboratory Report* 6763, (September 1968)

Blodgett, K. B. (1935). Films Built by Depositing Successive Monomolecular Layers on a Solid Surface. *Journal of American Chemical Society*, Vol. 57, No. 6, (June 1935), pp. 1007–1022, ISSN 1520-5126

Buckley, J. S. & Fan, T. (2007). Crude Oil/Brine Interfacial Tensions. *Petrophysics*, Vol. 48, No. 3, (June 2007), pp. 175–185

Chatzis, I.; Morrow, N. R. & Lim, H. T. (1983). Magnitude and Detailed Structure of Residual Oil Saturation. *Society of Petroleum Engineers Journal*, Vol. 23, No. 2, (April 1983), pp. 311–326, ISSN 1930-0220

Chow, R. S. & Takamura, K. (1988). Electrophoretic Mobilities of Bitumen and Conventional Crude-in-Water Emulsions Using the Laser Doppler Apparatus in the Presence of Multivalent Cations. *Journal of Colloid and Interface Science*, Vol. 125, No. 1, (September 1988), pp. 212–225, ISSN 0021-9797

Clint, J. H. & Wicks, A. C. (2001). Adhesion Under Water: Surface Energy Considerations. *International Journal of Adhesion & Adhesives*, Vol. 21, No. 4, pp. 267–273, ISSN 0143-7496

Cochran, R. A. & Scott, P. R. (1971). The Growth of Oil Slicks and Their Control by Surface Chemical Agents. *Journal of Petroleum Technology*, Vol. 23, No. 7, (July 1971), pp. 781–787, ISSN 0149-2136

Covington, E. J. (March 2011). Katharine B. Blodgett, October 11, 2011, Available from: http://home.frognet.net/~ejcov/blodgett2.html

Danielli, J. F. (1937). The Relations between Surface pH, Ion Concentrations and Interfacial Tension. *Proceedings of the Royal Society of London. Series B*, Vol. 122, (April 1937), pp. 155–174, ISSN 1471-2954

Davies, J. T., & Rideal, E. K. (1961). *Interfacial Phenomena*, Academic Press, 61-8494, London

Derrick, M. E. (1982). Profiles in Chemistry, Agnes Pockels, 1862-1935. *Journal of Chemical Education*, Vol. 59, No. 12, (December 1982), pp. 1030–1031, ISSN 0021-9584

Fowkes, F. M. (1964). Attractive Forces at Interfaces. *Industrial and Engineering Chemistry*, Vol. 56, No. 12, (December 1964), pp. 40–52

Garrett, W. D. & Barger, W. R. (1970). Factors Affecting the Use of Monomolecular Surface Films to Control Oil Pollution on Water. *Environmental Science & Technology*, Vol. 4, No. 2, (February 1970), pp. 123–127, ISSN 1520-5851

Girifalco, L. A. & Good, R. J. (1957). A Theory for the Estimation of Surface and Interfacial Energies. I. Derivation and Application to Interfacial Tension. *Journal of Physical Chemistry*, Vol. 61, No. 7, (July 1957), pp. 904–909

Harkins, W. D. & Feldman, A. (1922). Films. The Spreading of Liquids and the Spreading Coefficient. *Journal of American Chemical Society*, Vol. 44, No. 12, (December 1922), pp. 2665–2685, ISSN 1520-5126

Harkins, W. D. (1941). A General Thermodynamic Theory of the Spreading of Liquids to Form Duplex Films and of Liquids or Solids to Form Monolayers. *Journal of Chemical Physics*, Vol. 9, No. 7, (July 1941), pp. 552–568, ISSN 1089-7690

Hoffmann, H. (1990). Fascinating Phenomena in Surfactant Chemistry. *Advances in Colloid and Interface Science*, Vol. 32, No. 2-3, (August 1990), pp. 123–150, ISSN 0001-8686

Hunter, R. J. (1986). *Foundations of Colloid Science Vol. I*, pp. 219, Oxford University Press, ISBN 0-19-855187-8, Oxford

Johnson, C. E. (1976). Status of Caustic and Emulsion Methods. *Journal of Petroleum Technology*, Vol. 28, No. 1, (January 1976), pp. 85–92, ISSN 0149-2136

Kunieda, M.; Nakaoka, K.; Liang, Y.; Miranda, C. R.; Ueda, A.; Takahashi, S.; Okabe, H. & Matsuoka, T. (2010). Self-Accumulation of Aromatics at the Oil-Water Interface through Weak Hydrogen Bonding. *Journal of American Chemical Society*, Vol. 132, No. 51, (December 2010), pp. 18281–18286, ISSN 1520-5126

Kunieda, M.; Liang, Y.; Fukunaka, Y.; Matsuoka, T.; Takamura, K.; Loahardjo, N.; Winoto, W. & Morrow, N. R. (2012). Spreading of Multi-component Oils on Water. *Energy and Fuels*, in press, http://dx.doi.org/10.1021/ef201530k

Laidlaw, W. G. & Wardlaw, N. C. (1983). A Theoretical and Experimental Investigation of Trapping in Pore Doublets. *Canadian Journal of Chemical Engineering*, Vol. 61, No. 5, (October 1983), pp. 719–727

Langmuir, I. (1933). Oil Lenses on Water and the Nature of Monomolecular Expanded Films. *Journal of Chemical Physics*, Vol. 1, No. 11, (November 1933), pp. 756–776, ISSN 1089-7690

Langmuir, I. (1936). Two-Dimensional Gases, Liquids and Solids. *Science*, Vol. 84, pp. 379–383

Loahardjo, N. (December 2009). Improved Oil Recovery by Sequential Waterflooding and by Injection of Low Salinity Brine. *Ph.D. Dissertation*, University of Wyoming, Laramie, Wyoming

Mayer, E. H.; Berg, R. L.; Carmichael, J. D. & Weinbrandt, R. M. (1983). Alkaline Injection for Enhanced Oil Recovery - A Status Report. *Journal of Petroleum Technology*, Vol. 35, No. 1, (January 1983), pp. 209–221, ISSN 0149-2136

Owens, D. K. & Wendt, R. C. (1969). Estimation of the Surface Free Energy of Polymers. *Journal of Applied Polymer Science*, Vol. 13, No. 8, (August 1969), pp. 1741–1747, ISSN 1097-4628

Pomerantz, P.; Clinton, W. C. & Zisman, W. A. (1967). Spreading Pressures and Coefficients, Interfacial Tensions, and Adhesion Energies of the Lower Alkanes, Alkenes, and Alkyl Benzenes on Water. *Journal of Colloid and Interface Science*, Vol. 24, No. 1, (May 1967), pp. 16–28, ISSN 0021-9797

Qin, A.; Hoffman, D. E. & Munk, P. (1992). Excess Volume of Mixtures of Alkanes with Aromatic Hydrocarbons. *Journal of Chemical Engineering Data*, Vol. 37, No. 1, (January 1992), pp. 61–65, ISSN 1520-5134

Queimada, A. J.; Quinones-Cisneros, S. E.; Marrucho, I. M.; Coutinho, J. A. P. & Stenby, E. H. (2003). Viscosity and Liquid Density of Asymmetric Hydrocarbon Mixtures. *International Journal of Thermophysics*, Vol. 24, No. 5, (September 2003), pp. 1221–1239, ISSN 1572-9567

Queimada, A. J.; Caço, A. I.; Marrucho, I. M. & Coutinho, J. A. P. (2005). Surface Tension of Decane Binary and Ternary Mixtures with Eicosane, Docosane, and Tetracosane. *Journal of Chemical Engineering Data*, Vol. 50, No. 3, (May 2005), pp. 1043–1046, ISSN 1520-5134

Raschke, T. M. & Levitt, M. (2004). Detailed Hydration Maps of Benzene and Cyclohexane Reveal Distinct Water Structures. *Journal of Physical Chemistry B*, Vol. 108, No. 35, (September 2004), pp. 13492–13500, ISSN 1520-5207

Rayleigh & Pockels, A. (1891). Surface Tension. *Nature*, Vol. 43, (March, 1891), pp. 437–439

Rayleigh. (1899). Investigations in Capillarity. *Philosophical Magazine*, Vol. 48, pp. 321–337

Rolo, L. I.; Caço, A. I.; Queimada, A. J.; Marrucho, I. M. & Coutinho, J. A. P. (2002). Surface Tension of Heptane, Decane, Hexadecane, Eicosane, and Some of Their Binary Mixtures. *Journal of Chemical Engineering Data*, Vol. 47, No. 6, (November 2002), pp. 1442–1445, ISSN 1520-5134

Shewmaker, J. E.; Vogler, C. E. & Washburn, E. R. (1954). Spreading of Hydrocarbons and Related Compounds on Water. *Journal of Physical Chemistry*, Vol. 58, No. 11, (November 1954), pp. 945–948

Sjöblom, J.; Aske, N.; Auflen, I. H.; Brandal, Ø.; Havre, T. E.; Sæther, Ø., Westvik, A.; Johnsen, E. E. & Kallevik, H. (2003). Our Current Understanding of Water-in-Crude Oil Emulsions. Recent Characterization Techniques and High Pressure Performance. *Advances in Colloid and Interface Science*, Vol. 100–102, (February 2003), pp. 399–473, ISSN 0001-8686

Squires, F. (1921). Method of Recovering Oil and Gas. *U.S. Patent* No. 1,238,355

Takamura, K. & Chow, R. S. (1985). The Electric Properties of the Bitumen/Water Interface Part II. Application of the Ionizable Surface-Group Model. *Colloids and Surfaces*, Vol. 15, pp. 35–48, ISSN 0927-7757

Transue, L. F.; Washburn, E. R. & Kahler, F. H. (1942). The Direct Measurement of the Spreading Pressures of Volatile Organic Liquids on Water. *Journal of American Chemical Society*, Vol. 64, No. 2, (February 1942), pp. 274–276, ISSN 1520-5126

Yu, L. & Wardlaw, N. C. (1986). The Influence of Wettability and Critical Pore-Throat Size Ratio on Snap-off. *Journal of Colloid and Interface Science*, Vol. 109, No. 2, (February 1986), pp. 461–472, ISSN 0021-9797

van Oss, C. J.; Good, R. J. & Chaudhury, M. K. (1988). Additive and Nonadditive Surface Tension Components and the Interpretation of Contact Angles. *Langmuir*, Vol. 4, No. 4, (July 1988), pp. 884–891, ISSN 1520-5827

Zapf, A.; Beck, R.; Platz, G. & Hoffmann, H. (2003). Calcium Surfactants: a Review. *Advances in Colloid and Interface Science,* Vol. 100–102, (February 2003), pp. 349–380, ISSN 0001-8686

Zisman, W. A. (1941a). The Spreading of Oils on Water Part I. Ionized Molecules Having Only One Polar Group. *Journal of Chemical Physics*, Vol. 9, No. 7, (July 1941), pp. 534–551, ISSN 1089-7690

Zisman, W. A. (1941b). The Spreading of Oils on Water Part II. Non-Ionized Molecules Having Only One Polar Group. *Journal of Chemical Physics*, Vol. 9, No. 10, (October 1941), pp. 729–741, ISSN 1089-7690

Zisman, W. A. (1941c). The Spreading of Oils on Water Part III. Spreading Pressures and the Gibbs Adsorption Relation. *Journal of Chemical Physics*, Vol. 9, No. 11, (November 1941), pp. 789–793, ISSN 1089-7690

Zisman, W. A. (1943). Spreading Agents for Clearing Water Surface of Oil Film. *Naval Research Laboratory Report* P-1984, (January 1943)

7

Fate of Subsurface Migration of Crude Oil Spill: A Review

P. O. Youdeowei
Institute of Geosciences and Space Technology,
Rivers State University of Science and Technology,
Port Harcourt,
Nigeria

1. Introduction

The sensitivity of crude oil operational areas and the devastating effect of crude oil spillage on the lithosphere, biosphere, hydrosphere and atmosphere is established fact. The impacts of these spills on the ecosystem leave indelible imprints. While literature abounds on studies involving the collection of baseline data and the prediction of effects using expert judgments, experiments, interviews and models, mitigation and control measures to minimize the impact has recorded limited success over the years.

With research still in progress on ways and means to handle ecological crude oil pollution when they occur, including the search for alternative energy resources, reviews on the knowledge and understanding of the interaction of crude oil spill and the environment must remain essential. This will help to foster and evolve the much needed solution to better effective mitigation and control strategies and hence ensure a sound and sustainable environment.

Pollution occurs when the concentration of various chemical or biological constituents exceed a level at which a negative impact on amenities, the ecosystem, resources and human health can occur. Pollution results primarily from human activities (Awobajo, 1981).

The sources of pollution include sewage, urban run-off, industrial processing wastes and effluents, coastal developments, shipping activities and atmospheric dust and fall out. The chemical or biological constituents creating pollution are known as contaminants. Contaminants degrade the natural quality of a substance or medium. Inorganic contaminants include zinc, lead and pesticides. Organic contaminants consist of petroleum hydrocarbons and biological pollutants (coliform bacteria and pathogens).

Petroleum as the contaminant or pollutant of interest here is a complex mixture of naturally occurring hydrocarbons in the solid (Asphalt, pitchblende, tar), liquid (crude oil) and gaseous state (natural gas). It is a mixture of hundreds of hydrocarbons highly variable in composition whose individual chemical properties vary widely. Its properties depend on the properties of the individual constituents and is made up basically of paraffin, aromatics, asphalt and naphthenes.

Petroleum deposits contain similar elements, usually 11-15% Hydrogen and 82-87% Carbon. Some of the compounds are more amenable to volatilization than to dissolution, and vice versa. Also, other components within petroleum are not particularly prone to either and will tend to persist in the subsurface (Webb, 1985).

Petroleum products fall within the class of liquids that do not readily dissolve in water and can exist as a separate fluid phase known as non-aqueous phase liquids (NAPLs). It is also known to be lighter that water and sub-grouped as LNAPLs. Those with a density greater than water (DNAPLs) are chlorinated hydrocarbons such as carbon tetrachloride and chlorophenols (Palmer and Johnson,1989).

Non-aqueous phase liquids (NAPLs), such as petroleum, chlorinated solvents and polychlorinated biphenyl (PCB) oils, are a common cause of groundwater contamination in many industrialized countries (Keely, 1989).

Although these liquids exist as a separate fluid phase in the subsurface, they typically have solubilities orders of magnitude greater than drinking water standards.

Predicting the fate of these chemicals in the subsurface is a challenging problem that needs to be addressed at many sites, before an effective remediation result can be achieved.

2. Evaluation of physical transport parameters

Access to and the utilization of knowledge about contaminant transport and fate is difficult because of the complexity of the sub-surface environment. Furthermore transport and fate assessments require inter-disciplinary analyses and interpretations because the process involved in these activities are naturally intertwined (Keely, 1989).

The integration of information on geologic, hydrologic, chemical and biological processes into an effective contaminant transport evaluation requires data that are accurate, precise, and appropriate. Even though a given parameter, such as hydraulic conductivity, can be measured correctly and with great reproducibility, it is difficult to know how closely an observation actually represents the vertical and horizontal distribution of conductivities found at a site.

It is important to appreciate the processes involved in the transport of contaminants in both porous and fractured media under saturated or unsaturated conditions. This information will assist in the design of efficient and cost-effective monitoring networks and remediation strategies of ground and surface water resources.

The severity of soil contamination tends to be a function of the properties of the soil. Knowledge of the infiltration process is prerequisite for managing contaminant transport in the unsaturated or vadose zone (Fig.1). The vadose zone is the area between the surface of the land and the aquifer water table in which the moisture content is less than the saturation point and the pressure is less than atmospheric. Depth of the vadose zone varies greatly, depending on the region of the site. Because the vadose zone overlies the saturated zone, chemical releases at or near the land surface must pass through the vadose zone before reaching the water table. The depth to the water table which is equivalent to the thickness of the unsaturated zone, is one of the parameters that determine whether or not a pollutant will reach the water table from a surface spill. Therefore, at many contaminated sites, often both the vadose zone and the saturated zone need to be characterized and remediated (Mercer and Spalding,1989). The unsaturated zone is an integral component of the hydrological cycle, which directly influences infiltration. Infiltration is defined as the initial process of water (or contaminant) movement into unsaturated zone through the soil surface (Tombul, 2003). The maximum rate at which fluid can move into the soil is called the infiltration capacity or potential infiltration rate (Bouwer, 1978).

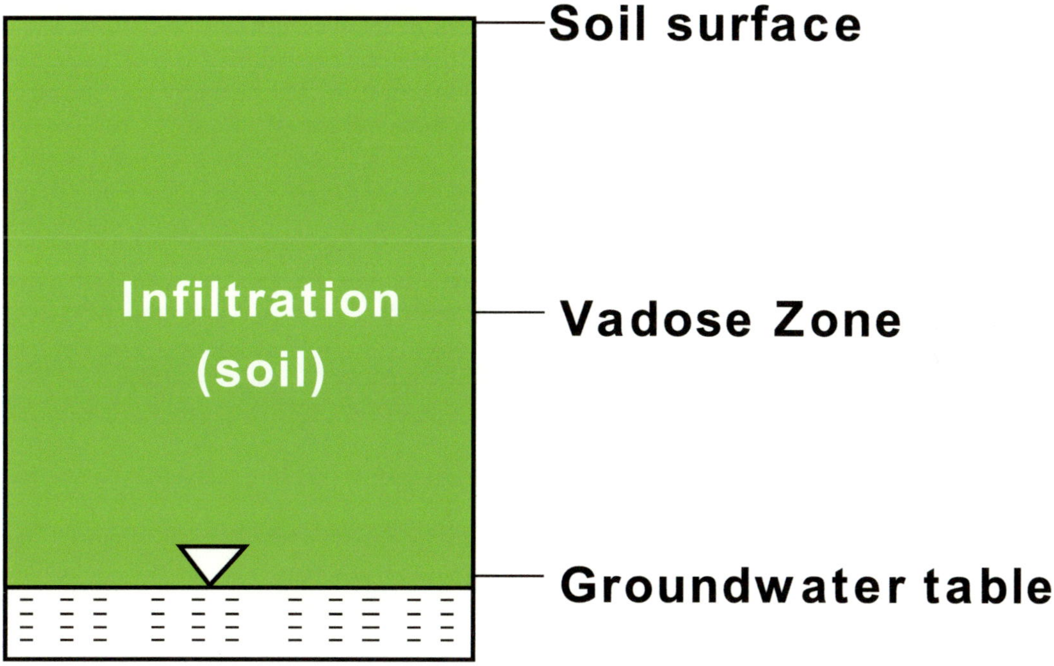

Fig. 1. Typifying the Vadose Zone

The ability to transmit fluid or the hydraulic conductivity of the soil is a highly variable quantity. If the soil is composed of well-sorted sand or gravel, the conductivity will be high and will vary only slightly with time. Most natural sediments, however, develop a stratified structure. As a result of their depositional history, the hydraulic properties (permeability, porosity etc.) in the Formations are mostly heterogeneous and anisotropic. In a homogeneous sediment, porosity and permeability are equal everywhere.

Geologic heterogeneities within the Formations make it difficult to precisely quantify the hydrogeologic system and the resulting properties affecting contaminant transport. Groundwater flow or solute (pollutant) transport at a given location depends on the permeability of the subsoil and the potential or hydraulic gradient. Effective hydraulic conductivity or permeability is an important parameter for the prediction of infiltration and run-off volume of fluids. The ability of sediments to hold and transmit fluids is determined by their porosity and permeability. The porosity of a soil or rock material is the percentage of the total volume of the material that is occupied by pores or interstices; or simply the percentage of pore spaces in the material. These pores may be filled with water if the material is saturated or with air and water if it is unsaturated.

The facility of fluid flow through any porous medium is termed permeability. Any material with voids is porous and, if the voids are interconnected, possesses permeability. Permeability or hydraulic conductivity is also the ability of a soil to conduct or discharge water under a hydraulic gradient. It depends on soil density, degree of saturation, viscosity and particle size. Materials with larger void spaces generally have larger void ratios, and so, even the densest soils are more permeable than materials such as rocks and concrete. Materials such as clays and silts in natural deposits have large values of porosity (or void ratio) but are nearly impermeable, primarily because of their very small void sizes. Since natural subsurface geology is very heterogeneous, there can be various sizes of hydraulic conductivities in a relatively small area (Miller and Hogan, 1996). Permeability vary strongly for different sediments, but porosities vary only from values of 0.2 (20%) for coarse, unsorted sands to 0.65 (65%) for clay (Bowles,1985).

3. Contamination characterization

The fate of hydrocarbons in the subsurface depends on the processes of transport, multiphase flow, volatilization, dissolution, geochemical reactions, biodegradation, and sorption (Figure 2). An interdisciplinary investigation of these processes is critical to successfully evaluate the potential for migration of hydrocarbons in the subsurface.

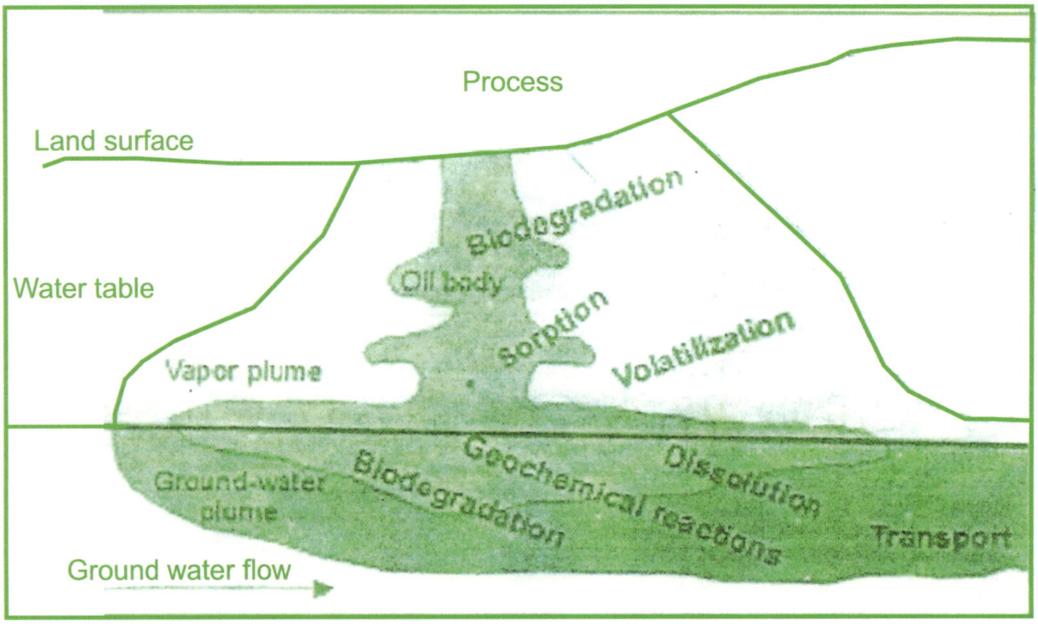

Fig. 2. Processes critical to understanding the fate and transport of hydrocarbons In the subsurface.

Contaminant transport in the subsurface is affected by different processes. They include advection, dispersion, diffusion, adsorption and decay. These processes can work together or separate in groundwater flow (Miller and Hogan, 1996).

The process by which contaminants are transported by the bulk motion of the flowing groundwater is known as advection. Non-reactive solutes are carried at an average rate equal to the average linear velocity, V, of the water (Freeze and Cherry, 1979). The average linear velocity, V, at which groundwater flows through a porous aquifer is given from the modified Darcy relations:

$$V = -\frac{K}{n}\frac{dh}{dL} \tag{1}$$

Where K is the hydraulic conductivity of the formation in the direction of groundwater flow, n is the porosity of the formation and $^{dh}/_{dl}$ is the hydraulic gradient in the direction of groundwater flow.

Dispersion is a mixing process. According to the advective hydraulics of flow system, the solute has the tendency to spread out from its flow path by a spreading phenomenon called hydrodynamic dispersion, which causes dilution of the solute. Hydrodynamic dispersion is the combined effect of mechanical dispersion and molecular diffusion (Mercer and Spalding, 1989). Freeze and Cherry (1979) states that it occurs because of mechanical mixing during fluid advection and because of molecular diffusion due to the thermal-kinetic energy of the solute particles. Mechanical dispersion is the mixing of the contaminant resulting from movement through complex pore structures (Greenkorn, 1983) and is due mainly to the porous medium. Mechanical dispersion is predominant at high groundwater velocities while, molecular diffusion is significant at low velocities.

When a contaminant is diluted in the groundwater it follows the path of normal flow and is called lateral dispersion. Due to heterogeneities in soil particle size and pore size, dispersion will occur.

Freeze and Cherry (1979) defines longitudinal dispersion as the spreading of solute in the direction of bulk flow. Spreading in directions perpendicular to the flow is called transverse dispersion. Longitudinal dispersion is normally much stronger than lateral dispersion and is the process whereby some of the water molecules and solute molecules travel more rapidly than the average linear velocity and some travel more slowly. The solute therefore spreads out in the direction of flow and declines in concentration. Dispersion coefficient varies with the groundwater velocity and is relatively constant at low velocities, but increases linearly with velocity as the groundwater velocities increase.

The study of dispersion phenomenon is important for predicting the time when a concentration limit used in regulations such as drinking water standards, will be reached and for determining optimal, cost-effective strategies for aquifer remediation (Palmer and Johnson, 1989).

Molecular diffusion is due to concentration gradients and the random motion of molecules (Miller and Hogan, 1996). Diffusion does not need advective velocity to occur. It is a process due to the contaminant alone. The larger the amount of pollutants the greater and farther the effects of diffusion can be. It is usually found to have effects at low velocities or long time periods of travel. Diffusion is a dispersion process of importance only at low velocities. Formation that is dependent on diffusion would be a low hydraulic conductivity layer. These layers will have such low velocities regular dispersion will not be noticed.

According to Mercer and Spalding (1987), adsorption is the transfer of contaminants from the groundwater to the soil. Desorption is the transfer of contaminants from the soil to the groundwater. These processes involve mass transfer of contaminants. Adsorption is also the ability of a compound to "attach" itself to the soil (Miller and Hogan, 1996). It is dependent on the properties of the soil and the compound. Differences in solubility and reaction with organic materials help make up a wide range of adsorption strengths.

Decay or degradation is the biological decomposition or chemical alteration of dissolved compounds. Miller and Hogan (1996) stated that decay does not affect how fast or how far pollutants will travel. Biological or chemical processes will reduce the amount of compounds traveling through the system but the effects of advection will stay the same. Processes that involve mass transfer of contaminants by chemical reactions will include precipitation and dissolution, oxidation and reduction while biological transformation may remove contaminants from the systems by biological degradation, or transform contaminants to other toxic compounds that are subject to mass transfer by the other processes earlier discussed.

The processes of adsorption-disorption, chemical reactions, and biological transformation play important roles in controlling the migration rate as well as concentration distributions. These processes tend to retard the rate of contaminant migration and act as mechanisms to reduce concentrations (Mercer and Spalding, 1989). Crude oil that is fairly viscous, with a high wax content and high pour point will have slow rates of spreading and dispersion (Webb, 1985). Generally, the rate of hydrocarbon loss due to evaporation is high during the early states of a spill, which renders the oil less toxic and less flammable. Evaporation results in an increase in oil viscosity and density, which further retards the spreading rate.

4. Fate of crude oil spill pathways

Naturally formed soil profiles are rarely homogenous with depth, rather they contain distinct layers, or horizons with specific hydraulic and physical characteristics. Because migration depends on subsurface lithology, the presence of these layers in the soil profile will generally retard water and contaminant movement during infiltration (Tombul 2003). Clay layers will impede flow due to their lower saturated hydraulic conductivity. However, when these layers are near the surface and initially very dry, the initial infiltration rate may be much higher and then drop off rapidly. Hult and Grabbe (1985) observe that as crude oil moves from a spill site, it contaminants soil in the vadose zone since its components are largely water insoluble and less dense than water, hydrocarbon free product tends to reside and spread along the water table boundary. Hydrocarbon free product can easily pollute wells within the zone of contamination, and also can sorb to and contaminate those soil areas influenced by water table fluctuations. This sorbed material tends to be another more subtle source of secondary contamination. In addition, free product can contaminate surface waters, and hydrocarbon vapors can collect in basements of buildings and create inhalation or explosion risks.

It was observed that vertical hydraulic gradients exist between some zones at some study sites. These gradients would physically provide the vertical hydraulic force necessary to transport light-non-aqueous phase liquids (LNAPLs) such as crude oil through the various water-bearing zones (Mercer and Spalding, 1989). Vertical migration proceeds until the interface between the unsaturated zone and saturated zone is met. At this point, free phase solvent would either spread laterally, continue through the water table vertically as DNAPL, or a combination of both. Because advection and dispersion are the primary transport mechanisms for LNAPLs (Miller and Hogan, 1996), horizontal (lateral) transport of LNAPLs within the Formation may likely occur through advection and dispersion within the shallow, intermediate and deep water-bearing zones at the study sites. Movement will be predominantly in the direction of groundwater flow through advection.

Density and solubility are among the primary physical properties affecting the transport of separate phase liquids in the soil and water. The density of a slightly soluble compound will determine whether it will sink or float in the saturated zone (Tombul, 2003). In some cases, dissolved concentrations are large enough that the density of the contaminant plume may contribute to the direction of solute transport. The contribution of density to the vertical component of groundwater V_g, can be calculated using the concept of equivalent freshwater head (Frind, 1982) by:

$$V_g = -\frac{K_v}{n}\left(\frac{p-1}{p_0}\right) \tag{2}$$

Where K_v is the conductivity in the vertical direction, n is porosity, p is the density of the contaminated water and p_o is the density of the native groundwater. According to Freeze and Cherry (1979), except for small amounts of hydrocarbons that go into solution, oil does not penetrate below the water table (oil is immiscible in water and is less dense). As oil accumulates on the water table, the oil zone spreads laterally, initially under the influence of gradients caused by gravity and later in response mainly to capillary forces. Capillary spreading becomes very slow and eventually a relatively stable condition is attained.

In theory, stability occurs when a condition known as residual oil saturation or immobile saturation is reached. Below a certain degree of saturation, oil is held in a relatively immobile state in the pore spaces. If the percent oil saturation is reduced further, isolated islands or globules of oil become the dominant mode of oil occurrence. Over the range of pressure gradients that can occur, these islands are stable. As the mass of oil spreads laterally due to capillary forces, the residual oil saturation condition must eventually be attained, provided that the influx of oil from the source ceases. This is referred to as the stable stage.

Degradation is the biological decomposition or chemical alteration of dissolved compounds (Awobajo,1981). Biodegradation plays an important role in the fate of crude oil. Many components of petroleum are readily degraded by subsurface micro-organisms (Hult and Grabbe, 1985). In the saturated zone, biodegradation frequently makes the aquifer anaerobic, resulting in much slower rates of degradation. In the unsaturated zone, vapor-phase molecular diffusion can maintain an oxygen supply at distances below the ground surface (Hult and Grabbe, 1985). Palmer and Johnson (1989) note that the unsaturated or vadose zone often contains greater amounts of organic matter and metal oxides than the saturated zone. Contaminants can adsorb onto these materials, making their rate of movement substantially less than in the saturated zone. Materials adhering to these absorbents can act as a source of contaminants to the saturated zone even after remediation.

The activity of the micro-organisms in the vadose zone generally is considered to be much greater than below the water table.

Furthermore, the unsaturated portion of the vadose zone can be a pathway for the transport of gases and volatile organics. These characteristics of the vadose zone can be important when predicting the transport of contaminants and designing systems for remediation. Hydrogeological studies are essential to underscore the fate of contaminant transport in the subsurface and the processes that govern them, such as advection, dispersion, diffusion, adsorption and decay. Its goal is also to determine the directions of groundwater flow at the polluted sites which information is essential to any groundwater remediation or monitoring program. The determined depth-to-water table (thickness of the unsaturated zone) is one of the hydrogeological parameters that determines whether or not a pollutant will reach the water table from a surface spill. High water table conditions will ensure that contaminant motion will be more of a lateral than vertical flow pattern. After the pollutants may have impacted the groundwater, it will tend to move laterally along these pathways, with its concentration decreasing as it moves away from the point or line sources of pollution due to dispersion and other attenuation effects, such as biological decomposition of organic compounds and precipitation of dissolved chemicals (Bouwer, 1978).

5. Ecologic-lithospheric sensitivity of crude oil spill

Sensitivity index is essentially a geomorphological classification and relates to the sensitivity of a particular area. Geomorphology and related physical processes govern the deposition and persistence of oil in the environment. The Mangrove swamp is a peculiar sensitive environment; its soils consist mainly of clays, shales, clayey shales, organic silts, organic silty clays and frequently peat materials. The soils contain abundant organic matter and iron compounds, and bacterial decay is active (Hutchings and Saenger, 1987). The most commonly occurring soil type in the mangrove environment is the peaty silty clayey mud which is noted for its very high water absorbing capacity. The dense network of rivers and creeks, low ground elevation and negligible gravity drainage ensure shallow depth to groundwater in this environment. There is also dense vegetation typical of a brackish environment. These characteristics of the mangrove zone tend to enable it retain oil pollution in its system so that pollutants experience longer residence period in the environment (Saenger et al., 1983). As a result, mangrove swamps have been found to be the most sensitive environment that could be affected by an oil spill over time (Hutchings and Saenger, 1987). The saturated, anaerobic nature of the mangrove zone will also result in slower rates of crude oil degradation (Hult and Grabbe, 1985).

The mangrove swamp zone, by virtue of its characteristic geomorphology, lithology, drainage condition and vegetation is identified as having a peculiar sensitivity to crude oil pollution, as a result of the long residence period it exhibits over time to contaminants.

6. Discussion

The information revealed by this review is further pointer to the need for all remediation technologies to consider where the contaminants reside in the subsurface. It is important to appreciate the processes involved in the transport of contaminants in both permeable and non-permeable media under saturated conditions. The environmental sensitivity ranking of the polluted sites will also serve as guide to identifying the geomorphologic zones most susceptible to crude oil pollution. This information will assist in the design of efficient and cost-effective monitoring networks and remediation strategies of the soil, ground and surface water resources.

7. References

Awobajo, A. (1981) An analysis of oil spill incidents in Nigeria: 1976-1980. *Proc. Int. Sem.* Lagos. pp. 51-62.

Bouwer, H. (1978) *Groundwater Hydrology*. McGraw-Hill Book Co. 480pp.

Bowles, J.E. (1985) *Physical and Geotechnical Properties of Soils*. 2nd Ed. McGraw-Hill Book Company, New York. 449pp.

Freeze, R.A; Cherry, J.A.(1979) *Groundwater*. Prentice-Hall, Inc. Englewood Cliffs, N. J. 604pp.

Frind, E.O. (1982) Simulation of long-term Transient Density-dependent Transport in Groundwater. *Advances in Water Resources*, Vol.5, pp. 73-88.

Frind, E.O. and Hokkanen, G.E. (1987) Simulation of the Borden Plume using the Alternating Direction Galerkin Technique; *Water Resources Research*, Vol. 23, No.5, pp. 918-930.

Greenkorn, R.A. (1983) *Flow Phenomena in Porous Media*. New York: Marcel Dekker.

Hult, M.F. and Grabbe, R.R. (1985) Permanent Gases and Hydrocarbon Vapors in the Unsaturated Zone. In: *Proceedings, U.S. Geological Survey Second Toxic-Waste Technical Meeting*, Cape Cod, MA, October, 1985.

Hutchings, P. and Saenger, P. (1987) *Ecology of Mangroves*. University of Queensland Press New York.

Keely, J.F. (1989) Introduction In *transport and fate of contaminants in the subsurface*.EPA/625/4-89/019. Cincinnati OH 45268. 4pp.

Mercer, J.W. and Spalding C.P. (1989) Characterization of the vadose zone. In *transport and fate of contaminants in the subsurface*.EPA/625/4-89/019. Cincinnati OH 45268.

Miller, J. and Hogan (1996) Dispersion. In *Groundwater Pollution Primer*. Civil Engineering Dept., Virginia Tech. 6pp.

Palmer, C.D. and Johnson, R.L. (1989) Physical processes controlling the transport of non-aqueous phase liquids in the subsurface. *In Seminar Publication: Transport and fate of contaminants in the subsurface*. EPA/625/4-89/019. Cincinnati 0H 45268, pp.23-27.

Saenger, P., Hegerl, E.J. and Davis, J.D.S. (1983) Global Station Mangrove Ecosystems. *Environmentalist* 3 (Sup. No. 3: 1-88)

Tombul, M. (2003) Relationship between infiltration rate and contaminant transport in unsaturated zone. *Proceedings of the first International Conference on Environmental Research and Assessment*; Bucharest, Romania, March 23-27, 2003. pp.355-362.

Webb, C.L.F. (1985) Offshore Oil Production in the Baltic Sea: a coastal sensitivity study. *Proc. 1985 Oil Spill Conf.* (Vol.1). California.

Crude Oil Transportation: Nigerian Niger Delta Waxy Crude

Elijah Taiwo[1], John Otolorin[1] and Tinuade Afolabi[2]
[1]*Obafemi Awolowo University, Ile-Ife, Department of Chemical Engineering,*
[2]*Ladoke Akintola University of Technology, Ogbomoso,*
Department of Chemical Engineering,
Nigeria

1. Introduction

Crude oil is one actively traded commodity globally. Its demand has been growing steadily over the decades, from 60 million barrels per day to 84 million barrels per day (Hasan et al , 2010). In Nigeria, crude oil production has grown from a little above 1000 barrels per day in 1970 to over 3000 barrels per day in 2010 (Fig. 1.1). The world production was conservatively 73 million barrels per day in the year 2005 and within the range of 72 and 75 million barrel per day between 2005 and 2010.

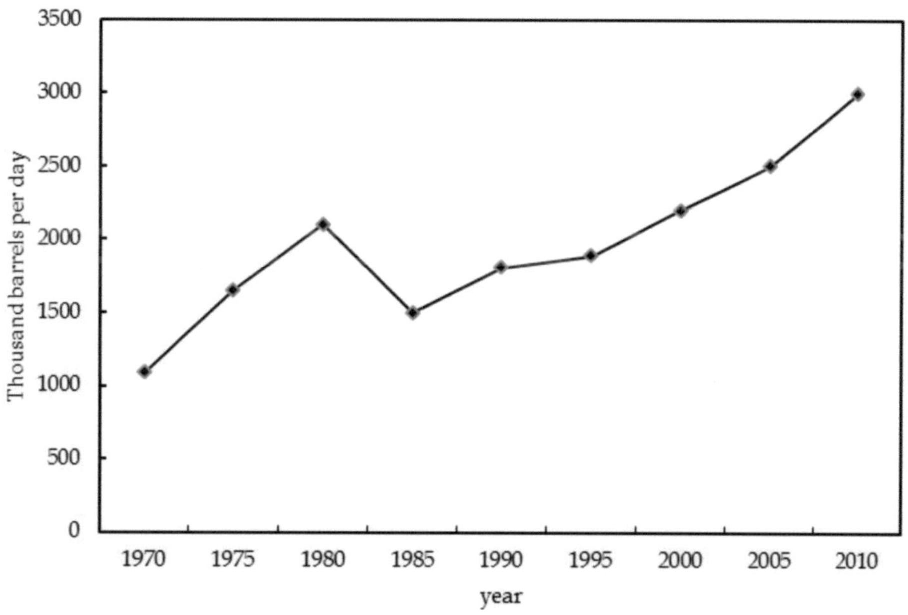

Fig. 1.1. Nigerian crude oil production.

Nigeria is the largest oil producer in Africa and the eleventh largest in the world. ChevronTexaco, ExxonMobil, Total, Agip, and ConocoPhillips are the major multinationals involve in Nigeria oil sector. The main production activity in Nigeria is in the Niger Delta region, which according to master plan, extends over an area of about 70,000 square kilometers which amounts 7.5% of Nigeria's land mass. It lies between latitude 3°N and 6°N and longitude 5°E and 8°E (Fig 1.2). The Niger delta is world's third largest wetland after Holland and Missisippi. It covers a coastline of 560 km, which is about two-thirds of the entire coastline of Nigeria (Fawehinmi, 2007).

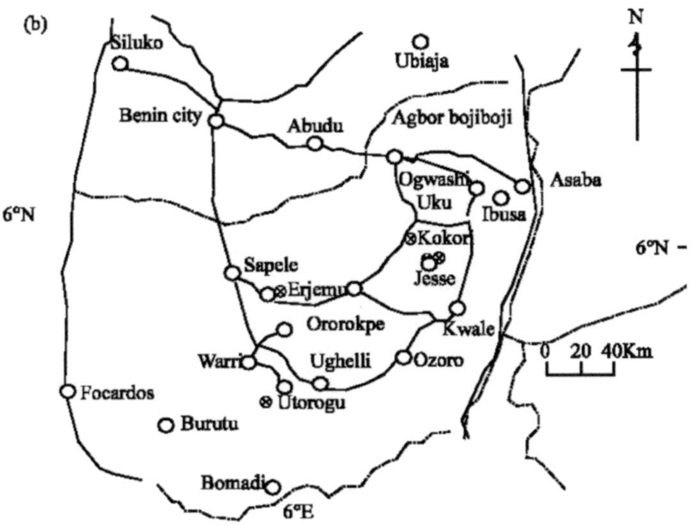

Fig. 1.2. Niger Delta region, Nigeria.

Nigeria has a substantial reserve of paraffinic crude oils (Ajienka and Ikoku, 1997), known for their good quality (low sulphur, high API gravity), and containing moderate to high contents of paraffinic waxes. The data correlated for light, medium and heavy crude oil samples from different sites in Nigeria show densities ranging from 0.813-0.849 g/ml, 0.866-0.886 g/ml, and 0.925-.935 g/ml at 15°C respectively. Characteristically, waxy crude oils have undesirably high pour points and are difficult to handle where the flowing and ambient temperatures are about or less than the pour-point. They exhibit non-Newtonian flow behaviour at temperature below the cloud point due to wax crystallization. Consequently, the pipeline transportation of petroleum crude oil from the production wells to the refineries is threatened.

The Nigerian Niger Delta crude oil, which is the mainstay of Nigerian economy, exhibits waxiness, with deposits in the range of 30-45 % (Adewusi 1997; Fasesan and Adewumi, 2003; Taiwo et al., 2009 and Oladiipo et al., 2009). In fact, pipelines have been known to wax up beyond recovery in Nigeria. Production tubing has also been known to wax up, necessitating frequent wax cutting, using scrapers conveyed by wireline, which is an expensive practice. Billions of dollars has been lost to its prevention and remediation (Oladiipo et. al., 2009). The resultant effect on the petroleum industries include among others, reduced or deferred production, well shut-in, pipeline replacements and/or abandonment. For efficient operation of a pipeline system, steady and continuous flow without any interruption is desirable (Chang et al., 1999). The difficulties in pipeline transportation are due to this complex nature of crude oil, which cause a variety of difficulties during the production, separation, transportation and refining of oil (Al-Besharah et al., 1987). For example, formation of asphaltic sludge after shutting in a production well temporarily and after stimulation treatment by acid has resulted in partial or complete plugging of the well (Ayala et al., 2007and Escobedo et al., 1997). Relative amount and molecular distribution of wax, resin and asphaltene as well as thermal and shear history of high pour point waxy crude sample directly affects the rheological properties of crude oil (Ajienka and Ikoku, 1997).

Phase changes in petroleum fluids during production, transportation and processing, constitute a challenging and an industrially important phenomenon. Polydisperse nature of hydrocarbons and other organic molecules in petroleum fluids accounts for the complexity of their phase behavior (Mansoori, 2009), which could be reversible or irreversible (Abedi et al., 1998). Generally, heavy fractions have little or no effect on the liquid-vapour phase behaviour of the majority of petroleum fluids. Their main contribution is in solid separation from petroleum fluids, due to changes in the composition, temperature and pressure (Mansoori, 2009; Escobedo and Mansoori, 2010). The main components of the heavy fraction, which participate in the solid phase formation include asphaltenes, diamondoids, petroleum resins and wax. Petroleum wax consist mainly saturated paraffin hydrocarbons with number of carbon atoms in the range of 18–36. Wax may also contain small amounts of naphthenic hydrocarbons with their number of carbon atoms in the range of 30–60. Wax usually exists in intermediate crudes, heavy oils, tar sands and oil shales. The distribution of n-alkanes as a function of the number of carbon-atom in a paraffin wax sample is given in figure 1.3 below.

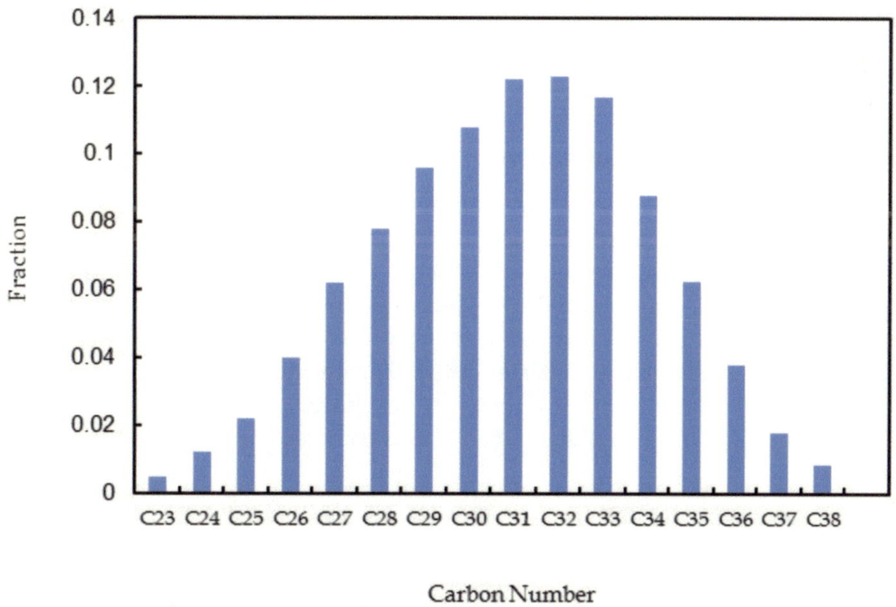

Fig. 1.3. Carbon number distribution of paraffin wax.

Phase equilibrium deal with the various situations in which two or more phases (or state of aggregation) coexist in thermodynamic equilibrium with each other. Reservoir oil contains paraffin wax in solution. The average temperature of the oil in production well is somewhat higher than the atmospheric temperature. When a waxy crude oil is cooled, the heavier paraffinic constituents begin to separate as solid crystals once the solubility limit is exceeded (Karan et al., 2000). Thermodynamically, the solid-liquid phase boundary temperature, that is the maximum temperature at which the solid and liquid phases co-exist in equilibrium at a fixed pressure, is the wax appearance temperature (WAT). This depends on wax concentration, the crystallization habit of wax, and the shear stability of different wax structures (Holder and Winkler, 1965; Hussain et al., 1999).

This contribution therefore present the various phase-transitions, which may occur in petroleum fluids, and a unified perspective of their phase behaviors. Experimental determination of the rheological properties and the characteristics of the crude oil were carried out and the data tested on different established rheological models for ease of simulating the flow behavior of the crude oil. The wax deposition tendencies of crude oil in the pipelines and its influences on the transportation capacity were determined. The various methods of mitigating flow assurance problems and wax deposition inhibition highlighted towards adequate crude oil production.

2. Experimental methods

2.1 Materials

All crude oil samples used in this study were from Niger Delta Oil field in Nigeria, having density and API gravity in the range of 847 - 869 kg/m^3 and 24.4 – 36.5 ^{o} at 15 ^{o}C, respectively. For physical properties measurements, the crude samples were shaken vigorously for one hour to homogenize and presents good representation of samples. Standard methods were employed in determination of the physical properties.

2.2 Rheological properties measurements

The rheological behavior of the crude oil samples were studied using Haake Rheo Stress model RS100 rheometer having several operating test modes. The test sample charged into the rheometer cup was allowed to equilibrate at a particular temperature. The unit was set to desired shear rates and operated at 10 revolution per minute, then the temperature of the test, the shear rate and the shear stress were recorded. The procedure was repeated at various other set temperatures and shear rates. The resulting sample deformation was detected using digital encoder with high impulse resolution. Thus, it allows measurements of small yield values, strains, or shear rates. The rheometer is equipped with a cone and plate sensor.

2.3 Determination of the wax content

Wax content was determined by precipitation method (Coto et al., 2008). It involves samples dissolution in n-pentane, precipitation with acetone:n-pentane mixture in ratio 3:1 and separating by filtration in Buchner funnel using glass microfiber Whatman filter N934.

2.4 Pipeline crude oil transportation simulation

The pipeline transportation system simulated for crude oil experimental measurement comprises the experimental pipe system, the circulating water system and crude oil reservoir system (Fig. 2.1.).

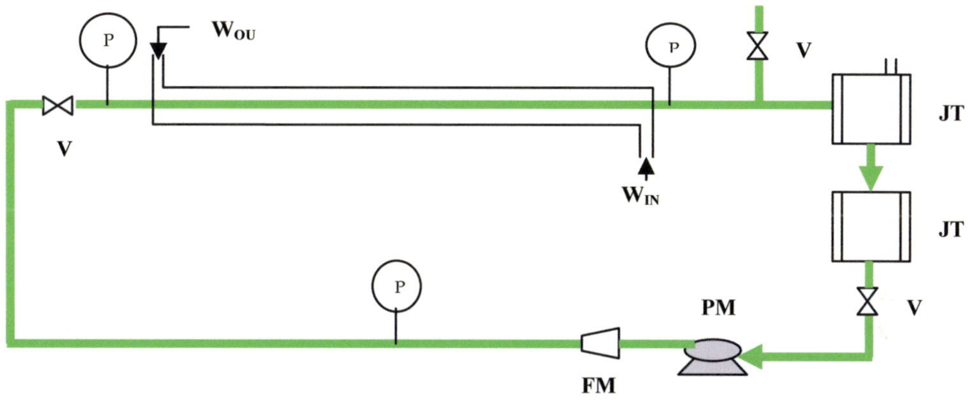

V- Valves; P - Pressure gauge; FM - Flow Meter; PM - Pump;
W_{IN} & W_{OUT} - Heating/cooling Water IN and OUT; JT- jacketed Tank

Fig. 2.1. Simulated pipeline transportation system.

The experimental pipe system and water circulation system are arranged in shell and tube mode. The water jacket around the reservoir tank can control the oil temperature in the pipe system while water circulation system controls the wall temperature of the test section. The internal diameter of pipe at both the test and reference section is 20.47 mm with the total length of 2.75 m.

2.4.1 Wax deposition in simulated transportation flow line

The differential pressure of the test section and the reference section were measured during the wax transportation experiment. The wall temperature of the test section was varied and the differential pressure measured. The data collected were used to calculate the extent of the wax deposition and to elucidate on the mechanism of deposition. Carbon number distribution of paraffin wax determined by gas chromatography technique using the IP 372/ 85 methods.

2.5 Evaluation tests of flow improvers

Effectiveness of some flow improvers for the waxy crude oils was determined through the pour point test (using Herzog MC 850 pour point test equipment) according to the ASTM-97 and kinematic viscosity by the IP 71 procedures. The dynamic inhibitive strength of these modifiers was equally experimented using the simulated crude oil pipeline transportation.

3. Results and discussions

3.1 Rheological behaviour and modelling

Rheology is the science of the deformation of matter. It involves the study of the change in form and flow of matter in term of elasticity, viscosity and plasticity under applied stresses or strains. The rheological behavior of waxy oil is crucial in the design of pipeline, flow handling equipment and processing purposes in the oil industry. The study of the rheological characteristics of crude oil is significant to lowering the energy consumption and ensuring safety and cost effectiveness in pipeline transportation of waxy crudes. Figure 3.1 shows a typical rheogram in terms of shear stress and shear rate. Shear stress increases asymptotically and significantly with shear rate. Several researchers have investigated models to describe the rheological properties of waxy oils (Davenport and Somper, 1971; Matveenko et al., 1995 and Remizov et al., 2000). Generally the behaviour can be broadly grouped into two; Newtonian and non-Newtonian fluids. Whereas the Newtonian fluids exhibit a linear relationship between shear stress and shear rate, the non-Newtonian fluids do not. The non-Newtonian fluids have yield stress, which is the upper limit of stress before flow occurs. At this point, the range of reversible elastic deformation ends and range of irreversible deformation or visco-elastic viscous flow occur (Chang et al., 2000). The existence of yield stress behavior depends upon the degree on interlocking structure developed by waxing and fragility of the network (Philip et al., 2011).

Modeling analysis carried out by fitting the rheological data to three models are reported in Table 3.1. The models include

Bingham plastic,

$$\tau = \tau_o + \mu_P \gamma \tag{3.1}$$

Casson,

$$\tau^{1/2} = \tau_o + \mu \gamma^{1/2} \tag{3.2}$$

and Power law,

$$\tau = m\gamma^n \tag{3.3}$$

where γ is the applied shear rate (s^{-1}) and τ is the corresponding shear stress in Pa, m and n are consistency index in Pa s and flow behavior index, the τ_o and μ are the apparent yield stress in Pa and the apparent viscosity in Pa s. Bingham model predicted very adequately the flow behavior of the crude oil over the tested range of shear rates. This observation suggests Nigerian Niger Delta waxy crude oil exhibits plastic behavior. Table 3.1 reports the highest regression correlation coefficient, R^2, of 0.992 for the Bingham model. This is similar to an earlier work on waxy crudes from this area (Taiwo et al., 2009), with good prediction of the yield stress within percentage deviation of 2.88 ± 0.15.

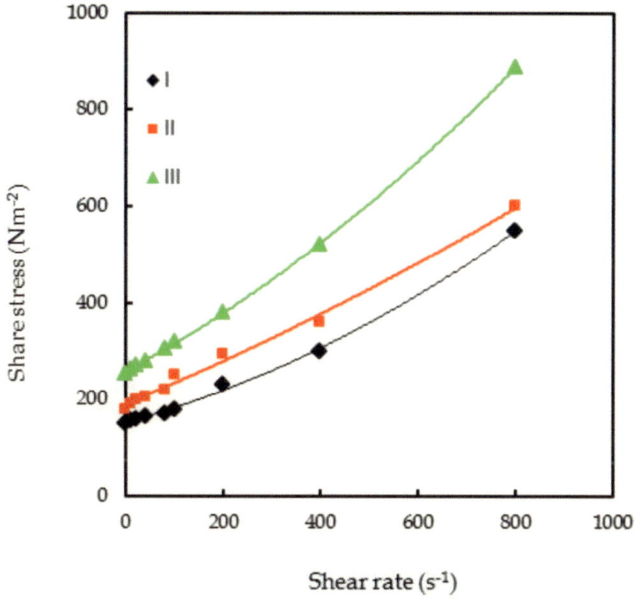

Fig. 3.1. Rheogram of the waxy crude samples.

Sample	Bingham Plastic			Casson			Power Law		
	τ_o	μ_P	R^2	τ_o	μ	R^2	m	n	R^2
I	139.4	0.487	0.983	113.0	0.387	0.898	66.09	0.259	0.766
II	184.9	0.505	0.991	149.8	0.385	0.944	91.20	0.240	0.836
III	244.0	0.779	0.992	197.4	0.485	0.919	501.19	0.253	0.800

Table 3.1. Rheological Parameters and Correlation Coefficients of Models.

According to Bogne and Doughty (1966), the rheological characteristics of materials form a continuous spectrum of behavior, ranging from the perfectly elastic Hookean solid at one extreme, to that of purely viscous Newtonian fluid at the other. Between these idealized extremes is the behavior of real materials that include, among others, non-Hookean solids, non-Newtonian fluids, and viscoelastic substances. The waxy crude oil generally belongs to the non-Newtonian fluids while the Nigerian Niger delta waxy crude showed plastic behavior, which is time dependent non-Newtonian fluid.

3.2 Temperature effect

Temperature has a strong effect on viscosity and viscous behavior. This effect provides the flow behavior curve in terms of the viscosity-shear rate or viscosity-shear stress relationships. Fig. 3.2 shows the effect of temperature on shear stress-shear rate behavior over the temperature range of 25 - 55 °C experimented. The crude oil shows non-Newtonian shear thinning (viscosity reduction) behavior over the range of shear rates studied. By definition, yield stress, τ_o, is the limiting stress below which a sample behaves as a solid. At low stress, the elastic deformation takes place, which disappears when the applied stress is released (Guozhong and Gang , 2010). Chang et al., (1998) described the

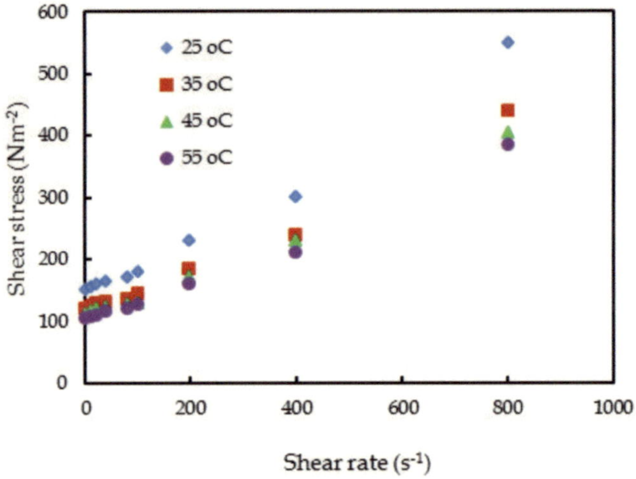

Fig. 3.2. Shear stress as a function of shear rate with varying temperature (Sample I).

yielding behavior of a waxy crude oil as having three distinct characteristics—an elastic response, a creep and a fracture. The shear stress at the point of fracture is the value of practical importance and is usually taken as the yield stress. Above the yield stress point, the applied stress leads to unlimited deformation which causes the sample to start flow (Ghannam and Esmail, 2005). The yield point, which is required to start the flow, decreases with temperature from 150 Nm^{-2} at room temperature of 25 °C to 104 Nm^{-2} at the temperature of 55 °C. Similar trend ensued for the other samples. At a higher cooling rate, the rate of wax precipitation is higher. Hence, a higher stress is necessary either to aggregate the crystals, or to breakdown the structure.

In addition, the apparent viscosity decreases considerably with increasing temperature (Fig 3.3).

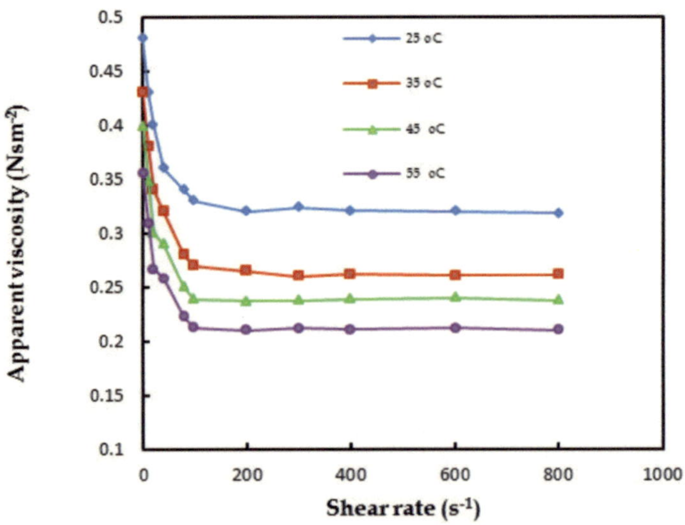

Fig. 3.3. Effect of temperature on viscosity of the crude.

Viscosity differences are relatively larger at low shear rates than at high shear rates. At high temperature, wax in the crude oil could not agglomerate and form aggregates, and hence reducing the oil viscosity. At high shear rate, (≥ 100 s^{-1}), an almost constant viscosity was observed with increasing share rate. This could result from effective dispersion of wax agglomerates in the continuous phase originally immobilized within the agglomerate, after being completely broken down into the basic particles.

The observed variation with temperature is attributable to the strong effect of temperature on the viscosity of wax and asphaltene components in crude oil. At high temperature, the ordered structure of these chemical components are destroyed, and hence reducing the oil viscosity (Khan, 1996). As the shear rate increases, the chain type molecules disentangled, stretched, and reoriented parallel to the driving force, and hence reduced the heavy crude oil viscosity (Ghannam and Esmail,1998, 2006).

3.3 Wax deposition characteristics

Fig. 3.4 shows the wax deposited as a function of temperature for the three Nigerian crude oils samples. Sample II has the highest percentage wax content of 33.5% with wax appearance temperature (WAT) of 43.5 °C while for sample I the wax content is 10.5 % with WAT of 35 °C.

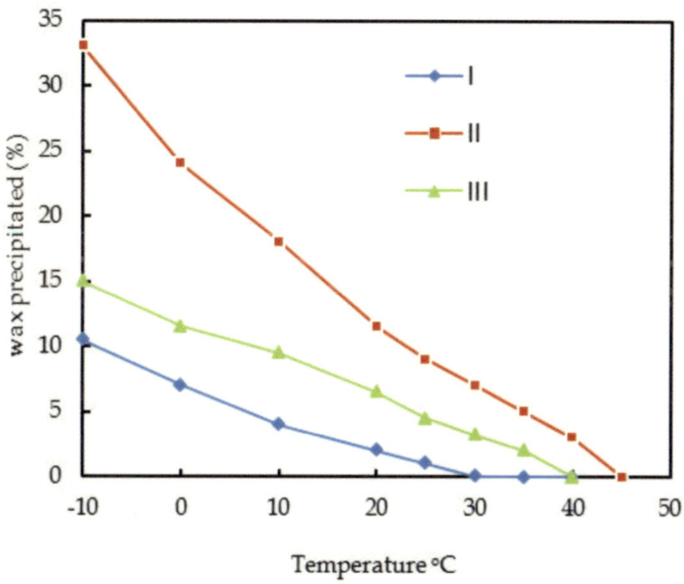

Fig. 3.4. Wax precpitation as a function of temperature.

The wax appearance temperature (WAT) and the pour point temperature (PPT) are good pointers to the temperature regime, in which a waxy crude oil is expected to start to show non-Newtonian behavior. Their determination is an important step in the study of rheological behavior of such systems. In fact, WAT is probably the most important flow assurance measurement for waxy crudes. Pour point represents the lowest temperature at which the crude oil can be stored or handled without congealing in the tanks or pipelines. Usually, it is 10-20 °C lower than the cloud point (Ajienka, 1983). The pour point of samples experimented were in the range of 18 and 32 °C. As the fluid cools below WAT, crystal size tends to increase and crystal aggregation is usual, particularly under quiescent and low shear condition, such that the solid-like behavior of the waxy suspension increases. At some point between the WAT and pour point, a transition can be determined depending on wax development and is thus strongly affected by thermal and shear conditions. This is considered an indication of initial development of interlocking network (Lopes-da-Silva and Coutinho, 2004).

3.4 Wax deposition rate in flow system

As trot to understanding the deposition pattern and hence the flow assurance of the oil samples, simulated transportation flow pipe described in section 2.4 was developed. It is a modification to earlier work (Taiwo et al., 2009), with consideration for the idea of Guozhong and Gang (2010). As expected, the wax deposit thickness showed inverse proportionality with wall temperature and direct relation with temperature difference between the oil and pipe wall (Fig 3.5). In addition, a monotonic increase with time was observed. The low thickness observed at 42 °C wall temperature is significant of closeness to the WAT of oil sample which is 43.5 °C. In addition, the average wax deposition rate (mm/d) increases with increased difference in temperature as shown in Fig. 3.6.

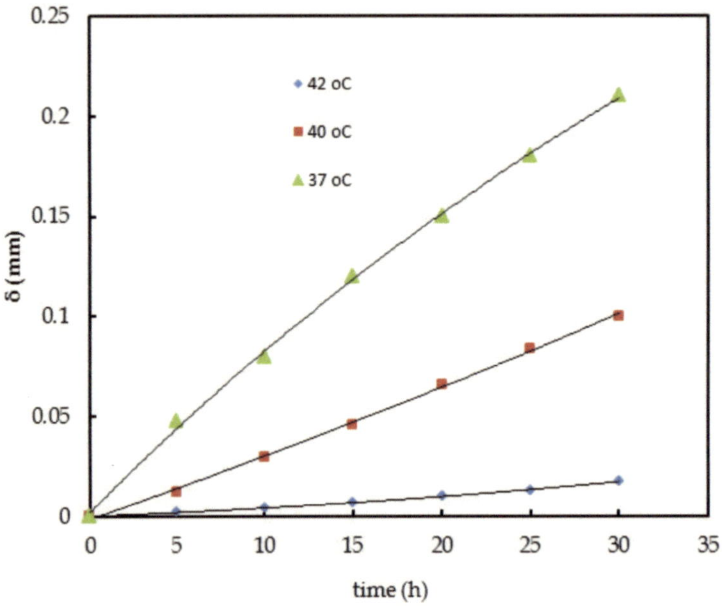

Fig. 3.5. Wax deposit thickness variation with time as a function of wall temperature.

The average wax deposition rates reported in Fig. 3.6 were at temperature differences of 3 °C, 5 °C and 8 °C respectively, at 30 h deposition time. Generally, gradual increase in deposition rate with increasing temperature difference was observed. However, at zero °C temperature difference, experimental results showed that no deposition occur because there is no driving force for lateral diffusion of the wax molecules. When the oil temperature is 50 °C, the average wax deposition rate was zero at temperature difference of 3 °C, corresponding to the wall temperature of 47 °C, which is slightly above the WAT of the crude sample. Under this condition, the paraffin molecule super saturation concentration is very low, thus, its wax deposition power tend to zero. However, at 5 °C temperature difference, the deposition rate was 0.025 mm/d. This observation was due to lateral diffusion of the wax in crude oil. When temperature difference increased to 8 °C, the rate increased to 0.10 mm/d. It reflects wall temperature of 42 °C which is lower than the WAT of the crude oil sample.

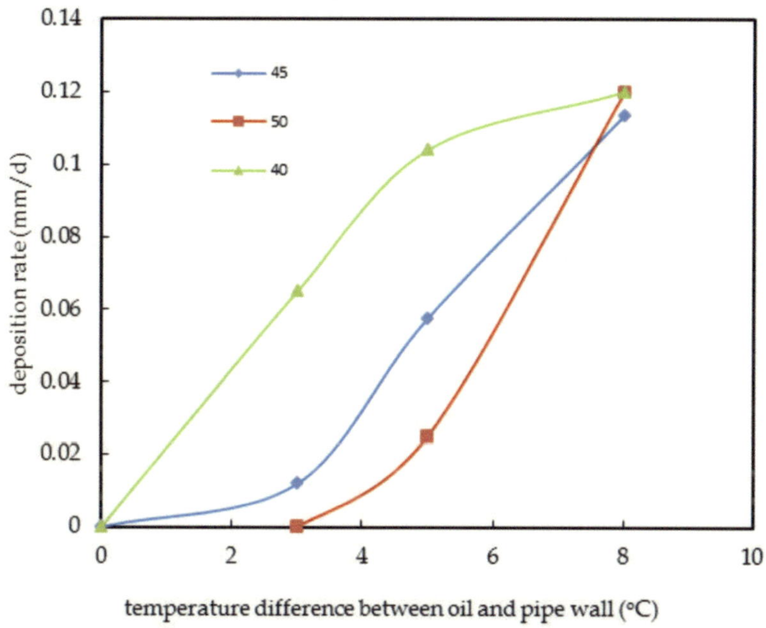

Fig. 3.6. Variation of deposition rate with temperature difference.

At the oil temperature of 45 °C, the average wax deposition rate increases with the fall in the wall temperature. When the temperature difference between the oil and the wall is 8 ° C (the wall temperature is 37 °C), the average wax deposition rate is 0.12 mm/d and higher than that experienced for 50 °C oil temperature. For the oil temperature of 40 °C, deposition rate is generally higher than other temperature conditions. However, at higher temperature difference, the deposition rate almost assumed same value. This may be due to wax content in the crude.

The theory of the molecular diffusion suggests gradual wax deposition in the pipe as the wall temperature decreases. However, at low wall temperature of the test section, the differential pressure of the test section increased sharply. This rapid increase of the differential pressure should therefore not be the resultant effect of the wax deposition, since a gradual increase in wax deposition was observed (Figs. 3.4 and 3.5). A fast increase of the oil viscosity near the wall could have resulted from the drop in wall temperature, thereby distorting the flow regime in the test section due to the radial differential temperature; this makes the differential pressure of the test section increase suddenly. Mansoori, (2009) attributes this observation to phase transition which influence the Gibbs energy of the system hence sudden change in pressure drop.

3.5 Doping of waxy crude

Wax build-up in tubing and flow lines is a typical flow assurance problem. This makes transportation of such waxy crudes un-economically feasible. To improve mobility of the oils, viscosity reduction is imminent. This can be done by heating, blending of the oils with lighter oils or other solvents having pour-point depressing effect (Hemant et al., 2008). In addition there are applications of microwaves, ultrasound irradiation, magnetic fields, lining and coating pipelines with fiber-reinforced plastics to reduce the wettability of paraffin with the wall of the pipe, and covering the inner wall surfaces with polypropylene (Hemant et al., 2008). Pipeline heating is usually deployed for small distances (with insulation) and moving the oil as quickly as possible. This passive insulation becomes in effective when longer pipe length oil transport is required. High backpressure will set in requiring expensive booster pumps. Electrically heated subsea pipeline is considered an alternative (Langner and Brass, 2001). Even then, the complexity of the pipeline design required and need to heat whole pipe length makes the configuration costly to deploy and operate (Martinez-Paloua et al., 2011). Doping with solvent, which keeps the wax in solution, is therefore essential in ensuring oil mobility. Based on evaluation of preliminary studies (Adewusi, 1998; Fasesan and Adewumi, 2003; Bello et al., 2005a, 2005b, 2006 and Taiwo et al., 2009), xylene based pour point depression solvents were considered. Table 3.2 reports some solvents for improving the transportation of Niger Delta waxy crudes. Xylene (X) and tri chloro ethylene (TCE) are good pour point depressant based on the work of Adewusi (1998) and Bello et al., (2005a), while Taiwo et al., (2009) showed tri ethanol amine (TEA) to be very good wax deposition inhibitor (Table 3.2). These three solvents further confirm their wax inhibition strength from the report of the present experimentation (Table 3.3). Xylene reduced the pour point from 32 °C to 18 °C for sample II that has the highest wax content of 30.5 percent, while TCE and TEA reduced it to 19.5 and 17 °C, respectively. Pour point depressants (PPDs) accomplish this task by modifying the size and shape of wax crystals and inhibit the formation of large wax crystal lattices (Wang et al., 2003). Generally, the PPDs create barrier to the formation of the interlocking crystal wax network (Wang et al. 1999). As a result, the altered shape and smaller size of the wax crystals reduce the formation of the interlocking networks and reduces the pour point (Hemant et al., 2008) by preventing wax agglomeration (Hafiz and Khidr, 2007). In some cases pendant chains in the additive co-crystallize with the wax and the polar end groups disrupts the orthorhombic crystal structure into a compact pyramidal form (Hemant et al., 2008).

Trichloroethylene, xylene and tri ethanolamine actively decreased the pour point of the samples and their wax deposition potentials on doping. The oxygen containing groups in the doping agents take the role of inhibiting the growth of waxes and poisoning them by adsorptive surface poisoning mechanism. The PPDs on adsorption on the surface of the wax renders its nuclei inactive for further growth. The waxes then occur in small sized particles distributed within the crude oil samples, thus cannot form net-like structure required for solidification and deposition.

Pour Point Depressant	Adewusi, 1998 Bello et. al., 2005		Taiwo et al 2009					
	EB		FL		BL		EL	
	Wax deposition (g/g)	Pour point (°F)	Wax deposition (g/g)	Pour point (°C)	Wax deposition (g/g)	Pour point (°C)	Wax deposition (g/g)	Pour point (°C)
No additive	0.05	36	0.25	28	0.28	33	0.335	36
Methyl Ethyl Ketone	0.037	22	0.15	22	0.185	18	0.188	25
Tri Ethanol Amine	-	-	0.085	18	0.106	23	0.107	16
Xylene	0.013	17	0.115	17	0.143	17	0.154	18
Cyclo-hexanone	0.028	18	-	-	-	-	-	-
Tri Chloro Ethylene	0.009	16	-	-	-	-	-	-

Table 3.2 Evaluation of various solvents for flow improver.

Pour Point Depressant	I		II		III	
	Wax deposition (g/g)	Pour point (°C)	Wax deposition (g/g)	Pour point (°C)	Wax deposition (g/g)	Pour point (°C)
No additive	0.105	18	0.305	32	0.150	27
Methyl Ethyl Ketone	0.069	14	0.164	23.5	0.084	21.5
Tri Ethanol Amine	0.040	12	0.092	17	0.053	18.5
Xylene	0.054	11.5	0.130	18	0.067	17.5
Cyclo-hexanone	0.052	12.5	0.182	22	0.076	18
Tri Chloro Ethylene	0.048	11	0.171	19.5	0.062	17

Table 3.3 Screening of common solvents for flow improver.

It is suggested that the reduction of the measured viscosity is due to the interactions between the hydroxyl functions or the delocalized unpaired electron over the π-orbital especially in the phenyl group of PPD and some functionalities of the wax (Gateau et al., 2004). This could have premised the performance of TEAX and TCEX. With 0.1 volume fraction of TEA or TCE in TEAX or TCEX binary, the shear thinning behavior is enhanced, and the viscosity as well as the pour point decreases as the temperature increases (Figs 3.7 and 3.8). TEAX improved the rheological characteristic behavior better than TCEX. The reduction of viscosity on the addition of the solvents is due to the dissolution of paraffin wax, the effect xylene on the effectiveness of these dopants is significant and play a major role in reducing the viscosity. This observation gave credence to the report of Chanda et al., (1998).

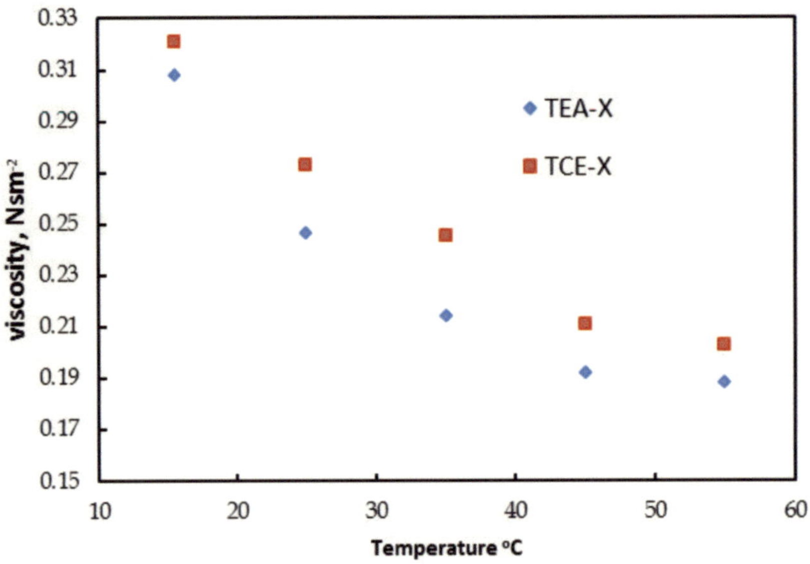

Fig. 3.7. Viscosity variation for doped sample II with temperature.

3.6 Hydrocarbon distribution in the crude oil wax

The crude oil samples wax had similar hydrocarbon distributions. The carbon number distribution of the paraffin wax is shown in Fig. 3.9. Samples I and III had very identical carbon number distribution, and all samples showed the same peak carbon number. This ranged between C31 and C33 indicative of paraffinic wax, although tailing to C38 an occurrence of naphthenic crude.

The dynamic inhibitive strength of the flow improvers tested by applying the doping agents into the crude samples in the reservoir tanks of the simulated pipeline transportation set-up, showed a marked reduction relative to that obtained for undoped samples. This showed that the PPD apart from lowering the pour point of the waxy crude caused reduction in pressure build-up by keeping the wax particles in solution and hence reduced viscosity of crude samples. Wax deposition was not feasible in the test pipe surfaces even at 8 °C temperature difference between the oil temperature and the wall temperature. The moderately higher-pressure drop with TCEX over the TEAX modified crude samples further confirms the strength of TEAX in wax crystal modification (Fig 3.10). Plate 3.1 showed the wax dispersed in crude oil sample upon modification. This evidently supports the view of Azevedo and Teixeira (2003) that shear dispersion play significant role in wax deposit removal thus affecting the rate of wax accumulation.

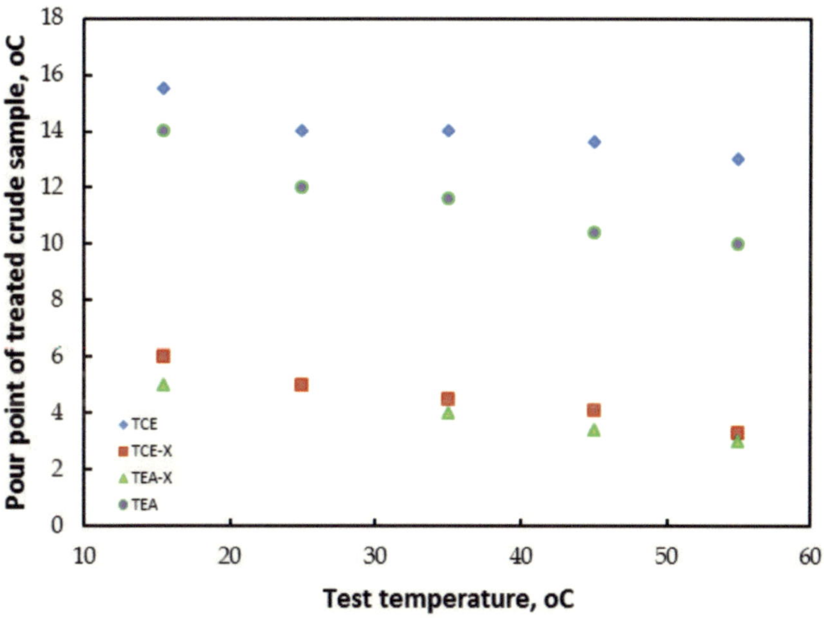

Fig. 3.8. Pour point depression test for sample II at varying test temperature.

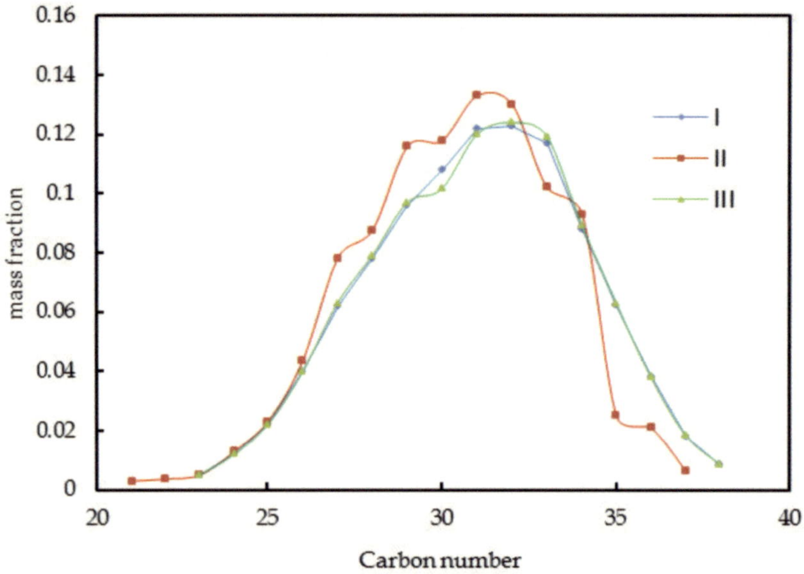

Fig. 3.9. Hydrocarbon distribution in wax samples.

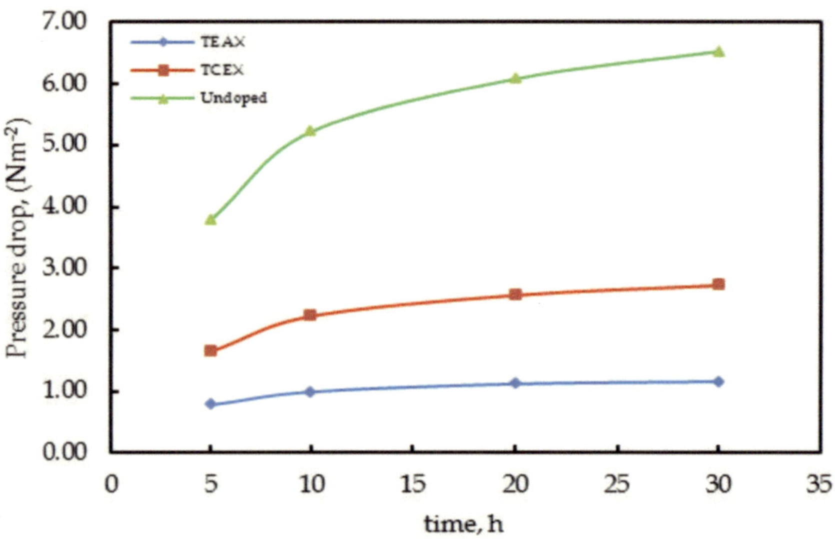

Fig. 3.10. Variation of pressure drop with time for doped and undoped sample in dynamic test.

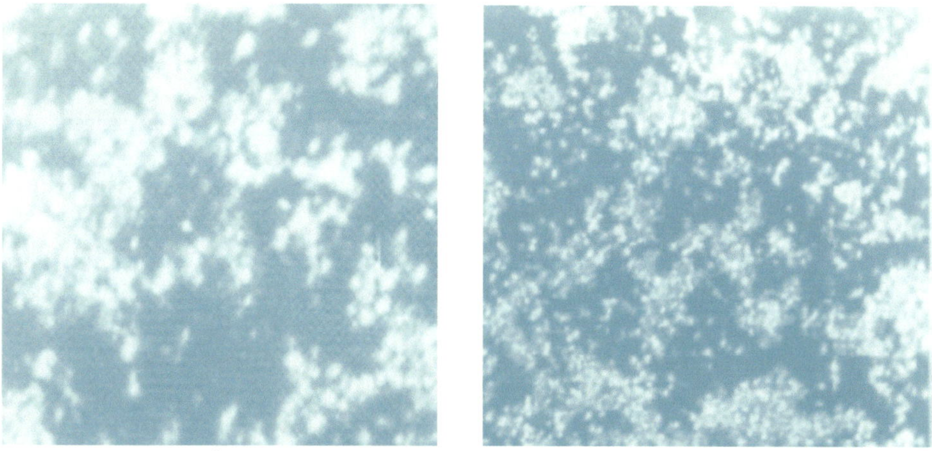

A B

Plate 3.1. Micrograph of crude sample at X 500 magnification (A: undoped B: doped)

4. Conclusion

A study of transportation of waxy Nigerian Niger Delta crude oil using flow improver was performed. This study has been able to highlight the significance of TEA in improving the flow of this crude oil by significantly lowering the pour point and wax deposition potentials of the crude sample. The crude oil had been established to show plastic behaviour with Bingham non-newtonian model adequately predicts its rheological behaviour with occurrence of yield stress. The simulated pipeline transportation system for experimentation revealed mass diffusion mechanism for transport and deposition of waxes on the tube wall. The characteristics and behaviour of solvent modified crude oils should help in addressing cost-effective solution to flow assurance problems in the industry.

5. Acknowledgment

The authors appreciate the technical staff of Chemical Engineering Department, Obafemi Awolowo University, Ile-Ife, Nigeria for the fabrication works, Department of Petroleum Engineering, University of Ibadan, Nigeria for making their facility available for this study, and the Department of Petroleum Resources (DPR), Nigeria, for assistance during the sample collection.

6. References

Abedi, S.J., Seyfaie, S. and Shaw, J.M. (1998). Unusual retrograde condensation and asphaltene precipitation in a model heavy oil system, *Journal of Petroleum Science and Technology*. 16 (3/4), 209–226.

Adewusi V. A. (1998). An improved inhibition of paraffin deposition from waxy crudes. *Petroleum Science and Technology* 16(9 & 10), 953 - 970

Adewusi, V. A. (1997). Prediction of wax deposition potential of hydrocarbon systems for viscosity-pressure correlations. *Fuel* 76:1079–1083.

Ajienka, J. A. (1983) The effect of Temperature on rheology of Waxy crude oils and its implication in production operation, Ph. D Dissertation, University of Port Harcourt, Nigeria.

Ajienka, J.A. and Ikoku, C.V, (1997). Waxy crude oil handling in Nigeria: practices, problems and prospects, *Energy sources* , Vol. 12. No. 4. pp. 463-478.

Al-Besharah, J. M.; Salman, O. A. and Akashah, S. A. (1987) Viscosity of crude oil blends. *Industrial and Engineering Chemistry Research*, 26, 2445–2449.

Ayala, O. F.; Ayala, L. F. *and* Ayala, O. M. (2007) Multi-phase Flow Analysis in Oil and Gas Engineering Systems and its Modeling. *Hydrocarbon World*, 2007, 57-60

Azevedo L. F. A. and Teixeira A. M. (2003) A critical review of the modeling of wax deposition mechanism. *Petroleum Science & Technology*, 21, 393-408

Bello O. O, Ademodi B. T. and Akinyemi P. O. (2005a). Xylene-based inhibitor solves crude oil wax problems in Niger Delta pipeline. *Oil & Gas Journal*, 103, 10-56.

Bello, O. O.; Fasesan, S. O.; Akinyemi, P. O.; Macaulay, S. R. A. and Latinwo, G. K. (2005b) Study of the influence of xylene-based chemical additive on crude oil flow properties and paraffin deposition inhibition, *Engineering Journal of the University of Qatar*, 18, 15-28

Bello, O. O.; Fasesan, S. O.; Teodoriu, C. and Reicicke, K. M. (2006). An evaluation of the performance of selected wax inhibitors on paraffin deposition of Nigerian crude oils. *Petroleum Science and Technology* 24, 195-206

Chanda, D.; Sarmah, A.; Borthaku, R. A.; Rao, K.V.; Subrahmanyam, B. and Das, H. C. (1998). Combined effect of asphaltenes and flow improvers on the rheological behaviour of Indian waxy crude oil. *Fuel* 77, 1163 –1167.

Chang C.; Nguyen Q. D. and Rønningsen H. P. (1999) Isothermal start-up of pipeline transporting waxy crude oil. *Journal of Non-Newtonian Fluid Mechanics*. 87 (1999) 127–154

Chang, C.; Boger, D. V. and Nguyen Q. D. (2000). Influence of thermal history on the waxy structure of statically cooled waxy crude oil. *SPE Journal*. (2000) Vol. 5:148-157.

Chang, C.; Boger, D. V. and Nguyen, Q. D. (1998). The yielding of waxy crude oils. Industrial & Engineering Chemistry Research 37, 1551–1559.

Coto, B.; Martos, C.; Pena, J. L.; Espada, J. J. and Robustillo, M. D. (2008). A new method for the determination of wax precipitation from non-diluted crude oils by fractional precipitation. *Fuel*, 87, 2090-2094

Davenport, T.C. and Somper, R.S.H. (1971). The yield value and breakdown of crude oil gels. Journal of the Institute of Petroleum 55,86–105.

Escobedo, J. and Mansoori, G. A. (2010). Heavy-organic particle deposition from petroleum fluid flow in oil wells and pipelines. *Petroleum Science*, 7:502-508;

Escobedo, J.; Mansoori, G. A.; Balderas-Joers, C.; Carranza-Becerra, L. J. and Mendez-Garcia, M. A. (1997). Heavy organic deposit during oil production from a hot deep reservoir: A field experience. *Proceeding of the 5th Latin America and Caribbean Society for Petroleum Engineering conference and Exhibition*, SPE Paper 38989, 9p, 1997

Fasesan, S. O., and Adewunmi, O. O. (2003). Wax variation in some Nigerian oil wells in Delta Field. *Petroleum Science and Technology*, 21:91–111.

Fawehinmi, A. (2007). Oil Companies' Corporate responsibility and sustainable infrastructure provision in the Niger Delta, *Petroleum Training Journal*, 4(1),22-29.

Gateau, P.; Henaut, I.; Barre, L. and Argillier, J. F. (2004). Heavy oil dilution. *Oil and Gas Science and Technology Review* IFP, 59(5):503–9.

Ghannam, M. T. and Esmail, N. (2006). Flow enhancement of medium-viscosity crude oil. *Journal of Petroleum Science and Technology,* 24(8):985–99.

Ghannam, M. T. and Esmail, N. (1998) Rheological properties of carboxymethyl cellulose. *Journal of Applied Polymer Science,* 64(2):289–301.

Ghannam, M. T. and Esmail, N. (2005) Yield stress behavior for crude oil–polymer emulsions. *Journal of Petroleum Science and Engineering,* 47:105–15.

Guozhong, Z. and Gang, L. (2010). Study on the wax deposition of waxy crude in pipelines and its application. *Journal of Petroleum Science and Engineering,* 70, 1–9

Hafiz, A. A. and Khidr, T. T. (2007). Hexa-triethanolamine oleate esters as pour point depressant for waxy crude oils. *Journal of Petroleum Science and Engineering,* 56, 296–302

Hasan, S. W.; Ghannam, M. T. and Esmail, N. (2010). Heavy crude oil viscosity reduction and rheology for pipeline transportation. *Fuel, 89 (2010) 1095–1100*

Hemant, P. S.; Kiranbala, A. and Bharambe D. P. (2008). Performance based designing of wax crystal growth inhibitors. *Energy and Fuels, 22 (6), 3930-3938*

Holder, G.A. and Winkler, J. (1965) Wax crystallization from distillate fuels." Part 1 and 2 *Journal of Institute of Petroleum* 51(499), 238-252.

Hussain, M.; Mansoori, G. A. and Ghotbi, S. (1999)Phase behavior prediction of petroleum fluids with minimum characterization data. *Journal of Petroleum Science and Engineering,* 22(1-3), 1999, 67-93

IP The Institute of Petroleum, London, (1993b)

Karan, K. Ratulowski, J. and German P. (2000). Measurement of waxy crude properties using novel laboratory techniques. SPE Annual conference and Exhibition, Dallas , Texas, 1-4, October 2000. SPE 62945

Khan, M. R. (1996). Rheological properties of heavy oils and heavy oil emulsions, *Energy Source,* 18, 385–91.

Langner, C. C. and Bass, R, M (2001). Method for enhancing the flow of heavy crude through subsea pipelines. US Patent 6264401

Lopes-da-Silva, J. A. and Coutinho, J. A. P., (2004). Dynamic rheological analysis of the gelation behavior of waxy crude oils. *Rheol. Acta,* 43 (5) 433-441.

Mansoori, G.A. (2009) A unified perspective on the phase behaviour of petroleum fluids, *Internaional Journal of Oil, Gas and Coal Technology,* Vol. 2, No. 2, pp.141–167.

Martinez-Paloua, R.; Mosquirra, M. L.; Zapata-Rendonb, B.; Mar-Juarezc, E.; Bernal-Huicochead, C.; Clavel-Lopezo, J. and Aburtob, J. (2011). Transportation of heavy and extra heavy crude oil by pipeline: A review. *Journal of Petroleum Science and Engineering,* 75 (3-4), 274-282

Matveenko, V. N.; Kirsanov, E. A. and Remizov, S. V. (1995). Colloids and Surfaces A: Physicochemical and Engineering Aspects 101, 1–7.

Oladiipo A., Bankole A. and Taiwo E (2009) Artificial Neural Network Modeling of Viscosity and Wax Deposition Potential of Nigerian Crude Oil and Gas

Condensates. *33rd Annual SPE International Technical Conference and Exhibition in Abuja, Nigeria,* August 3-5, 2009. SPE 128600

Philip, D. A.; Forsdyke, I. N.; McCracken I. R. and Ravenscroft P. D. (2011) Novel approach to waxy crude restart: Part 2: An investigation of flow events following shut down. *Journal of Petroleum Science and Engineering* 77(3-4), 286-304

Remizov S. V., Kirsanov E. A. Matveenko N. V. (2000), Colloids and Surfaces A. Physicochemical and Engineering Aspects 175, 271-275

Taiwo, E. A., Fasesan, S. O. and Akinyemi, O. P.(2009) Rheology of Doped Nigerian Niger-Delta Waxy Crude Oil. *Petroleum Science and Technology,* 27(13),1381 – 1393

Wang J., Brower K. And Buckley J. (1999) Advances in observation of asphaltene destabilisation. *Proceeding of SPE on Oil field Chemistry,* Houston, Texas. February, 16-19, 1999. SPE Paper 50745

Wang, K. S.; Wu, C. H.; Creek, J. L.; Shuler P. J. and Tang Y. (2003). Evaluation of effects of selected wax inhibitors on paraffin deposition. *Petroleum Science and Technology.* 21, 369-379

9

Degradation of Petroleum Fractions in Soil Under Natural Environment: A Gravimetric and Gas Chromatographic Analysis

Prahash Chandra Sarma
Cotton College, Guwahati,
Assam,
India

1. Introduction

The modern civilization depends on exploration, distribution and use of petroleum and thousands of products derived from it. The environment all throughout the world has been very badly damaged due to these chemicals. While pollution of the atmosphere has been caused almost totally by burning of the petroleum products, the leaks, spillages and accidental fallout of petroleum products and the crude petroleum itself have greatly affected land and water resources. The oil field areas always receive a large amount of effluents rich in crude oil and land degradation is a common phenomenon. While using different fractions of petroleum (gasoline, kerosene, diesel etc.) spills and leakages cannot be avoided and in industrial areas, garages and other places, a large amount of it flows to the nearby areas. All types of oil have amazing spreading power and once a leak occurs, the oil may spread horizontally as well as vertically depending upon soil conditions, moisture, temperature etc. The lighter fraction being volatile are easily removed by evaporation and other physical processes, but the heavy components such as aromatics - simple and polycyclic, Hopanes etc. remain in the soil for a very long time unless biodegraded by soil micro organisms. The detection of presence of some carcinogenic hydrocarbons viz. PAH has made the situation a matter of serious concern. In most cases; the pollutant oil exerts its detrimental effects before it degrades into harmless and simple compounds. Hence a study on the type of degradation mechanism, the measures which can expedite the process of bioremediation and how the soil parameters are influenced by oil pollutants and their degradation is demanded by situation. The present investigation intends to gains some true knowledge on the matter based on experimental findings.

2. Soil is a depleting natural resource

Human race is dependent on a number of gifts of the nature. One such gift of nature is the Soil resource. It has been defined as the thin layer of the earth's crust in which biological activities take place. The beneficial activities of soil are multidimensional. Animal life is absolutely dependent on plant kingdom for food. Plants absorb nutrients from soil and convert it into a form acceptable to animal kingdom. Thus it is one of the best supporters of life. Soil is made of weathered rock material and organic matter. Weathering, the physical and chemical breakdown of rocks is the first step in the soil forming process. The process of soil formation is a very long and slow geological process. The process is cyclic, because the rocks are changed to unconsolidated particles, which may be eventually be cemented together by other chemical and physical mechanisms to yield new rocks. The slowness of this soil producing process indicates that soil is a depleting natural resource when subjected to erosion, loss of fertility or pollution. Soil fertility, the ability of the soil to supply the plants with their essential nutrient elements, directly determines what crops can be grown on a soil and the nutritional value of these crops for man and animals.

3. Land degradation due to hydrocarbon pollutants

By virtue of the properties such as a portable, dense energy source and as the base of many industrial chemicals; Petroleum is one of the world's most important commodities. Today about 90% of fuel needs are met by oil. The presence of the oil industry has significant environmental impacts. Accidents and routine activities such as seismic exploration, drilling, and generation of polluting wastes affect the society. Oil extraction is often environmentally damaging. Leaks, spillages and accidental fallouts of petroleum and its various fractions are of frequent occurrence. Offshore exploration and extraction of oil disturbs the surrounding marine environment. Crude oil and refined fuel spills from tanker, underground pipelines, and ship accidents and have damaged soil resource, ecosystem in many parts of the world.

Land degradation due to this is a common phenomenon in the modern world. Since the early part of the twentieth century, the damaging effect of the hydrocarbons from oil has been known. Many workers reported the damaging effect of petroleum and its various fractions on soil and water resources (Young, 1935). It was reported that crude oil can sterilise soils and prevent crop growth for various period of time. The duration of the damaging effect depends largely on the degree and depth to which the soil is saturated with the oil. The damage that the oil does is due mostly to the prevention of the plant from obtaining sufficient moisture and from ramifying its roots very little is due to toxicity, as such (Plice, 1948).

Out of various processes of disappearance of oils from polluted soil, microbial degradation of pollutants is one of the most important ones (Davies & Westlaki, 1979). Crude petroleum and its fractions are converted into soil organic matter by bacteria and fungi. During the conversion, the organisms, which are free lives, fix fairly large amount of atmosphere nitrogen in the soil. Later this nitrogen becomes available for plant growth and the organic matter improves soil physical conditions. Based on general principle of hydrocarbon degradation, a programme of rehabilitation such as liming, fertiliser addition, and frequent tilling is considered to be broadly applicable for all types of mineral oils and experiments showed that these are really effective measures (Dibble & Bartha, 1979). Dry micro organisms contain approximately 14% N, 3% P and 1% S in the form of proteins, nucleic acids, polysaccharides and low molecular weight compounds. On the other hand, Petroleum products are composed of hydrogen and sulphur and essentially no phosphorous in their environment in order to grow on hydrocarbons as their carbon and energy sources (Rosenberg, 1993). This application of nutrients especially that of Phosphorous seems to be an effective means of rehabilitation and experiment showed that this is really so (Reynolds & Walworth, 2000). On the contrary, studies showed that the presence of Polychlorinated

Biphenyls negatively impact fossil fuel degradation (Hoeppel et.al, 1995).Out of several methods to expedite the process of degradation of Petroleum Hydrocarbons and to rehabilitate an oil inundated soil such as Bioremediation, Polyencapsulation, Vapour Transport, Land Farming, Alcohol Flooding etc. Bioremediation is considered effective and environmentally benign. Before taking up remediation measures emergency clean up operation is an important activity.

4. Methodology

Soil sample having no history of oil pollution were taken, polluted by known quantity of oil using emulsifier- some with nutrients, some with oxidising agents and some without any additive and the stated experiment was done by placing indoor in polyethene bags and maintaining the conditions necessary for microbial activities (Table-1). Series with B, C, D and E are similar to series A except that the oil is replaced by Kerosene oil, Diesel oil, Lubricating oil and Residual oil respectively. Gravimetric determination was done by withdrawing 20 g polluted soil sample from each bag and by recovering the oil by soxhlet extraction method using Petroleum ether as solvent after definite time interval. Gas Liquid Chromatography (GLC/GC) is the method of choice for rapidly and accurately analysing the volatile substances. It is also applicable to non volatile ones due to availability of higher column temperatures. Different compounds maintain similar sequence in the chromatogram analyzed under same system of column and identical conditions.

Sample No.	Mass of Crude Oil	Mass of Nutrient	Mass of Soil	Total mass	Concentration of Crude Oil	Volume of Emulsifier	Volume of H_2O_2 added
-	g	g	g	g	ppm	mL	mL
Set-1							
1A0	0	0	3000	3000	0	2	0
1A1	3	0	2997	3000	1000	2	0
1A2	15	0	2985	3000	5000	2	0
1A3	30	0	2970	3000	10000	2	0
1A4	45	0	2955	3000	15000	2	0
1A5	60	0	2940	3000	20000	2	0
Set-2							
2A0	0	10	2990	3000	0	2	0
2A1	3	10	2987	3000	1000	2	0
2A2	15	10	2975	3000	5000	2	0
2A3	30	10	2960	3000	10000	2	0
2A4	45	10	2945	3000	15000	2	0
2A5	60	10	2930	3000	20000	2	0
Set-3							
3A0	0	0	3000	3000	0	2	10
3A1	3	0	2997	3000	1000	2	10
3A2	15	0	2985	3000	5000	2	10
3A3	30	0	2970	3000	10000	2	10
3A4	45	0	2955	3000	15000	2	10
3A5	60	0	2940	3000	20000	2	10

Table 1. Sample Numbers and their Contents possessing Crude Oil.

5. Results and discussions

A few important parameters of the soil such as Texture, pH, Hydraulic Conductivity, Bulk Density, Water Holding Capacity, Soil Organic Carbon, NPK nutrients etc. were determined before the start, in between, and at the end of the experiment(Sing et al, 2000) (Table-2).

Parameter	Value	Unit	Parameter	Value	Unit
Texture	Silty clay	-	pH	6.86	-
Bulk Density	1.082	g cm^{-3}	WHC	41.80	%
Electrical Conductivity	0.43	$mScm^{-1}$	SOC	1.00	%
Hydraulic Conductivity	0.073	cm min^{-1}	SOM	1.724	%
Particle Density	2.650	g cm^{-3}	Porosity	59.17	%

Table 2. Values of Physico-Chemical Parameters of Soil.

Remains of hydrocarbons and their oxygenated derivatives in the above mentioned soil bags were extracted by soxhlet extraction using Petroleum Ether as solvent and gravimetric determination was done (Table-3). It was found that the NPK supplementation is an effective measure of rehabilitation of oil inundated soil. Supply of nutrients to the soil helps micro organisms to grow abundantly, consequently the degradation becomes faster. Hydrogen peroxide decomposes some of the organic matter present in the polluted soil and it also contributes towards the degradation to some extent. It has been found that more than 50% of the applied mass undergoes degradation in average in all the three sets of crude oil experiment during the first two months of placement. The activities of micro organisms are highly dependent on temperature. Ambient temperature was also recorded during the period of the experiment. The monthly average of the ambient temperature during the

Sample	Oil Added	Amount of Oil Recovered (g) after				
No.	g	1 month	2 month	3 month	4 month	5 month
1A1	3	01.975	01.650	01.111	00.142	00.030
1A2	15	12.797	09.375	07.979	05.634	03.832
1A3	30	17.715	14.200	13.339	10.842	06.327
1A4	45	34.567	20.051	19.018	17.878	10.001
1A5	60	41.400	30.645	23.243	19.835	10.770
2A1	3	01.321	01.200	00.915	00.098	00.022
2A2	15	11.850	08.476	07.700	04.545	02.875
2A3	30	16.715	13.100	11.990	08.895	05.500
2A4	45	31.922	18.025	14.019	13.950	07.290
2A5	60	37.950	27.409	20.890	14.160	09.110
3A1	3	01.613	01.400	00.965	00.110	00.000
3A2	15	12.040	08.734	07.800	05.312	03.005
3A3	30	16.973	13.362	12.319	09.312	05.780
3A4	45	33.727	19.632	16.737	14.499	08.560
3A5	60	38.419	29.319	21.783	15.322	09.234

Table 3. Balance Amount of Crude Oil Recovered against Months after subtracting the mass found in Controls.

period of the experiment is given (Table 4). The experiment was done during summer in the city of Guwahati, where the average of ambient temperature was 28.24 ± 0.40^{0}C in the range of 26.89-29.19^{0}C with negligible decreasing trend (correlation coefficient, r = -0.20). These conditions along with other physico-chemical parameters were favourable for microbial growth (Khan & Anjaneyulu, 2005).

Temp.	At 1st month	At 2nd	At 3rd	At 4th	At 5th	At 6th	Average	Standard Error	Correlation Coefficient
^{0}C	27.28	29.19	28.65	28.81	28.63	26.89	28.24	±0.40	-0.20

Table 4. Average Ambient Temperature during the Period of Experiment.

Many agencies working on Hydrocarbon pollution use GC technique to estimate hydrocarbon pollution. G.C. analysis of the oil recovered from three samples viz. 1A5, 2A5 and 3A5 (Plate-1) from three sets after two months since placement have shown that number of components become 22, 37 and 39 respectively including solvent (Sarma et al, 2005a). Hydrogen Peroxide decomposes organic compounds from soil. Hence a good number of volatile components under GC conditions have been found in the 3A5 sample. Later these components escape from soil as carbon dioxide and a fraction fixes with soil. In the mass determination of the remaining hydrocarbons, it has been found that degradation in the second set i.e. set having applied NPK is highest. It means that it has already degraded and given away more compounds from the polluted soil sample. It has been seen from the chromatograms that there are no components with retention time in the range of 16.315 -43.247 minutes, 17.682 - 40.729 minutes and 11.589 - 42.677 minutes respectively in1A5, 2A5 and 3A5. A large number of components with minute differences in their retention time are seen in the chromatograms. Table- 5 represents some of such peaks between 1A5 and 2A5. The corresponding peaks in 2A5 appeared with an advance retention time of 0.452 minute in average belonging to the range of 0.571 to 0.392 minute. There are no corresponding peaks in 2A5 for five compounds present in 1A5 (Table-6). Probably these are the compounds degraded completely as a result of profuse microbial activities due to application of nutrients. Table (Table-7) represents the list of peaks which are found in 2A5 but not in 1A5. These are the peaks of compounds formed due to higher microbial activities as a result of application of nutrients. Number of such components are 20 and these compounds occupied a minute area % (average 0.63 %) of the chromatogram in the range

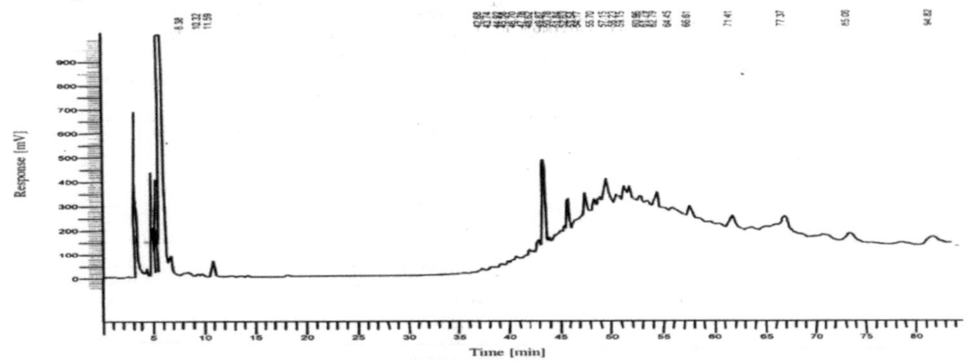

Plate 1. GC Chromatogram of oil recovered after 2 months of placement of sample 3A5.

Sl. No.	Serial Number As in 1A5 Chromatogram	1A5 Retention Time	1A5 Area % Report	2A5 Retention Time	2A5 Area % Report	Minute Advance in Ret. time
1	6	12.271	19.38	11.700	45.15	0.571
2	7	12.810	1.27	12.363	3.54	0.447
3	8	13.288	0.39	12.839	1.92	0.449
4	9	14.538	6.36	14.080	12.32	0.458
5	10	15.286	2.11	14.835	6.17	0.451
6	11	15.823	0.12	15.339	2.17	0.484
7	12	16.315	3.09	15.860	5.38	0.455
8	13	43.247	0.68	42.845	0.30	0.402
9	14	48.772	1.85	48.380	0.26	0.392
10	15	49.805	2.47	49.363	0.78	0.442
11	16	50.389	23.34	49.993	5.26	0.396
12	17	51.724	0.08	51.275	0.16	0.449
13	18	52.736	0.27	52.320	1.03	0.416
14	19	53.152	4.70	52.744	1.55	0.408
15	20	55.236	2.48	54.832	0.83	0.404
16	21	56.304	1.06	55.814	0.46	0.490
17	22	59.828	0.21	59.261	0.10	0.567

Table 5. List of Similar Peaks in GC Analysis Report of 1A5 and 2A5.

Sl No in Chromatogram	1	2	3	4	5
Retention Time in minute	09.053	09.355	09.774	10.497	10.992
Area %	05.33	07.42	03.87	02.13	11.41

Table 6. List of peaks found in 1A5 giving no corresponding peak in 2A5.

Sl. No.	Sl. No. of peak	Retention Time in minute	Area %	Sl. No.	Sl. No. of peak	Retention Time in minute	Area %
1	4	13.220	0.30	11	25	53.630	0.07
2	6	14.513	2.56	12	28	56.367	0.34
3	10	17.682	0.65	13	29	57.210	2.31
4	11	40.729	0.09	14	30	58.320	0.16
5	12	42.359	0.04	15	32	59.782	0.25
6	14	44.121	0.73	16	33	61.080	0.04
7	15	45.072	0.07	17	34	62.848	1.74
8	16	45.714	0.16	18	35	66.697	0.61
9	17	46.270	0.15	19	36	71.440	1.52
10	18	46.840	0.66	20	37	77.554	0.18

Table 7. List of peaks found in 2A5 but not in 1A5.

of 0.04 to 2.56%. Table (Table-8) represents 14 peaks due to same compounds between 1A5 and 3A5 with their time of appearance. The corresponding peaks in 3A5 appeared after an average advancement of retention time of 0.589 minute in the range of 0.502 to 0.682 minute. These common components are the examples of some of the compounds which are resistant to degradation during this period. There are no corresponding peaks for eight compounds present in 1A5 (Table-9).These have already disappeared in 3A5. Probably these are the compounds degraded completely as a result of degradation activities due to application of hydrogen peroxide. Table-10 represents the list of peaks which are found in 3A5 but not in 1A5. These are the peaks of compounds formed due to higher microbial activities that occurred on the compounds. Number of such components are 25 and these compounds occupied a minute area (average 1.84%) of the chromatogram in the range of 0.06 to15.19%.

-	Sl. No.	1A5		3A5		
Sl No	as in 1A5 Chromatogram	Retention Time	Area % Report	Retention Time	Area% Report	Advance in Retention Time
1	1	9.053	5.33	8.383	0.68	0.670
2	2	9.355	7.42	8.726	0.52	0.629
3	5	10.992	11.41	10.325	0.46	0.667
4	6	12.271	19.38	11.589	3.68	0.682
5	13	43.247	0.68	42.677	0.22	0.570
6	14	48.772	1.85	48.249	1.15	0.523
7	15	49.805	2.47	49.228	2.97	0.577
8	16	50.389	23.34	49.868	20.99	0.521
9	17	51.724	0.08	51.132	1.19	0.592
10	18	52.736	0.27	52.229	1.50	0.507
11	19	53.152	4.70	52.630	7.23	0.522
12	20	55.236	2.48	54.734	4.93	0.502
13	21	56.304	1.06	55.695	3.11	0.609
14	22	59.828	0.21	59.154	5.48	0.674

Table 8. List of Similar Peaks in GC Analysis Report of 1A5 and 3A5.

Sl. No.	1	2	3	4	5	6	7	8
Peak No	3	4	7	8	9	10	11	12
RT in minute	9.774	10.497	12.810	13.288	14.538	15.286	15.823	16.315
Area %	3.87	2.13	1.27	0.39	6.36	2.11	0.12	3.09

Table 9. List of peaks found in 1A5 giving no corresponding peak in 3A5.

GC analysis of the oil, recovered after 1 month and 2 months in the sample 2A3 gave a total of 17 and 24 peaks respectively. There is no peaks during12.217 to 47.277 minutes and during 15.303 to 44.055 minutes in the samples after 1 month and after 2 months respectively. Out of the 17 compounds obtained after I month, it appears that 14 compounds are still present in oil recovered after two months. It means that these components could resist degradation in the second month. These are placed in table (Table-11). These compounds appear 0.232 to 0.387 minute (average 0.275 minute) advance in the sample obtained after 2 months. This change in retention time is probably due to minor change of experimental conditions during transit. The three components, which appeared at 12.217,

Sl. No.	Sl. No. of peak	RT in minute	Area %	Sl. No.	Sl. No. of peak	RT in minute	Area %
1	6	43.743	0.10	14	27	57.153	15.19
2	7	44.919	0.13	15	28	58.234	0.95
3	8	45.577	0.43	16	30	59.705	3.71
4	9	46.125	0.40	17	31	60.961	0.29
5	10	46.702	0.84	18	32	61.788	0.09
6	11	47.778	0.14	19	33	62.792	4.38
7	13	48.620	0.59	20	34	64.446	0.09
8	16	50.340	0.66	21	35	66.609	3.74
9	17	50.776	0.06	22	36	71.413	6.76
10	19	51.836	1.66	23	37	77.370	0.35
11	22	53.537	0.40	24	38	85.062	0.44
12	23	54.169	0.17	25	39	94.824	0.54
13	26	56.245	3.78	-			

Table 10. List of peaks found in 3A5 but not in 1A5.

-	Sl. No.	After 1 m		After 2 m		Minute
Sl No	1 month report	Retention Time	Area %	Retention Time	Area %	Advance in RT
1	1	10.941	0.90	10.686	9.58	0.255
2	3	47.277	0.71	47.001	2.21	0.276
3	4	49.850	0.44	49.571	3.13	0.279
4	5	50.406	9.43	50.159	11.01	0.247
5	6	51.771	0.27	51.465	0.25	0.306
6	7	52.736	7.70	52.496	7.66	0.240
7	8	53.171	3.43	52.928	4.94	0.243
8	9	55.244	1.65	55.012	2.77	0.232
9	10	56.314	4.17	56.037	3.29	0.277
10	11	56.893	0.55	56.594	0.29	0.299
11	12	57.621	18.27	57.365	17.62	0.256
12	13	58.132	3.12	57.860	2.74	0.272
13	14	59.848	0.81	59.561	4.05	0.287
14	15	63.516	20.07	63.129	13.55	0.387

Table 11. List of Similar Peaks in GC Analysis Report of 2A3 after 1Month and 2Months.

72.525 and 81.026 minutes with area percent report of 25.47, 1.85 and 1.17 respectively in the sample after 1 month got disappeared in the second month. It has also been seen that another 10 peaks due to 10 new compounds appear in the sample after 2 months (Table- 12). These compounds seem to be the degradation products of various hydrocarbons. These peaks have area percent in the range of 0.11 to 11.00 with an average of 1.69%.

In the GC analysis of the sample 3A3 after 1 month and after 2 months, it has been seen that there are 11 and 33 components respectively including that for the solvent. There are no peaks appear in between 20.082 to 46.835 and in between 11.662 to 42.728 minutes in the two samples respectively. As many as 8 numbers of peaks of components present after 1 month in 3A3 can be linked to same number of components present in the extract obtained after 2 months. Table (Table-13) represents these peaks. The corresponding peaks appear in the sample after 2 months with an advance retention time average of 0.06 minute in the range of 0.011 to 0.126 minute. Corresponding peaks for the rests three viz. at retention time 20.082, 62.788 and 71.406 minutes are not present in the second month. It implies that these components disappear during the month. The other 25 components which appear in the sample after 2 months are probably due to formation of derivatives as a result of degradation activities. These components have an average area percent of 1.42 in the range of 0.05 to 6.33 (Table-14).

Sl. No.	Sl. No. of peak	Retention Time in minute	Area %	Sl. No.	Sl. No. of peak	Retention Time in minute	Area %
1	1	6.823	0.86	6	7	44.055	0.11
2	2	7.256	0.17	7	14	53.817	0.38
3	3	9.225	2.24	8	15	54.476	0.39
4	4	13.335	0.61	9	21	58.546	0.24
5	5	15.303	0.92	10	24	71.991	11.00

Table 12. List of peaks found in the sample 2A3 after 2 months but not in after 1 month.

-	Sl. No.	After 1 month		After 2 months		Minute
Sl No	1 m report	Retention .Time in minute	Area %	Ret. Time in minute	Area %	Advance in RT (minute)
1	1	11.732	36.21	11.662	2.83	0.070
2	3	46.835	1.00	46.709	0.31	0.126
3	4	49.952	2.45	49.912	25.59	0.040
4	5	52.303	5.93	52.275	1.28	0.028
5	6	52.734	1.43	52.681	7.05	0.053
6	7	54.822	0.89	54.777	4.20	0.045
7	8	55.863	3.19	55.760	4.85	0.103
8	9	57.207	16.13	57.196	18.27	0.011

Table 13. List of Similar Peaks in GC analysis report of 3A3 extracted after 1 and 2 months.

The 17 and 11 peaks found respectively in the GC chromatogram of samples 2A3 and 3A3 after 1 month of placement, possess 9 peaks (Table-15) having a very narrow range of 0.414 to 0.454 minute difference in the retention time except one which appeared at a difference of 0.728 minute. The average difference in retention time is 0.469 minute. It has been seen that larger number of components disappeared from 3A3 at the same time compared to 2A3. Out of 24 and 33 peaks respectively found in the GC chromatogram of samples 2A3 and 3A3 after 2 months 15 similar peaks have been identified and presented in table (Table- 16). Compared to number of peaks found in 2A3, the corresponding peaks in 3A3 appear in advance by a narrow range of 0.169 to 0.365 minutes where the average value obtained is0.261 minute. The components responsible for these peaks are resistant to degradation at least up to two months under the experimental conditions.

Sl. No.	Sl. No. of peak	Retention Time (minute)	Area %	Sl. No.	Sl. No. of peak	Retention Time (minute)	Area %
1	1	8.258	0.10	14	17	51.891	2.70
2	2	8.807	0.07	15	20	53.620	0.71
3	4	42.728	0.33	16	21	54.246	0.05
4	5	44.020	0.48	17	24	56.266	3.55
5	6	44.956	0.26	18	26	58.276	1.02
6	7	45.629	0.65	19	27	59.196	6.33
7	8	46.177	0.62	20	28	59.742	2.19
8	10	47.826	0.21	21	29	61.125	1.71
9	11	48.299	1.18	22	30	62.843	1.70
10	12	48.682	0.86	23	31	66.631	1.29
11	13	49.288	5.85	24	32	71.433	0.26
12	15	50.393	1.20	25	33	77.479	0.49
13	16	51.195	1.81	-	-	-	-

Table 14. List of peaks found in the sample 3A3 after 2 months but not in after 1 month.

Sl. No.	Sl. No. 2A3 report	2A3-1M Retention Time	2A3-1M Area %	3A3-1M Retention Time(min)	3A3-1M Area %	Minute Advance in Retention time
1	2	12.217	25.47	11.772	36.21	0.440
2	3	47.277	0.71	46.835	1.00	0.442
3	5	50.406	9.43	49.952	2.45	0.454
4	7	52.736	7.70	52.303	5.93	0.433
5	8	53.171	3.43	52.734	1.43	0.437
6	9	55.244	1.65	54.822	0.89	0.422
7	10	56.314	0.417	55.863	3.19	0.451
8	12	57.621	18.27	57.207	16.13	0.414
9	15	63.516	20.07	62.788	21.04	0.728

Table 15. List of Similar Peaks in GC Analysis Report of 2A3and 3A3 after one month.

Gas Chromatographic analysis of recovered oil from kerosene oil polluted soil (B-set) after 2 months of placement showed that number of components at 1B5 and 3B5 (Plate-2) became 17 and 35 respectively. Gravimetric determination shows that mass of oil recovered in the samples having applied Hydrogen Peroxide is less, whereas number of components in the same is more than that of the sample without Hydrogen Peroxide (Table-17). It indicates that Hydrogen Peroxide removes the pollutants by degrading into smaller compounds. There are no peaks appeared in between the retention time ranges 4.838 to 32.566 minute in1B5 and in the range of 9.611 to 32.669 minute in 3B5. It appears that for every peak in the 1B5 sample, there is a corresponding peak in the 3B5 sample (Table-18) with an average delay of 0.115 minute in the range of 0.095 to 0.190 minute. All peaks present in 1B5 have been found in 3B5. Excluding peak at serial number 9, all other similar peaks in the table have higher peak area at 1B5. Besides these, there are about 18 peaks in 3B5 but not in 1B5, out of which the peaks at 2.716 appears to be for the solvent. These 18 peaks (Table-19) seem to be for some degradation products.

.	Sl. No.	2A3	2A3	3A3	3A3	Minute
Sl. No	As in 2A3	Retention Time	Area % Report	Retention Time (minute)	Area % Report	Advance in Retention time
1	8	47.001	2.21	46.709	0.31	0.292
2	9	49.571	3.13	49.288	5.85	0.283
3	10	50.159	11.01	49.912	25.59	0.247
4	11	51.465	0.25	51.195	1.81	0.270
5	12	52.496	7.66	52.275	1.28	0.221
6	13	52.928	4.94	52.681	7.05	0.247
7	14	53.817	0.38	53.620	0.71	0.197
8	15	54.476	0.39	54.246	0.05	0.230
9	16	55.012	2.77	54.777	4.20	0.235
10	17	56.037	3.29	55.760	4.85	0.277
11	18	56.594	0.27	56.266	3.55	0.328
12	19	57.365	17.62	57.196	18.27	0.169
13	21	58.546	0.24	58.276	1.02	0.270
14	22	59.561	4.05	59.196	6.33	0.365
15	23	63.129	13.55	62.843	1.70	0.286

Table 16. List of Similar Peaks in GC chromatogram of 2A3 and 3A3 extracted after two months.

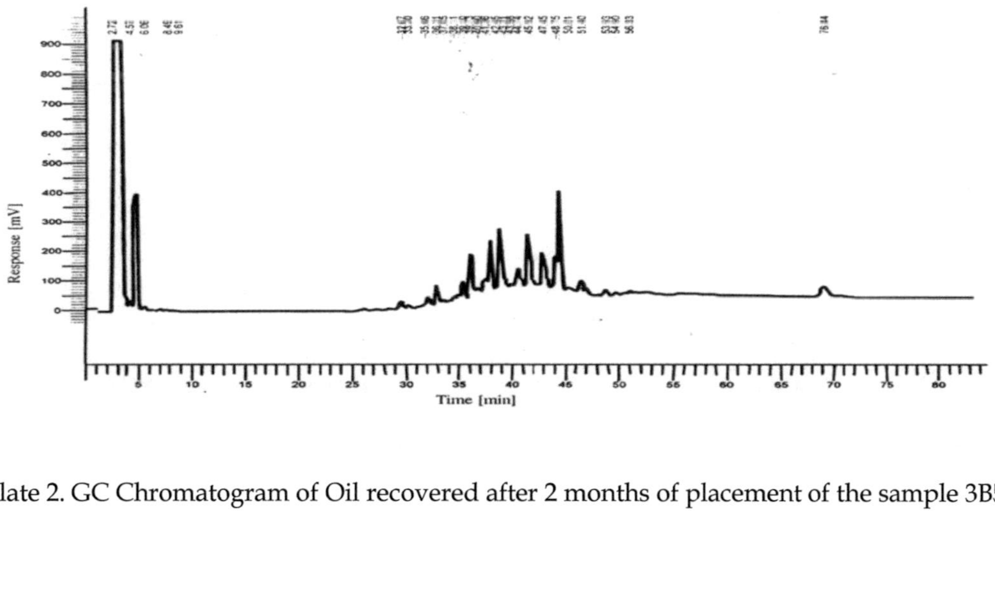

Plate 2. GC Chromatogram of Oil recovered after 2 months of placement of the sample 3B5.

Sample Number	Oil added g	Amount of degraded oil (g) after 1 month	2 month	3 month	4 month	5 month
1B5	60	21.126	30.355	47.322	53.832	54.753
3B5	60	25.751	32.790	48.048	54.132	55.052

Table 17. Amount of Kerosene Oil Degraded against time in Months.

Gas Chromatographic Analysis of the extracted oil after 3 months of mixing in the Diesel oil polluted samples of 1C3, 2C3(Plate-3) and 3C3 gave 50, 42 and 27 peaks respectively (Sarma ,2010). It has been seen that the peaks at 1C3 starts at retention time 29.592 minute onwards, which for the 2C3 and 3C3 are 31.873 and 40.645 minute respectively.

Sl. No.	Sl No As in 1B5 Chromatogram	1B5 Retention Time (Minute)	Area %	3B5 Retention Time (Minute)	Area %	Minute Delay in Ret. time
01	01	04.387	01.45	04.567	0.27	0.180
02	02	04.838	19.73	05.028	7.09	0.190
03	03	32.566	01.23	32.669	0.34	0.103
04	04	35.389	02.07	35.490	0.55	0.101
05	05	36.207	05.41	36.309	1.28	0.102
06	06	38.435	00.43	38.541	0.18	0.106
07	07	38.993	03.59	39.096	0.89	0.103
08	08	39.623	10.59	39.726	2.51	0.103
09	09	41.251	00.30	41.360	1.09	0.109
10	10	41.760	09.89	41.867	3.94	0.107
11	11	42.813	11.15	42.923	4.41	0.110
12	12	44.627	01.28	44.743	0.93	0.116
13	13	45.813	07.65	45.921	3.03	0.108
14	14	47.341	04.21	47.446	3.05	0.105
15	15	48.645	04.00	48.748	2.07	0.103
16	16	48.962	15.74	49.073	6.84	0.111
17	17	51.309	01.27	51.404	0.75	0.095

Table 18. List of Similar Peaks in GC Analysis Report of 1B5 and 3B5.

Sl. No.	Sl. No. of peak	Retention Time in minute	Area %	Sl. No.	Sl. No. of peak	Retention Time in minute	Area %
1	1	2.716	57.97	10	20	42.454	0.10
2	4	6.061	0.10	11	22	43.307	0.50
3	5	8.486	0.00	12	23	43.976	0.02
4	6	9.611	0.00	13	29	50.011	0.13
5	8	33.300	0.18	14	31	51.815	0.44
6	9	35.062	0.05	15	32	53.927	0.28
7	12	37.053	0.49	16	33	54.903	0.01
8	13	38.110	0.16	17	34	56.332	0.01
9	17	40.604	0.22	18	35	76.439	0.12

Table 19. List of peaks found in the sample 3B5 but not in 1B5.

Against 50 compounds present in 1C3, there are 42 and 27 compounds present in the samples 2C3 and 3C3, where nutrients and Hydrogen Peroxide were applied respectively. It implies that nutrients and Hydrogen Peroxide have a positive role in degradation of higher hydrocarbons into lower ones and subsequent removal of them from soil. A list of similar peaks in GC analysis report of 1C3, 2C3 and 3C3 are given in table (Table-20). It has been seen that the peaks due to same compounds appear in the chromatogram of 2C3 and 3C3 after an average delay in retention time of 0.078 and 0.123 minutes compared to the peaks of 1C3 in a range of 0.018 to 0.194 and 0.087 to 0.232 minutes respectively. This minute difference in retention time is probably due to minor change in experimental conditions during transit. There two peaks in the chromatogram of 2C3 and 3C3 at 54.466 and 54.552 minute with area % 0.53 and 0.22 respectively are probably due to same compound, which is a degradation product. The other two such compounds appear at retention time 61.654 and 71.422 minute in the 2C3 and 3C3 reports are also due to degradation products.

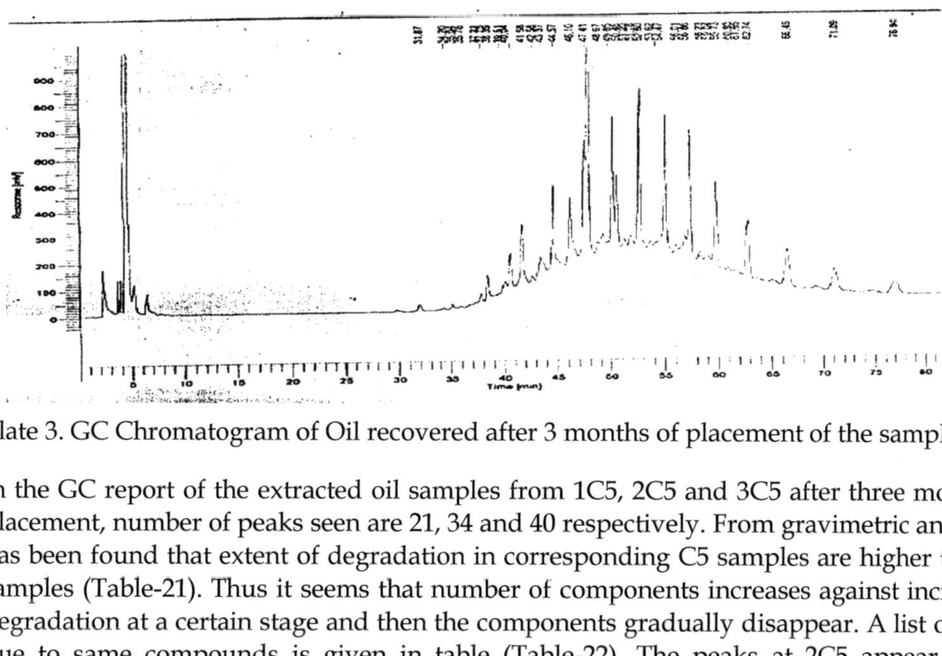

Plate 3. GC Chromatogram of Oil recovered after 3 months of placement of the sample 2C3.

In the GC report of the extracted oil samples from 1C5, 2C5 and 3C5 after three months of placement, number of peaks seen are 21, 34 and 40 respectively. From gravimetric analysis it has been found that extent of degradation in corresponding C5 samples are higher than C3 samples (Table-21). Thus it seems that number of components increases against increase of degradation at a certain stage and then the components gradually disappear. A list of peaks due to same compounds is given in table (Table-22). The peaks at 2C5 appear with a negligible average delay of 0.012 minute in retention time in the range of -0.036 to 0.055. The same for 3C5 peaks is 0.007 minute in the range of -0.070 to 0.034 minute. The existence of these compounds in all the samples indicates that these

compounds could resist degradation during the initial period of three months. The other compounds present in 2C5 and 3C5 are some of the degradation products. In the same way similar peaks have been found in 1C3 and 1C5, 2C3 and 2C5, and between 3C3 and 3C5.

The chromatograms obtained from the extracts of Lubricating oil polluted samples 1D1, 1D2 and 1D3 (Plate-4) are represented in plates. Number of components which persists degradation and number of components produced as a result of microbial degradation taken together has been found to be 64, 53 and 67 respectively, including solvent peak after lapse of 3 months (Sarma & Bhattacharyya ,2010). From the gravimetric analysis, it appears

-1- SlNo. (as per GC 1C3)	-2- 1C3 Retention Time (RT)	-3- 1C3 Area % of the peak	-4- 2C3 RT	-5- 2C3 Area % of the peak	-6- 3C3 RT	-7- 3C3 Area % of the peak	-8- Minute Delay in RT of 2C3 compared to 1C3	-9- Minute Delay in R T of 3C3 compared to 1C3
1	29.952	0.03	-	-	-	-	-	-
2	31.792	0.44	31.873	0.33	-	-	0.081	-
3	34.112	0.11	34.195	0.06	-	-	0.083	-
4	34.996	0.03	35.042	0.21	-	-	0.046	-
5	35.693	0.27	35.776	0.01	-	-	0.083	-
6	36.533	0.24	-	-	-	-	-	-
7	37.100	0.08	37.214	0.02	-	-	0.114	-
8	37.662	0.96	37.753	0.41	-	-	0.091	-
9	38.287	1.60	38.392	1.32	-	-	0.105	-
10	38.806	0.15	-	-	-	-	-	-
11	39.415	0.39	39.508	0.01	-	-	0.093	-
12	39.849	2.57	40.043	0.93	-	-	0.194	-
13	40.449	3.17	40.514	2.21	40.645	0.53	0.065	0.131
14	41.504	2.19	41.574	4.41	41.722	0.90	0.070	0.148
15	42.504	1.35	42.576	0.32	-	-	0.072	-
16	43.276	2.95	43.368	0.75	43.495	0.05	0.092	0.127
17	43.651	1.73	-	-	-	-	-	-
18	44.190	1.35	-	-	-	-	-	-
19	44.513	3.53	44.574	4.03	44.690	2.37	0.061	0.116
20	44.974	0.83	-	-	-	-	-	-
21	45.364	0.03	-	-	-	-	-	-
22	46.046	5.28	46.101	5.23	46.195	2.19	0.055	0.094
23	47.365	3.74	47.414	5.96	47.509	5.33	0.049	0.095
24	47.556	22.91	47.634	15.65	47.810	19.82	0.078	0.176
25	48.608	0.51	48.674	0.06	48.786	0.10	0.066	0.112
26	49.142	1.12	49.160	0.81	49.265	0.57	0.018	0.105
27	49.583	0.02	49.649	0.20	-	-	0.066	-
28	50.016	3.99	50.073	6.62	50.163	9.04	0.057	0.090
29	50.425	6.06	50.476	3.81	50.563	4.54	0.051	0.087
30	51.141	0.76	51.200	0.44	51.317	0.07	0.059	0.117
31	51.755	0.10	51.779	0.09	51.980	0.23	0.024	0.201
32	52.532	6.49	52.597	8.59	52.687	12.80	0.065	0.090
33	53.474	0.58	53.623	0.54	53.747	0.37	0.149	0.124
34	54.007	0.03	54.136	0.32	54.237	0.04	0.129	0.101
35	54.938	5.55	54.993	7.34	55.087	10.74	0.055	0.094
36	55.954	0.88	56.006	0.68	56.110	0.08	0.052	0.104
37	56.442	0.43	56.476	0.43	56.578	0.03	0.034	0.102
38	56.834	1.97	56.859	1.32	56.972	0.86	0.025	0.113
39	57.213	4.15	57.273	6.59	57.370	9.76	0.060	0.097
40	58.169	0.86	58.229	0.61	58.341	0.06	0.060	0.112
41	58.738	0.10	-	-	-	-	-	-
42	58.927	0.41	59.042	0.37	-	-	0.115	-
43	59.664	3.21	59.735	5.39	59.869	7.94	0.071	0.134
44	60.389	0.04	-	-	-	-	-	-
45	60.865	0.23	60.930	0.02	-	-	0.065	-
46	62.645	2.45	62.735	4.64	62.911	6.41	0.090	0.176
47	66.342	2.18	66.447	3.90	66.679	4.70	0.105	0.232
48	70.973	1.54	71.094	2.92	-	-	0.121	-

49	72.360	0.33	-	-	-	-	-	-
50	76.800	0.08	76.942	1.88	-	-	0.142	-

Table 20. List of Similar Peaks in GC Analysis Report of 1C3, 2C3 and 3C3.

Sample	Oil Added	Percent of the applied oil Degraded after				
No.	g	1 month	2 months	3 months	4 months	5 months
1C3	30	27.84	43.18	45.20	65.53	67.66
1C5	60	24.25	46.03	56.04	72.43	75.50
2C3	30	33.19	49.88	51.89	69.97	70.98
2C5	60	38.33	53.13	59.03	73.78	77.19
3C3	30	30.91	45.38	47.93	69.09	69.99
3C5	60	36.30	49.52	57.45	73.33	76.45

Table 21. Percent of Applied Diesel Oil Degraded against time in Months.

-1- Sl No. (as per GC report of 1C5)	-2- 1C5 Retention Time (RT)	-3- 1C5 Area % of the peak	-4- 2C5 RT	-5- 2C5 Area % of the peak	-6- 3C5 RT	-7- 3C5 Area % of the peak	-8- Minute Delay in RT of 2C5peaks compared to peaks of 1C5	-9- Minute Delay in R T of 3C5peaks compared to peaks of 1C5
1	0.986	0.00	-	-	-	-	-	-
2	35.087	0.77	35.051	0.81	35.017	0.65	-0.036	-0.070
3	37.761	0.83	37.759	0.71	37.753	0.63	-0.002	-0.008
4	38.426	3.84	38.411	3.22	38.395	2.60	-0.015	-0.031
5	40.075	0.36	40.059	1.48	40.034	1.33	-0.016	-0.041
6	40.525	2.94	40.530	2.76	40.526	2.51	0.005	0.001
7	41.595	7.77	41.595	6.44	41.589	5.82	0.000	-0.006
8	43.364	0.72	43.379	1.03	43.381	0.97	0.015	0.017
9	44.581	6.46	44.589	5.64	44.586	5.05	0.008	0.005
10	46.083	2.79	46.106	2.33	46.106	2.30	0.023	0.023
11	47.399	6.71	47.411	6.56	47.415	6.16	0.012	0.016
12	47.699	19.61	47.724	16.63	47.729	16.09	0.025	0.030
13	50.055	7.50	50.073	6.96	50.075	6.44	0.018	0.020
14	50.448	3.35	50.478	2.91	50.479	2.84	0.030	0.031
15	52.578	8.30	52.597	8.43	52.598	8.28	0.019	0.020
16	54.982	6.70	54.995	6.87	54.999	6.67	0.013	0.017
17	57.263	6.27	57.275	6.37	57.280	6.32	0.012	0.017
18	59.730	5.87	59.751	5.53	59.748	5.65	0.021	0.018
19	62.729	5.07	62.755	4.83	62.752	5.13	0.026	0.023
20	66.451	3.95	66.477	4.06	66.479	4.44	0.026	0.028
21	71.096	0.21	71.151	3.02	71.130	3.44	0.055	0.034

Table 22. List of Similar peaks in GC Analysis Report of 1C5, 2C5and 3C5.

that the extent of degradation is highest in the samples with nutrients. The extent of degradation in presence of hydrogen peroxide is moderate. It seems that hydrogen peroxide increases number of components and later these degradation products disappear from soil. There are some peaks which seem to be due to same compounds (Table-23).

-1- Sl No.	-2- 1D1	-3- 1D1	-4- 1D2	-5- 1D2	-6- 1D3	-7- 1D3	-8- Minute	-9- Minute
(as per GC report of 1D1)	Retention Time (RT)	Area % of the peak	RT	Area % of the peak	RT	Area % of the peak	Delay in RT of 1D2 peaks compared to 1D1	Adv in R T of 1D3 peaks compared to peaks of 1D1
1	8.569	0.06	-	-	8.340	0.03	-	0.229
2	9.381	2.03	9.706	1.07	9.107	4.32	0.325	0.274
3	11.562	21.09	11.822	18.96	11.316	10.34	0.260	0.246
4	12.595	0.47	-	-	12.363	0.43	-	0.232
5	13.203	0.09	-	-	12.912	0.55	-	0.291
6	14.122	0.77	14.367	0.37	13.865	0.76	0.245	0.257
7	14.792	0.51	15.026	0.29	14.578	0.47	0.234	0.214
8	15.823	0.23	16.056	0.14	15.535	0.10	0.233	0.288
9	16.167	0.42	16.397	0.27	15.896	0.18	0.230	0.271
10	17.268	0.16	17.541	0.07	16.934	0.67	0.273	0.334
11	18.339	0.18	18.584	0.14	18.052	0.16	0.245	0.287
12	18.714	2.08	18.933	1.54	18.405	3.96	0.219	0.309
13	19.484	7.78	19.689	6.89	19.080	8.69	0.205	0.404
14	19.989	11.70	20.220	11.26	19.667	10.63	0.231	0.322
15	20.761	0.25	-	-	-	-	-	-
16	21.742	0.65	21.973	0.42	21.440	1.29	0.231	0.302
17	22.356	10.06	22.557	8.57	21.920	12.71	0.201	0.436
18	24.520	0.18	-	-	24.165	0.45	-	0.355
19	24.787	0.30	24.985	0.28	24.494	0.66	0.198	0.293
20	27.586	0.28	27.782	0.33	27.260	1.03	0.196	0.326
21	28.222	11.95	28.402	12.15	27.867	8.89	0.180	0.355
22	28.979	0.75	29.143	0.84	28.683	0.75	0.164	0.296
23	29.300	0.68	29.472	0.81	28.996	0.78	0.172	0.304
24	30.794	1.50	30.980	1.14	30.466	3.20	0.186	0.328
25	31.355	1.61	31.609	1.14	30.890	3.71	0.254	0.465
26	34.324	0.29	34.492	0.16	33.982	0.33	0.168	0.342
27	36.155	0.18	36.310	0.12	35.804	0.33	0.155	0.351
28	37.154	0.02	-	-	36.700	0.14	-	0.454
29	38.152	1.27	38.311	0.89	37.798	2.15	0.159	0.354
30	39.031	0.06	-	-	38.677	0.04	-	0.354
31	40.092	0.11	40.246	0.04	39.770	0.06	0.154	0.322
32	40.681	0.32	40.828	0.13	40.416	0.01	0.147	0.265
33	42.504	0.54	42.645	0.10	42.133	0.41	0.141	0.371
34	42.818	0.35	42.963	0.06	42.799	0.07	0.145	0.019
35	43.996	2.55	44.169	1.07	43.681	1.98	0.173	0.315
36	44.518	1.35	44.668	0.62	44.154	2.03	0.150	0.364

37	45.121	0.35	-	-	44.736	0.24	-	0.385
38	45.709	0.13	45.904	0.03	-	-	0.195	-
39	46.872	1.25	47.021	0.49	46.464	0.11	0.149	0.408
40	48.391	0.48	48.549	0.12	-	-	0.158	-
41	49.701	1.68	49.842	0.72	-	-	0.141	-
42	50.014	1.40	50.160	0.52	-	-	0.146	-
43	51.005	0.12	51.126	0.02	-	-	0.121	-
44	51.342	0.33	51.505	0.32	-	-	0.163	-
45	52.366	1.60	52.523	0.60	-	-	0.157	-
46	52.771	0.82	52.920	0.39	-	-	0.149	52.771
47	53.531	0.29	53.715	0.09	-	-	0.184	53.531
48	54.260	0.44	54.469	0.12	-	-	0.209	54.26
49	54.891	1.82	55.049	0.55	54.482	0.03	0.158	0.409
50	55.294	0.19	-	-	55.108	0.06	-	0.186
51	55.933	0.24	56.121	0.08	55.492	0.33	0.188	0.441
52	56.423	0.03	-	-	-	-	-	-
53	56.735	0.02	-	-	-	-	-	-
54	57.294	2.09	57.477	0.90	57.003	0.52	0.183	0.291
55	58.369	0.03	58.555	0.01	-	-	0.186	-
56	58.872	0.02	-	-	-	-	-	-
57	59.253	0.02	-	-	-	-	-	-
58	59.864	1.07	60.083	0.35	-	-	0.219	-
59	61.138	0.02	61.187	0.31	-	-	0.049	-
60	62.107	0.03	-	-	-	-	-	-
61	62.978	1.12	63.260	0.25	-	-	0.282	-
62	66.811	0.70	67.215	0.02	-	-	0.404	-
63	71.637	0.87	-	-	-	-	-	-
64	77.684	0.01	-	-	-	-	-	-

Table 23. List of Similar Peaks in GC Analysis Report of 1D1, 1D2 and 1D3.

It means that these common compounds are the compounds which are resistant to degradation. The number of such compounds in between 1D1 and 1D2 is 48, between 1D1

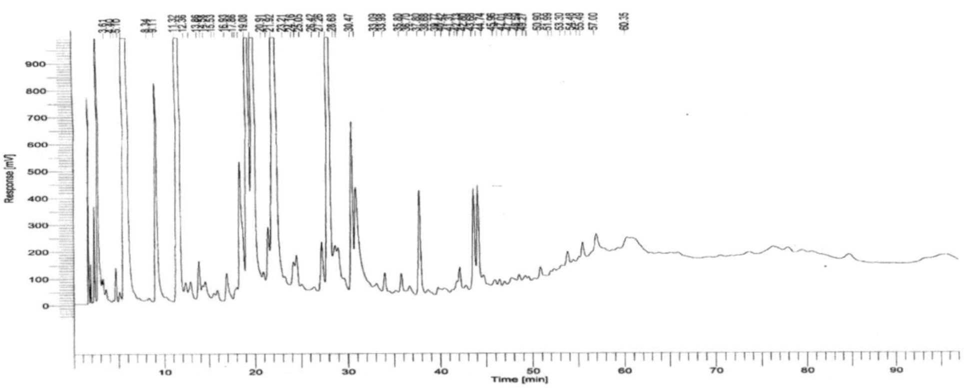

Plate 4. GC Chromatogram of oil recovered after 3 months of placement of sample 1D3.

and 1D3 is 41and among the three is 33 .The common compounds appear with a uniform difference in their retention time. 1D2 peaks appear with an average delay of 0.194 minute in the range of 0.049 to 0.404 minute and 1D3 peaks appear with an average advance retention time of 0.318 in the range of 0.019 to 0.441 minute. The other peaks of 1D2 and 1D3 are probably due to some degradation products. Number of such compounds in 1D2 is 5 and in 1D3 are 26. It indicates that Hydrogen Peroxide decomposes the higher hydrocarbons first and then the degraded components gradually disappear.

G.C. analysis of the recovered oil from residual oil polluted sample 1E5 after 1,2 (Plate-5) and 3 months gave 37, 65 and 52 peaks due to the undegraded components, derivatives and degradation products respectively. Out of the 37 components obtained after 1 month; 33 and 29 could resist degradation in the second and third months respectively, as revealed from their peaks appeared at a uniformly different retention time(Sarma et al,2004a) (Table-24). The stated 33 peaks appeared at a uniform advance retention time in the range of 0.914 to 1.247 minute with an average of 1.046. The stated 29 peaks appeared at a uniform advance retention time in the range of 0.914 to 1.247 minute with an average of 1.233. The remaining 4 and 4 components appeared in chromatograms obtained after 1 and 2 months respectively seems to be lost from the soil due to microbial activities. The number of peaks appeared in the chromatogram taken after 2 months but not in the previous chromatogram are 32. These peaks, and the corresponding peaks present in the last chromatogram are given in table (Table -25). Here only 17 peaks could resist complete degradation. These 17 peaks appeared at a uniform advance retention time in the range of 0.038 to 0.403 minutes with an average advance retention time of 0.196 minutes. The other 15 components seem to be lost from the soil due to microbial activities during the third month. Out of total 52 peaks appeared in the last chromatogram, only 6 peaks are due to new compounds (Table-26). These are the degradation products of the oil in the third month. Thus it can be concluded that degradation in the initial stage is more vigorous. The process slows down gradually. During degradation the higher compounds produce some fragments or some derivatives as a result of which number of components increases. Later on these compounds gradually disappear.

Table-27 shows number of components detected by GC analysis of a few samples of some of the mineral oils viz. Crude oil, Kerosene oil and Diesel oil–without agents, with nutrients, and with hydrogen peroxide. It appears that Hydrogen Peroxide increases the number of components in the sample and thus degraded the higher compounds. Application of NPK also increases the same, but lesser than Hydrogen Peroxide. For example, in the A5 samples, number of components after 3 months becomes 22 in 1A5 (without agent), 37 in 2A5 (with nutrients) and 39 in 3A5 (with hydrogen peroxide).The following table (Table 28) shows number of components detected by GC analysis of a few samples under identical conditions after one, two and three months. It appears that number of components increases initially and then decreases. It means that higher compounds give some smaller compounds and then escape from soil.

A good number of common components can be identified in the GC chromatogram of different samples (Table-29). These common components appear in the chromatogram after maintaining a uniform difference in their retention time. This difference is probably due to minor difference in the experimental conditions. In between the peaks at serial number 1and 2, the average difference in the retention time is 1.311 minute in the range of 1.304 to 1.316minute. Similarly the average difference in retention time between peaks at serial

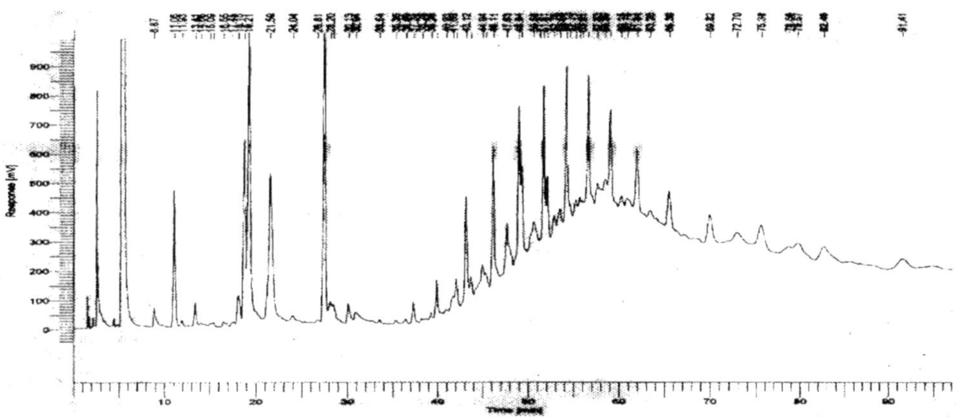

Plate 5. GC Chromatogram of Oil recovered after 2 months of placement of the sample 1E5.

-1- Sl No.	-2- After 1 month	-3- After1 month	-4- After 2 months	-5- After 2 months	-6- After3 month	-7- After 3 months	-8- Minute	-9- Minute
(as per GC report of analysis after 1 month)	Retention Time (RT)	Area % of the peak	RT	Area % of the peak	RT	Area % of the peak	Advance in RT of peaks after 2 months compared to peaks after 1 month.	Advance in Retention Time of peaks after 3 months compared to peaks after 1 months.
1	09.789	0.70	08.875	0.66	08.619	1.95	0.914	1.170
2	11.995	1.40	11.058	4.21	10.870	2.17	0.937	1.125
3	14.371	0.55	13.409	0.72	13.220	0.55	0.962	1.151
4	19.036	1.45	18.154	1.13	17.819	2.67	0.969	1.304
5	19.728	8.97	18.759	5.48	18.565	11.79	0.969	1.163
6	20.263	19.26	19.212	10.73	18.998	17.44	1.051	1.265
7	22.609	10.82	21.591	7.81	21.390	15.79	1.018	1.219
8	28.498	5.98	27.418	11.07	27.223	17.21	1.080	1.275
9	31.191	1.02	30.128	0.77	29.860	2.48	1.063	1.331
10	32.031	0.12	30.941	0.46	30.548	2.95	1.090	1.483
11	38.424	0.84	37.337	0.65	37.159	1.89	1.087	1.265
12	40.941	0.60	39.910	1.02	39.739	0.10	1.031	1.202
13	43.073	0.32	42.042	1.02	-	-	1.031	-
14	44.147	2.36	43.117	3.73	43.058	1.56	1.030	1.089
15	44.768	0.55	43.702	0.78	43.562	1.03	1.066	1.206
16	46.013	0.17	44.935	1.80	44.782	0.08	1.078	1.231
17	47.132	2.25	46.111	4.47	45.936	0.96	1.021	1.196
18	48.652	0.54	47.627	0.94	47.471	0.20	1.025	1.181
19	49.956	2.89	48.942	4.54	48.767	1.49	1.014	1.189
20	50.267	2.10	49.252	2.81	49.078	0.91	1.015	1.189

21	51.235	0.13	-	-	-	-		
22	51.630	0.47	50.560	0.95	50.339	0.40	1.070	1.291
23	52.626	2.78	51.610	4.47	51.439	1.42	1.016	1.187
24	53.023	1.26	52.005	1.81	51.829	0.62	1.018	1.194
25	53.820	0.19	52.760	0.75	52.632	0.33	1.060	1.188
26	54.538	0.34	53.444	0.99	53.248	0.53	1.094	1.290
27	55.148	2.61	54.134	4.84	53.964	1.42	1.014	1.184
28	56.205	0.41	55.165	0.42	54.955	0.82	1.040	1.250
29	56.695	0.02	55.642	0.32	-	-	1.053	-
30	57.554	2.85	56.534	3.53	56.375	1.46	1.020	1.179
31	58.662	0.04	57.531	0.36	57.353	0.26	1.131	1.309
32	60.198	1.80	58.951	2.97	58.747	1.18	1.247	1.451
33	61.305	0.58	60.145	0.42	-	-	1.160	-
34	63.399	1.66	-	-	-	-	-	-
35	67.374	1.17	-	-	-	-	-	-
36	72.364	0.10	-	-	-	-	-	-
37	80.707	20.73	79.568	0.03	-	-	1.139	-

Table 24. List of Similar Peaks in GC analysis report of 1E5 after 1,2 and 3 months.

number 2 and 3, 3 and 4, 4 and 5, and between 5 and 6 are 0.319 (0.300 to 0.351), 2.355 (2.347 to 2.369) , 0.397 (0.386 to 0.416), and 2.113 minute (in the range of 2.084 to 2.130 min) respectively. It clearly indicates that these are the compounds which are resistant to egradatdion till the period of their extraction from soil.

-1-	-2- Sl No.	-3- 2 months	-4- 2 months	-5- 3 months	-6- 3 months	-7- Minute
Sl No.	(as per GC report of analysis after 2 months)	Retention Time (RT)	Area % of the peak	Retention Time	Area % of the peak	Advance in RT of peaks after 3 month compared to peaks after 2 months
1	3	11.929	0.17	11.737	0.15	0.192
2	5	14.064	0.05	13.880	0.01	0.184
3	6	15.094	0.07	-	-	-
4	7	15.428	0.12	-	-	-
5	8	16.554	0.08	16.332	0.25	0.222
6	9	17.450	0.02	17.240	0.03	0.210
7	14	24.038	0.17	23.874	0.45	0.164
8	15	26.806	0.05	26.636	0.52	0.170
9	17	28.199	0.70	28.021	1.42	0.178
10	18	28.520	0.62	28.353	1.42	0.167
11	21	33.537	0.11	33.350	0.23	0.187
12	22	35.357	0.09	35.177	0.29	0.180
13	23	36.488	0.06	-	-	-

14	25	38.282	0.06	-	-	-
15	26	39.289	0.15	39.171	0.05	0.118
16	28	40.934	0.08	-	-	-
17	29	41.694	0.51	41.556	0.31	0.138
18	34	45.297	0.60	-	-	-
19	40	51.187	0.02	-	-	-
20	46	54.510	0.28	54.472	0.02	0.038
21	49	55.912	0.08	-	-	-
22	52	58.387	0.61	58.164	0.14	0.223
23	55	60.775	0.10	-	-	-
24	56	61.840	2.49	61.585	0.95	0.255
25	57	63.256	0.33	-	-	-
26	58	65.392	1.83	65.082	0.82	0.310
27	59	69.823	1.82	69.420	0.82	0.403
28	60	72.697	0.06	-	-	-
29	61	75.388	1.82	-	-	-
30	62	78.565	0.03	-	-	-
31	64	82.460	0.09	-	-	-
32	65	91.411	0.07	-	-	-

Table 25. List of Similar Peaks due to Degradation Products in GC Analysis Report of 1E5 after two and three months.

In order to make an attempt to identify different components those are persistent for a stipulated time in the soil samples, an experiment with some standard sample solutions were done (Sarma & Devi, 2009). Soil samples having no background of oil pollution were taken as per the following table (Table-30).

Sl No.	Sl No. (as per GC report)	After 3 months Retention Time (minute)	After 3 months Area % of the peak
1	8	17.471	0.03
2	22	36.086	0.07
3	43	56.823	0.07
4	47	59.522	0.16
5	51	74.861	0.04
6	52	76.556	0.08

Table 26. Peaks of compounds found after three months in 1E5 but not earlier.

Sl. No	Sample	Number of Components detected		
		Without Agent (first set)	With Nutrient (second set)	With H_2O_2 (third set)
1	A5(Crude Oil)	22	37	39 (plate-1)
2	B5(K Oil)	17	30	35(Plate-2)
3	C5(Diesel Oil)	21	34	40

Table 27. Number of Components detected after same time interval in Soil.

Extract of different samples by GC Analysis.

Sl. No.	Sample	Number of Components detected		
		After one month	After two months	After three months
1	2A3	17	24	23
2	3A3	11	33	32
3	1E5	37	65 (Plate-5)	52

Table 28. Number of Components detected after uniform time gap in Soil Extract of different samples by GC Analysis.

Laboratory temperatures during the experiment were in the range of 12.8 to 36.4^{0}C. The Physicochemical parameters of this soil sample are as given below (Table-31). These were suitable for the process of bioremediation to occur. The amount of recovered oil is 18046, 33638 and 51250 ppm in the samples S1, S2 and S4 respectively. It shows that the nutrients and Hydrogen Peroxide have expedited the process of oil degradation. The recovered oil from S2, S4 and some common known aromatic compounds in n-hexane were GLC analysed. The n-hexane soluble parts of the recovered oil exhibited peaks in the retention time of 39.006, 39.971, 40.118, 41.615, 41.656, 44.070, 44.151 and 62.596 minutes in the GC chromatograms. The GC peaks of the known compounds are as in Table 32A and B.

1-Naphthol and 2-Naphthol exhibited their peaks at retention time range of 37.095 to 37.238 minutes and 36.894 to 37.018 minutes averaging 37.184 ± 0.089 minutes and 36.969 ± 0.075 minutes respectively in different chromatograms. The lack of peaks before retention time of 39.066 minutes in the oil sample indicates that 2-Naphthol and 1-Naphthol are not present

Serial Number		Selection of common components from GC Chromatogram of Samples				
		1A5 -2 m	1B5- 2 m	1C5-3m	1D3-3m	1E5-2m
1	X	48.772	47.341	46.083	-	47.627
	Y	1.85	4.21	2.79	-	0.94
	Z	-	-	-	-	-
2	X	-	48.645	47.399	49.268	48.942
	Y	-	4.00	6.71	0.11	4.54
	Z(1.311)	-	1.304	1.316	-	1.315
3	X	50.389	48.962	47.699	49.619	49.252
	Y	23.34	15.74	19.61	0.07	2.81
	Z(0.319)	1.617*	0.317	0.300	0.351	0.310
4	X	52.736	51.309	50.055	51.988	51.610
	Y	0.27	1.27	7.50	0.04	4.47
	Z(2.355)	2.347	2.347	2.356	2.369	2.358
5	X	53.152	-	50.448	52.374	52.005
	Y	4.70	-	3.35	0.06	1.81
	Z(0.397)	0.416	-	0.393	0.386	0.395
6	X	55.236	-	52.578	54.482	54.134
	Y	2.48	-	8.30	0.03	4.84
	Z(2.113)	2.084	-	2.13	2.108	2.129

* = 1.307 +0.310 = 1.617

Table 29. List of Peaks due to Common Components in all the types of oil pollutants.

in here, X = Retention Time in minute in the chromatograms, Y = Area Percent in the Chromatograms, Z = Difference in Retention Time with the previous peak of the same sample in the chromatogram reported in the table. The figure within parenthesis ( ) indicates average difference in Retention Time of all the types of oil.

Sample No	Mass of Soil Taken(g)	Crude Oil Concentration (ppm)	Emulsifier added (mL)	NPK added (g)	H_2O_2 added (mL)	Water added (mL)
S 1	2910	20,000	10	30	-	100
S 2	2940	20,000	10	-	30	100
S 3	3000	0	10	-	-	100
S 4	2940	20,000	10	-	-	100

Table 30. Composition of Experimental Samples.

Physical Parameters			Chemical Parameters		
Name	Result	Unit	Name	Result	Unit
Texture	Sandy Loam	-	pH	6.87	-
Electrical Conductivity	0.0137	$mScm^{-1}$	Soil Org Carbon	1.41	%
Hydraulic Conductivity	$1.73X10^{-2}$	cm s^{-1}	Nitrogen	0.11	ppm
Water Holding Capacity	25.56	%	Phosphorous	0.09	ppm
Porosity	48.8	%	Potassium	0.92	ppm

Table 31. Physicochemical parameters of the soil sample.

Sl No	Compound	Sl No	Compound	Sl No	Compound
1	2-Naphthol	5	Naphthalene	9	Benzil
2	1- Naphthol	6	Anthracene	10	Phthalic Acid
3	Benzophenone	7	Benzoic Acid		
	Cinnamic Acid	8	Benzoin		

Table 32A. Compound Serial Numbers Used in the Table 32B.

Soln. No	Compounds Serial Number and their peaks with Retention Time ion Minutes									
	1	2	3	4	5	6	7	8	9	10
5				40.443						
6		37.173								51.856
7	37.018									
8							44.363			
9						40.786				
10	36.894	37.231			40.161		44.374			
11	36.996	37.095				41.243		44.910		
12			39.452				44.350		50.379	
13								44.998		
14			39.865					44.990		
15		37.238								

Table 32B. GC peaks of the known samples and their retention time.

On the other hand, lack of peaks around 44.9 minute, 50.3 minute and 51.8 minute in the oil sample chromatogram indicates that Benzoin, Benzil and Phthalic Acid are not present in the soil sample. The oil sample peaks at 39.971 minute, 40.118 minute, 41.615 minute, 41.656 minute, 44.070 minute and 44.157 minutes are very close to known sample peaks of Benzophenone, Cinnamic Acid, Anthracene and Benzoic Acid respectively. Out of these Anthracene and Naphthalene are tricyclic and bicyclic aromatic hydrocarbons respectively and others are oxygenated derivatives. The presence of these compounds in the polluted soil samples cannot be ruled out. That the polycyclic aromatic hydrocarbons, which are of great concern due to their toxicity and suspected carcinogenicity; are resistant to biodegradation was reported by many workers at different point of time.

The influence of applied phosphorous on bioremediation is positive. Since the phosphorous cycle is a sedimentary one, its fixation rate was studied. It has been found that a sample of sandy loam which possess 1.987μg g^{-1} available phosphorous fixes 99.85% phosphorous against addition of 50 g and 100 g of commercial single super phosphate fertiliser during progressive remediation from Lubricating oil pollution in a 120 days experiment(Sarma et al, 2008).

Bioremediation improves the soil physical conditions of petroleum polluted soil. During such remediation the amount of Soil Organic Carbon remarkably increases. A maximum of 111.51% and 65.20% increase of SOC was found in two samples of Sandy Clay and Sandy soil at an applied concentration of 20,000 ppm crude oil pollutant in an indoor experiment of 346 days. The increase of SOC in the samples where degradation was carried out in presence of added NPK was less and in those with added Hydrogen Peroxide was more than in the samples without NPK and Hydrogen Peroxide (Sarma & Sadhanidar , 2007).

It has been found that the pH and Electrical Conductivity of a remediating soil decreases. For example, in the first 6 months in the 1A5 sample pH decreases from 6.86 to 6.77; 6.73; 6.64; 6.59; 6.40 and 6.14 in regular monthly interval. Similarly, the EC decreases from 0.43 mS cm^{-1} to 0.41; 0.30; 0.27; 0.18; 0.14 and 0.11 mS cm^{-1} respectively in regular monthly intervals(Sarma et al, 2003a). Similar results were found when Lubricating oil is the pollutant (Sarma et al, 2003b, 2004b, 2005b). This is due to formation of some oxygenated derivatives from hydrocarbons, which are weakly acidic as the microbial degradation is a process of oxidation.

6. Conclusion

It has been found that under identical conditions a suitable soil sample degrades petroleum fractions to different extent. The extent of degradation of kerosene oil is highest, followed by crude oil, diesel oil, lubricating oil and residual oil. The disappearance of hydrocarbon is more in the initial stage and gradually it becomes a slow process. Complete recovery from hydrocarbon pollutants is not achieved during the experimental period of one year. The application of nutrients expedites the process of degradation. The action of hydrogen peroxide is moderate. The number of compounds in the recovered oil increases up to a period and then gradually decreases. In most cases the number of components generated is more in samples where hydrogen peroxide is applied. A good number of peaks formed probably due to same compounds can be pointed out and these appear in the chromatogram by maintaining an almost uniform difference in retention time. Parameters pH and Electrical

Conductivity show a decreasing trend on increase of degradation of the applied oil. Parameters Water Holding Capacity and Bulk Density become lower towards the side where presence of pollutants is higher. Parameters such as Hydraulic Conductivity and Organic Carbon become higher towards the side where presence of pollutants is higher. A good number of GC peaks in each of the samples are significant. The peaks so identified might be due to same components in the experimental oil. Since there is no probability of having such large number of common constituents in the oil samples; some of these peaks seem to be due to some degradation products.

7. Acknowledgement

The author is thankful to UGC, NER Office, Guwahati, Assam, India; ASTEC, Bigyan Bhawan, Guwahati, Assam, India; SAIF, CDRI, Lucknow, UP, India and Professor K G Bhattacharyya of Gauhati University, Assam, India.

8. References

Davies J S & Westlaki DWS, 1979- Crude Oil Utilisation by fungi, *Can J Microbiol* (25), pp146-56, (1979).

Dibble J T & Bartha R- Rehabilitation of oil Inundated Agricultural land: A case History, *J Soil Sci* pp56-60, (1979).

Hoeppel R E, Hinchee R E & Anderson D B,1995.*Bioremediation of Recalcitrant Organics* - Battelle Press, Columbus, OH.,pp 123-30.

Khan Z. and Anjaneyulu Y,2005.- Review on Application of Bioremediation Methods for Decontamination of Soils. *Res. J Chem. Environ.* Vol.9(2), pp. 75-79 , ISSN 0972-0626.

Plice M J, 1948. Some effects of Crude Petroleum on Soil fertility, *Soil Sci. Soc. Amer. Proc* Vol.13, pp. 413-416.

Reynolds C M & Walworth J L,2000- Bioremediation of a Petroleum Contaminated Cryic Soil: Effects of Phosphorous , Nitrogen, and Temperature-*Article Number 340074 – Internet.*

Rosenbeng E,1993- Exploiting Microbial growth on Hydrocarbons - new markets. *Review, Elsevier Science Publishers.* UK, Vol.11, pp.419-423.

Sarma P C , Bhattacharyya K G, andChoudhury S K, 2004a - Degradation of Residual Oil in Soil under Natural Environment: A Gravimetric and Gas Chromatographic Analysis, *Res. J of Cont. Concern,* . Vol.2, pp 86-93, ISSN 0972-7922

Sarma P C , Bhattacharyya K G, Choudhury S K, and Dutta U J, 2005b. Effect of Hydrogen Peroxide on hydrocarbon degradation in soil- a gas chromatographic analysis, *Res. J of Cont. Concern,* . Vol.3, pp30-35, ISSN 0972-7922.

Sarma P C and Bhattacharyya K G,2010. Degradation of Lubricating Oil in Soil under Natural Environment: A Gravimetric and Gas Chromatographic Analysis, *Res. J Chem. Environ.* Vol.14(3)(Sept.2010), pp12-16 (2010). ISSN 0972-0626.

Sarma P C and Devi U,2009. Gas Chromatographic Analysis of Persistent Hydrocarbon Components of Crude Oil and their Oxygenated Derivatives in Soil, *Res. J Chem. Environ.* Vol.13(3), pp 66-68. ISSN 0972-0626.

Sarma P C and Dutta U J, 2004b - Effect of Hydrocarbon Degradation on a few Physico-Chemical Parameters of a Soil Sample in Presence of an Oxidising Agent, *Proceedings, 49th Tech. Session of Assam Sc Soc.* Vol 5, pp 43-50.

Sarma P C, and Sadhanider U,2007, Effect of Hydrocarbon Degradation on Soil Organic Carbon, *Res. J of Cont. Concern,* Vol.5, pp08-12. ISSN 0972-7922.

Sarma P C, Bhattacharyya K G, and Choudhury S K, 2005a - Effect of Nutrients and Oxidising Agent on Degradation of Crude Oil Hydrocarbons in Soil under Natural Environment, *Ecological studies –New Horizons,* A Kumar, pp, 128-134, Daya Publishing House, ISBN 81-7035-384-X, Delhi.

Sarma P C, Bhattacharyya K G, and Choudhury S K,2003a, Effect on pH and Electrical Conductivities of Soil with respect to extent of degradation of Petroleum Hydrocarbons in Soil under Natural Environment, *Res. J Chem. Environ.* Vol.7(3), pp29-32 , ISSN 0972-0626.

Sarma P C, Bhattacharyya K G, and Choudhury S K,2003b - Effect on pH and Electrical Conductivities of Soil with respect to extent of degradation of Lubricating oil in Soil under Natural Environment, *Res. J of Cont. Concern,* Vol.1, pp24-28, ISSN 0972-7922.

Sarma P C, Momin M, and Sarma P,2008- Phosphate fixing Capacity of a Lubricating Oil Polluted Soil; A gravimetric and Spectrophotometric analysis, *Res. J of Cont. Concern,* Vol.6, pp23-26. ISSN 0972-7922.

Sarma P C,2010- Degradation of Diesel Oil Hydrocarbons in Soil in presence of a Green Reagent- a Gravimetric and Gas Chromatographic Analysis, *J of Ultra Chemistry,* Vol.6(2), (Aug, 2010). pp.247-251,ISSN 0973-3450.

Sing D, Chonkar P K & Pandey R N,2000, *Soil Plant Water Analysis - A Methods Mannual-* Indian Council of Agricultural Research,New Delhipp. 5-25.

Young PA,1935. Distribution and effect of petroleum oils and kerosene in potato, Cucumber, turnip, barley and onion *J.Agri Res.* V01.51,pp. 925-934.

10

To what Extent Do Oil Prices Depend on the Value of US Dollar: Theoretical Investigation and Empirical Evidence

Saleh Mothana Obadi
Institute of Economic Research, Slovak Academy of Sciences
Slovak Republic

1. Introduction

In majority, primary commodity prices are expressed in US Dollar, especially oil prices, not only in the commodity markets but also in many international organizations, for example, in the IMF International Financial Statistics, or in terms of indices based on dollar prices. As such, oil prices are obviously affected by inflation as well as real developments, and also by the value of the US dollar exchange rate. Therefore, a change of both variables affects the international trade of all economies. In the case of oil prices, any change of them affects prices of other primary commodities, products and services, and subsequently macroeconomic indicators of oil exporting and importing countries. There is therefore a definite link between monetary policies and exchange rates among other factors and oil prices. And this is the subject of analysis in this chapter.

The oil prices have signed a clearly fluctuated trend starting with the first through the second oil shock up to the present. According to Jalali-Naini &Manesh (2006) the crude oil price exhibits a high degree of volatility which varies significantly over time. Between 1987–2005, oil price volatility far exceeded that of other commodity prices. Behind that, there were many causes - the often mentioned one in the economic and energy-economic literature is the political (war conflicts) instability factor and subsequently the interruption of oil production or supply. It is clear that there are other factors influencing the oil prices - in the last decade, the increasing demand for oil in the emerging economies (China, India etc.), speculation in commodity markets and the weakening of the US Dollar.

The main objective of this chapter is to examine the correlation between oil prices and the value of the US dollar (USD), and to draw some conclusions about the oil market. Considering that one of the causes of raising the oil prices is a drop of the value of US dollar, to what the extent the US dollar declined we will see via its exchange rate against the main currencies like Japanese (yen), and other European currencies and the subsequent impact on the oil prices.

In addition to the qualitative analyses, as the research method we are using regression model, financial models, Granger causality and structural models to identify to what the extent the oil prices depend on the value of US Dollar, as one of the factors influencing the oil prices in the international markets, particularly in the last two decades.

The chapter is organized as follows. Section 2 presents some theoretical investigations and discusses the existing literature. Section 3 qualitatively and statistically analyzes the development of oil prices in nominal and real terms since 1970. Section 4 tries to analyze the development of the US Dollar exchange rate with relation to oil prices. The empirical analyses as well as econometric modeling of oil prices and its results are presented in Section 5.

2. Theoretical investigation and literature discussion

Traditionally, behind the increase or drop of oil prices is more than one factor. In addition to the fundamental market factors (Supply and demand) there are many others, such as, speculation in the crude oil markets, the less predictable factors (political instability hurricanes, tsunami, etc.), the US Dollar exchange rate as a more discussed factor in the energy-economic literature at least in the last two decades and other factors like capacity of the so-called downstream sector. In this chapter we will devote more attention to the role of movement of US Dollar exchange rate in the movement of oil prices.

Many studies related to this issue have been done. Part of them, theoretically and empirically examined the impact of oil prices on dollar real effective exchange rate, see (Coudert et al.,2008). They find that causality runs from oil prices to the exchange rate. "…, as we investigate the channels through which oil prices affect the dollar exchange rate, we find out that the link between the two variables is transmitted through the U.S. net foreign asset position" (Coudert, Mihnon & Penot, 2008). In the same link (Amano & van Norden,1998) examined whether a link exists between oil price shocks and the U.S. real effective exchange rate. They used the single-equation error correction and find that the two variables appear to be co-integrated and that causality runs from oil prices to the exchange rate and not vice versa. "The results suggest that oil prices may have been the dominant source of persistent real exchange rate shocks over the post–Bretton Woods period and that energy prices may have important implications for future work on exchange rate behavior" (Amano & van Norden,1998). According to Bénassy-Quéré et al.(2005), the relationship between oil price and USD exchange rate is clear. They provided evidence of a long-term relation (i.e. a cointegration relation) between the two series in real terms and of a causality running from oil to the dollar, over the period 1974-2004. Their estimation suggests that a 10% rise in the oil price leads to a 4.3% appreciation of the dollar in real effective terms in the long run. The estimation of an error correction model shows a slow adjustment speed of the dollar real effective exchange rate to its long-term target (with a half life of deviations of about 6.5 years). Although consistent with previous studies, our results are unable to explain the period 2002-2004 with a rising oil price and a depreciating dollar.

On the other hand, many other studies examined the impact of US dollar exchange rate on oil price, see e.g. (Alhajji, 2004; Cheng, 2008; Krichene, 2005; Yousefi & Wirjanto, 2004). According to Alhajji, "US Dollar depreciation reduces activities in upstream through different channels including lower return on investment, increasing cost, inflation, and purchasing power. Furthermore, US Dollar devaluation increases demand in countries with appreciated currencies because of increase in purchasing power and increases demand in the US as tourists prefer to spend their vacations in the US". In case of US Dollar depreciation, the revenues of oil exporting countries, at least those whose local currencies are tied to US Dollar, are more or less decreased. This leads to a deterioration of their terms of trade, because they must export more units of crude oil to get the same amounts of imported products for example from Europe, than they had to before US Dollar depreciation. Therefore, oil exporting countries are inclined to maintain oil prices high as much as appropriately in proportion to the US Dollar depreciation, and alleviate the loses in their oil revenues.

In the same link, Krichene (2005) examined the relation between oil prices, interest rates and US Dollar exchange rate NEER. In his study the attention was given to shocks arising from monetary policy—namely, shocks to interest and exchange rates. He used a vector autoregressive model (VAR) to analyze cointegration between crude oil prices, the dollar's NEER, and U.S. interest rates based on monthly, quarterly, and annual data. He reported that VAR analyses did not reject the existence of at least one cointegration relation. Although the cointegrating coefficients change both in sign and statistical significance according to frequency, sample period, and the number of lags, there is nevertheless a stylized relation during periods of large movements in interest and exchange rates. While an interest rate shock generally affects negatively and significantly oil prices for most of the sample periods, the effect of a NEER shock is significantly negative, essentially during periods of large movements in interest and exchange rates, (Krichene, 2005).

3. The development of nominal and real oil prices since 1970

The oil prices in the international oil markets have undergone tremendous changes since 1970. Starting from the renegotiation of the „posted price" - a reference price on which royalties to host countries were calculated - in 1970. Before that, this price was fixed (at 1.80 US dollar a barrel during the 1960s) by the major international oil companies that operated the oil concessions in these countries. Subsequent events culminating in the 1973 oil price shock and the eventual transfer of property rights to the host countries heralded the start of a new era in the oil industry (see Obadi, 1999). Middle Eastern countries - through their role in the Organization of the Petroleum Exporting Countries (OPEC) - were at the center of the transformation of the market since they owned the bulk of world proven crude oil reserves. In addition to transforming their societies through the inflow of substantial amounts of oil revenue, the Middle Eastern and North African (MENA) countries encountered new challenges in the area of economic policy and management, including how to cope with the adverse impact of the variability of oil prices on growth (see Bright E. Okogu, 2003).

When we compare the real (Nominal prices are measured in U.S. dollars per barrel. Real oil prices are calculated by Energy Information Agency-EIA, based on constant 1980 prices, and the deflator is consumer price index -CPI of USA based on data from the U.S. Department of Labor Bureau of Labor Statistic) and nominal oil prices (see Figure 1), we can say that the difference varies from one decade to another. For example in the 80s the difference between both average prices is about 4.47 US Dollar/barrel, while in the 90s it is 7.83 US Dollar/barrel. And in the first decade of the present century it is 26.32 US Dollar/barrel. That means that the value of US Dollar has sharply fallen against other world currencies at least in the last two decades. In other words, if the US Dollar had retained its value it had in 1980, the barrel of crude oil would have cost about 34 US Dollars in October 2011.

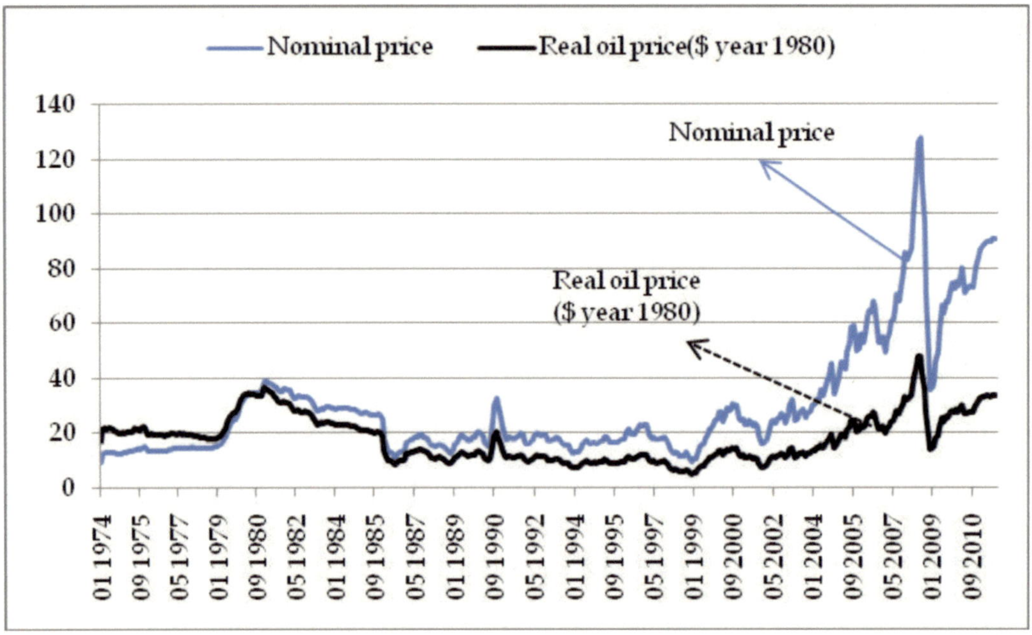

Source: author´s calculation based on EIA data, January 2011.

Fig. 1. The development of nominal and real oil prices, 1974-2010.

4. The historical exchange rate of US Dollar and oil prices

The devaluation of the US dollar began on December 18, 1971 continued through early 1972 and culminated in the global collapse of and heightened capital flight from the American unit, induced by the official burial of the two-tier gold market in November and the quadrupling of oil prices in 1973 (Historical Exchange Rate Regime of the United States of America. http://intl.econ.cuhk.edu.hk/exchange_rate_regime/index.php?cid=30) . Eventually, in April 1978, the par value of the US dollar in terms of gold and SDRs was repealed and the US dollar became a floating Effective Rate. The US dollar subsequently found strength in most exchange centers because of high interest rates; lower inflation and foreign capital flowed into the United States.

During 1986 and early 1987, America's unit was again battered in major exchange centres as Washington and the Group of Seven felt that a weak US dollar was the best course for world monetary peace. However, the free markets seemed to think otherwise, constantly pushing the US dollar up in 1988, 1989 and 1990, despite official efforts to keep it weak at a cost of billions of US dollars to the Federal Reserve (see also C K Liu, 2005).

US Dollar against world currencies

Since the 1970s, the two most important currencies, besides the dollar, have been the German mark and the Japanese yen (Mundell, 2003). The dollar has gyrated against other major currencies. Against the DEM, for example, the dollar was DEM 3.5 in 1975 and decreased in half to DEM 1.7 five years later, in 1980. Then the dollar doubled to DM 3.4 by early 1985, and then fell below DEM 1.35 in August 1992, at the peak of the ERM crisis in Europe. Since that time the dollar has risen far above DEM 2.0 This instability of the USD/DEM rate means that commodity prices in dollars and DEM would be for much of the period moving in opposite directions. In a period when the commodities prices were rising in dollars, they might have been falling in DEMs, and vice versa. The US dollar has lost also much value relative to the euro and the yen (see also Mundell, 2003).

What about the value of the US Dollar against the EUR? Looking to the movement of value of the two largest world reserve currencies during the last 12 years, we could say that it was a dramatic development (see figure 2).Since 2001 to 2011, the US dollar in annual averages lost 36.5% of its value relative to the euro (For the year 2011, the average of daily value of US dollar relative to euro for the period 3.1.2001 - 28.8.2011) when the value of the dollar has fallen from annual average 1.1169 euros to about 0.7095 euros per US dollar.

When the European currency has been launched in January 1999, 862 EUR would have been exchanged for about 1000 US Dollars. However, in February 2002 getting the same amount of US Dollars the Europeans have had to pay more than 1140 Euros - during that period the Euro has fallen sharply, about 33 %. What the EUR has lost during the aforementioned period, it gained again during the period March 2002 -January 2005. The race between these currencies has continued, but with less dramatic development. During the second half of 2005 and the first half of 2006 the US Dollar has gained about 11 % and returned to the approximately same rate it was in 1999. From this period on the US Dollar in average has registered only loses - during the period March 2006 -August 2011 the US Dollar has fallen against the Euro by about 16 %.

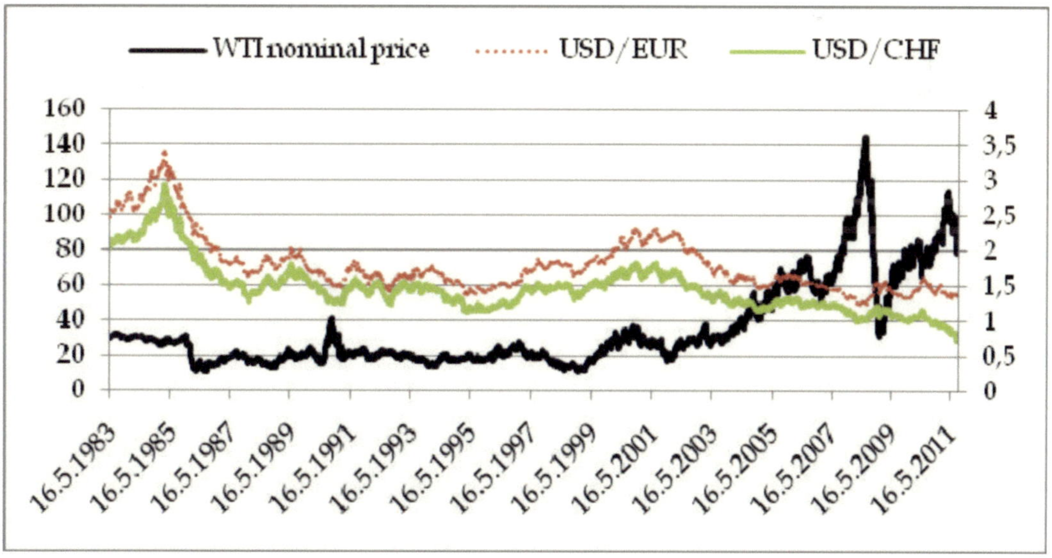

Left axis- oil price WTI in US Dollar and right axis - US Dollar exchange rate against EUR and CHF.
Source: Author's calculation based on data from http://www.economagic.com/ and EIA online data.

Fig. 2. US Dollar exchange rate against EUR and CHF and nominal oil price, 1983 - 2011

In figure 2, the US Dollar exchange rate against EUR is based on the historical US Dollar against Germany currency (DEM) up to the end of 1998. Because Germany is a largest economy in Europe and using a new currency (EUR), since 1.1.1999, it is used in this figure the official conversion rate of DEM/EUR (1 DEM = 0.5113 EUR)

As it is clearly seen in the above figure 2, the development of US Dollar exchange rate against CHF followed an identical trend such as against EUR. This explains the same reaction of the two European currencies to the change of the US Dollar value.

The development of the US Dollar against Japanese Yen was also interesting. Japan maintained a fixed exchange rate of 360.00 yen per US dollar until August 1971. In 1971, the yen was allowed to float above its fluctuation ceiling and an Effective Rate with a fluctuation range was established in December. With continuous devaluation of US dollar, it forced Bank of Japan to place a control on exchange rate, in a floating basis. Afterwards, the Effective Rate of yen was set to be floating freely. Since the introduction of a floating exchange rate system in February 1973, the Japanese economy has experienced large fluctuations in foreign exchange rates, with the yen on a long rising trend. In the same year, another rate called Interbank Rate was set up. The Yen was to be determined on the basis of underlying demand and supply conditions in the exchange markets. The Bank of Japan only intervened in the currency market when the yen fluctuated disorderly. The exchange rate regime was not changed much in Japan (Historical Exchange Rate Regime of Japan. http://intl.econ.cuhk.edu.hk/exchange_rate_regime/index.php?cid=9).

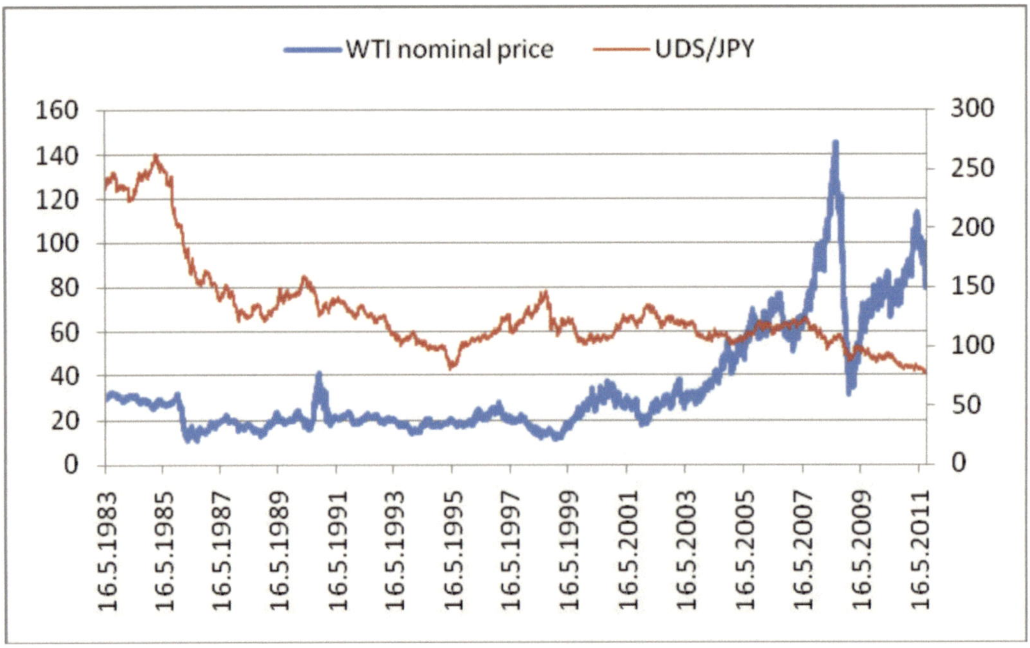

Left axis- oil price WTI in US Dollar and right axis – US Dollar exchange rate aginst JPY
Source: Author's calculation based on data from http://www.economagic.com/ and EIA online data.

Fig. 3. US Dollar exchange rate against JPY and nominal oil price, 1983 – 2011.

"Yen - dollar fluctuations have been just as extreme. In the hey-day of Bretton Woods, the yen - dollar rate was fixed at 360 yen to the dollar. After the 1970s this rate became flexible" (Mundell, 2003). Therefore, the Japanese yen started to rise against US Dollar - during the 1970s gained more than 14 %. During the 70s the US Dollar has lost in value against yen about 43%. During this period both economies have experienced economic slowdown as the consequences of two crude oil shocks. But the Japanese yen has gained a better position relative with US Dollar. During the first half of 80s the US Dollar gained in its value relative to yen again about 23 %. But it was the last time of its gaining in value against yen - by February 1985 the dollar was 260 yen. Ten years later, by April 1995, the dollar had fallen to 79 yen. In other words the yen had tripled in value against the dollar. "This was the period in which the balance sheets of Japanese companies were undermined, and Japanese banks ended up with the non-performing loans that persist in trillions of dollars to this day" (Mundell, 2003). As it is shown in Figure 3, the period from April 1995 up to August 2011, was the most fluctuated period between the dollar and the yen. In 1995, 10,000 Japanese Yen would have exchanged for about 107 USD, but in 2001 the same amount of yen exchanged for only about 82 USD - a close to 30% decline of the Yen (Historical Exchange Rate Regime of Japan. http://intl.econ.cuhk.edu.hk/exchange_rate_regime/index.php?cid=9). Since 1998, the US Dollar exchange rate against JPY has in average a decreasing trend. The US Dollar has lost against JPY during the period 11 August 1998 -1 Jun 2011 about 45 %.

4.1 The US dollar and the oil prices

When we go back to the history of crude oil invoicing by US Dollar, many sources agree that the formation of this idea dates back to 1945, when the US and Saudi Arabia had formed a cooperative partnership, following meetings between Franklin Delano Roosevelt and King Ibn Saud. US oil companies (Exxon, Mobil, Chevron, and Texaco) were already controlling Saudi discovery and production through a partnership with the Kingdom, the California Arabian Standard Oil Company [CASOC, the forerunner of the Arabian American Oil Company (Aramco, the forerunner of today's Saudi Aramco)]. In 1973, the Saudi Government increased its partner's share in the company to 25%, and then 60% the next year. In 1980, the Saudi government retroactively gained full ownership of Aramco with financial effect as of 1976. But, according to research by Spiro (1999), in 1974, the Nixon administration negotiated assurances from Saudi Arabia to setting the price of oil in dollars only, and investing their surplus oil proceeds in U.S. Treasury Bills. In return the U.S. would protect the Saudi regime. At about the same time this was happening (1975), the Saudis agreed to export their oil for US dollars exclusively. Soon OPEC as a whole adopted the rule. Now, as a result, the dollar was backed not by gold but, in effect, by oil. In 1973, with the dollar now floating freely, the Arab nations of OPEC embargoed oil exports to the US and other western countries in retaliation for American support for Israel in the Ramadan/Yom Kippur War. By this time it was clear that US oil production had peaked and was in permanent decline, and that America would become ever more dependent upon petroleum imports. As oil prices soared 400%, the US economy and other Western economies sharply turned down.

"In any case, the oil shock created enormously increased demand for the floating dollar. Oil importing countries, including Germany and Japan, were faced with the problem of how to earn or borrow dollars with which to pay their ballooning fuel bills. Meanwhile, OPEC oil countries were inundated with oil dollars. Many of these oil dollars ended up in accounts in London and New York banks, where a new process - which Henry Kissinger dubbed „recycling petrodollars" - was instituted" (Heinberg, 2004)..

OPEC countries were receiving billions of dollars they could not immediately use. When American and British banks took these dollars in deposit, they were thereby presented with the opportunity for writing more loans (banks make their profits primarily from loans, but they can only write loans if they have deposits to cover a certain percentage of the loan- usually 10% to 15% (Heinberg, 2004).

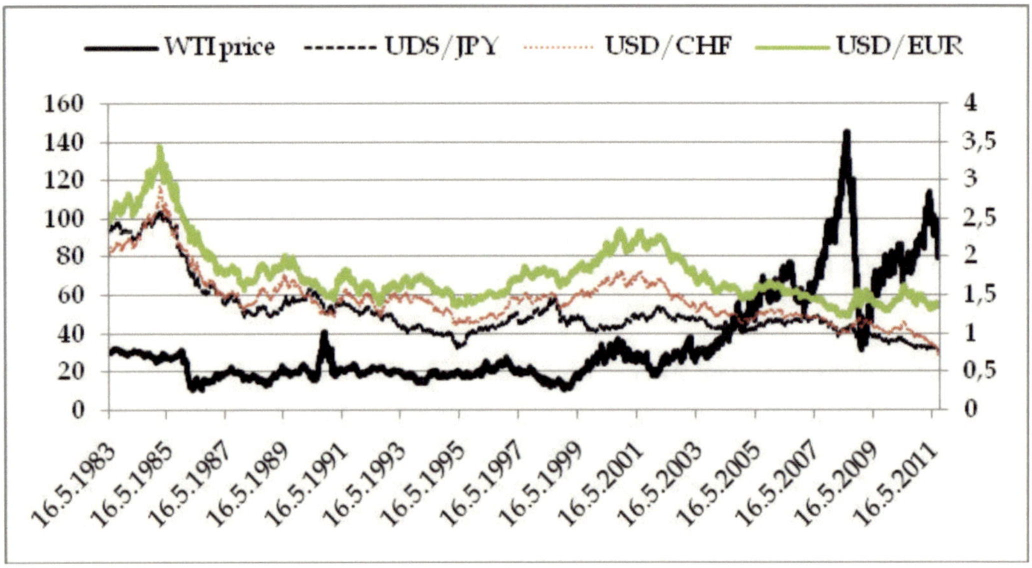

Left axis- oil price WTI and right axis - US Dollar exchange rate against the selected currencies (the value of JPY is divided by 100)
Source: Author's calculation based on data from http://www.economagic.com/ and EIA online data.

Fig. 4. US Dollar exchange rate against world currencies and nominal oil price, 1983 - 2011

Henry Kissinger, an advisor to David Rockefeller of Chase Manhattan Bank, suggested the bankers use OPEC dollars as a reserve base upon which to aggressively „sell" bonds or loans, not to US or British corporations and investors, but to Third World (developing) countries desperate to borrow dollars with which to pay for oil imports. By the late 1970s these petrodollar debts had laid the basis for the Third World debt crisis of the 1980s (after interest rates exploded). Most of that debt is still in place and is still strangling many of the poorer nations. Hundreds of billions of dollars were recycled in this fashion. (Incidentally, the borrowed money usually found its way back to Western corporations or banks in any event, either by way of contracts with Western construction companies or simple theft on the part of indigenous officials with foreign bank accounts) (Heinberg, 2004).

Also during the 1970s and '80s, the Saudis began using their petrodollar surpluses to buy huge inventories of unusable weaponry from US arms manufacturers. This was a hidden subsidy to the US economy, and especially to the so-called Defense Department. With the end of Bretton Woods in 1971, the US refused to continue supplying gold at 35 US dollars per ounce to other central banks. There was a December 1971 Smithsonian extension of the exchange rate peg system, with the final breakdown of the fixed exchange rates occurring in March 1973 (McCallum, 1999). The US inflation rate was left to increase dramatically since other countries were no longer locked into supporting the fixed US dollar gold price, or the exchange rate peg, and did not have to buy the excess US dollars. This provides a monetary explanation for the jump in inflation in 1974. Further jumps in the inflation rate in the 1970s occurred during the increasing deficits and money creation of the Carter Presidency, and before the US Federal Reserve tenure of Paul Volcker.

Some literature suggests that US-generated inflation was the main reason for the two „oil shocks" in 1974 and 1979. (Penrose, 1976) mentions that early attempts to raise the oil price were defended by OPEC with the explicit argument to offsetting the cumulative effect of US inflation, as well as to shelter against future erosion of revenues due to inflation (see Gillman & Nakov, 2001).

Similarly, Spero & Hart (1997) suggest that the increasing inflation rate and the devaluation of the dollar lowered the real value of earnings from oil production and led OPEC countries to demand a substantial increase in the price of oil. Barsky and Kilian (2000) question the extent to which oil shocks played a dominant role in triggering stagflation in the 1970s. They argue that there is little evidence that oil price rises actually raised the deflator, and suggest that monetary fluctuations can explain the variation in the price of oil and other commodities.

However, looking at the above figures and both the curve of oil prices and that of the US Dollar exchange rate, we can see a nearly identical trend of US Dollar exchange rate at least against the two selected European currencies, which have in the long term on average a decreasing trend. On the other hand, the oil prices also in the long term on average have an increasing trend (Obadi, 2006). This confirms the fact that the oil prices which are quoted in US dollar depend on the development of its value. The question needs to be asked whether the cycle of the value of the dollar against major currencies is related to the cycle of the dollar oil prices. A casual reading of the statistics suggests that this relationship is quite close. For further illustration, see the empirical section of this chapter, in which we tried to reveal the negative correlation between oil prices and the US dollar exchange rate.

The above figure shows the development of the value of dollar against selected world currencies (JPY, CHF and EUR) and the oil prices since 1983 to 2011, wherein an unambiguously reverse direction of oil prices and the exchange rate of US dollar against the selected world currencies is shown. Similar results had found (Weller & Lilly 2004) - oil prices have risen, while the dollar has simultaneously plummeted against the euro. The authors measured how much these two prices move in tandem, using a correlation coefficient. The correlation coefficient between oil and dollar is -0.7. That is, most of the time, when the dollar fell against the euro, oil prices rose.

OPEC is worried about the weakening value of the dollar: it has lost one-third of its value in just under two years (Rifkin, 2004). Since OPEC sells oil for dollars, the oil-producing countries are losing fortunes. The revenues just as the value of the dollar are diminishing (Obadi, 2006). And because oil-producing countries then turn around and purchase much of their goods and services from the EU and must pay in euros, their purchasing power continues to deteriorate. It is a big factor, perhaps bigger than the global demand and supply. In the late 1970s, the price of oil increased by 43% in USA but only 1% in Germany and 7% in Japan. That's despite the fact that Germany and Japan were more dependent on imported oil than USA.

5. Oil price in the econometric models

5.1 Methodology

Many empirical studies in this link found a negative linear relationship between oil prices and economic activity in oil importing countries. However, in the mid 80s it was indicated that the linear relationship between oil prices and economic activity began to lose statistical significance. Mork (1989), Lee et al. (1995) and Hamilton (1996) introduced the nonlinear transformation of oil prices to explain the negative relationship between the rise in oil prices and economic downturn and to prove Granger causality between these two variables.

Hamilton (2003) and Jiménez-Rodríguez (2004) also confirmed, on the example of USA, the existence of non-linear relationship between these variables. The importance of oil in the world economy explains why so much effort has been put into developing different types of econometric models to predict future price developments. In this section we describe the basic types of models. Financial models concentrate on the relationship between spot and futures prices. On the other hand, structural models explain the oil prices development by exogenous variables, which describe the physical oil market.

Financial models

Financial models are directly inspired by financial economic theory and based on the Market efficiency hypothesis (MEH). This theory is often attributed to Eugene F. Fama whose study "The Behavior of stock market prices" (Fama 1965) is considered crucial in the theory of market efficiency. He says that in the presence of full information and a large number of rational agents in the market, current prices reflect all available information and expectations for the future. In other words, the current prices are the best estimation of tomorrow's prices.

Generally, financial models examined the relationship between the spot price S_t and the future price F_t with maturity T. They examined, whether the future prices are unbiased and efficient an estimator of spot prices.

Reference Model looks like this:

$$S_{t+1} = \beta_0 + \beta_1 F_t + \varepsilon_{t+1} \quad (1)$$

In this equation F_t is an unbiased estimator of future prices S_{t+1}, when common hypothesis $\beta_0=0$ and $\beta_1=1$ is rejected and at the same time we do not indicate any autocorrelation between residues. Chernenko and Schwartz (2004) tested the validity of MEH, focused on the relationship between the difference of spot and future prices. Their model looks as follows:

$$S_t + S_{t+T} = \beta_0 + \beta_1(Ft - S_t) + \varepsilon_{t+1} \quad (2)$$

They analyzed monthly WTI oil prices in the period from April 1989 to December 2003. The/other authors have also compared the model with a random walk model and showed that both models exhibit nearly the same accuracy predicting future prices, and also confirmed the theory of market efficiency.

Structural models

Structural models concentrate on describing the development of oil prices market through explanatory variables that describe this market. Variables that are usually used to predict oil prices can be divided into two basic groups: variables that describe the role of OPEC in world oil market and the variables that capture current and future availability of oil.

Apart from the influence of OPEC, several authors emphasize the current and future physical availability of oil. With this in mind, many key variables are based on stock levels. Stocks are linkages between demand and production and consequently a good measure of variations in prices (Ye, Zyren and Shore 2005). Most authors distinguish two types of stocks: government stocks and industrial stocks. In terms of their origin, government stocks are not generated by real demand and supply. This explains the decision of many economists to put into models industrial stocks, which are changing within a short time and can capture the dynamics of oil prices. In the same link, Zamani (2004) has presented short-term forecasting models for WTI crude oil prices, where he incorporated both groups of variables, the physical availability of oil and the role of OPEC.

His model looks as follows:

$$S = \beta_1 + \beta_2 OQ + \beta_3 OV + \beta_4 RIS + \beta_5 RGS + \beta_6 DN + \beta_7 D90 + \varepsilon \quad (3)$$

Where:

OQ is fixed production quota issued by OPEC, the OV is overrun of this quota, RIS is relative industrial stocks given by equation: RIS = IS - ISN, where IS is value of industrial stocks and ISN is normal level of industrial stocks. RGS is relative government stocks given by equation: RGS = GS - GSN. Variable DN is a demand in countries outside the OECD and D90 is a dummy variable expressing the war in Iraq during the third and fourth quarter of 1990. Zamani used quarterly data for the period from 1988 to 2004. He suggested that an increase in all explanatory variables leads to a reduction of the oil prices, while the dummy variable and demand in non-OECD positively affect the price.

How OPEC decisions influence the oil price development?

The role of OPEC in the world oil markets has been examined by the academic community several years. Many studies brought evidence that OPEC has an ability to influence real oil prices, although in the recent years the demonstrated impact of the organization decreases. Many econometric analyses show that there was a statistically significant relationship between real oil prices and the following variables: capacity utilization by OPEC, OPEC quotas, exceeding the amount of these quotas (overproduction) and oil reserves in countries OECD (Organization for Economic Co-operation and Development). Between the abovementioned variables the existence of a Granger causality is demonstrated These variables causally affected the real oil prices, but oil prices did not affect these variables. Cointegration relationship between the real oil prices, OPEC capacity utilization, quotas, and maintaining these quotas shows that OPEC plays an important role in determining the world oil prices.

The impact of exchange rate of US dollar on oil prices

In this empirical part we will focus on the impact of US dollar exchange rate against main currencies on the oil prices. From the world currencies we selected Japanese yen, Euro, and Swiss franc, using the daily frequency data for the period 1986:1:1 - 2011:8:11. In case of the European Union currency, we used German mark value (DEM) multiple by the conversion rate which was approved on value 0.5113 DEM/EUR. This conversion rate is valid from January 1st 1999. Thereby, we obtained a consistent time series for the entire period.

	WTI	USD/EUR	USD/JPY	USD/CHF
Mean	35.86751	0.797079	118.0403	1.375751
Median	22.49	0.764905	117.27	1.3873
Maximum	145.31	1.2089	202.7	2.094
Minimum	10.25	0.570355	76.55	0.7208
Std. Deviation	26.12138	0.135684	19.52577	0.218743

Table 1. Basic descriptive statistics of the selected variables.

Overall, we have 6689 observations for this time period. As the dependent variable we have selected WTI crude oil price. The value of the variable WTI represents the average spot price of WTI crude oil with Incoterms standard FOB (Free on Board), expressed in U.S. dollars for barrel. Correlation between oil prices and the selected exchange rates will be examined through correlation matrix; - the results are shown in Table 2.

	WTI	USD/EUR	USD/JPY	USD/CHF
WTI	1	-0.1916	-0.5053	-0.6960
USDEUR	-0.1916	1	0.2630	0.6813
USDJPY	-0.5053	0.2630	1	0.7233
USDCHF	-0.6960	0.6813	0.7233	1

Table 2. Correlation matrix.

According to this results we can say, that exchange rates have negative impact on oil price development. So when the exchange rate is increasing, the oil prices will decrease. The real impact on crude oil price we examine through regression analyses of each explanatory variable on the dependent variable.

Before we started an analysis, we have tested each variable for the presence of unit root. The incidence of unit root indicates that time series are nonstationary, and because of these results, the regression could be spurious. Integration of time series has been tested using the Augmented Dickey-Fuller test (ADF test). The null hypothesis of this test is that the series has a unit root. The null hypothesis of nonstationarity is rejected if the t-statistic is less than the critical value. Critical ADF statistic values are considerably larger (in absolute value) than critical values used in standard regression. (Critical values used in ADF test: Equation with no intercept and no trend: -2.58 (1%), -1.94 (5%), -1.62 (10%). Equation with intercept: -3.46 (1%), -2.88 (5%), -2.58 (10%). Equation with intercept and trend: -4.01 (1%), -3.43 (5%), -3.14 (10%).) Lag structure of the ADF test were determined using the Akaike Information criteria. The ADF test confirmed a presence of a unit root in each variable. The results of the test we show in table below:

Variable	Intercept/Trend	Lag length	ADF statistic	Significance[1]
WTI	None	33	-0.274031	
D(WTI)	None	32	-12.97032	***
USD/EUR	None	0	-1.095342	
D(USD/EUR)	None	0	-82.28953	***
USD/JPY	None	15	-1.754621	*
D(USD/JPY)	None	14	-19.54761	***
USD/CHF	None	1	-1.873186	*
D(USD/CHF)	None	0	-83.14117	***

[1]Rejecting the null hypothesis on significance level: 10% - * ; 5% - ** ; 1% - ***

Table 3. Results of ADF test.

The ADF test has proved that all variables are nonstationary, because they contain stochastic trend, which can be clearly seen from the time series development. Therefore, to describe long term relations between variables, we are going to use cointegration analyses. Engle and Granger (1987) pointed out that a linear combination of two or more nonstationary series may be stationary. If such a stationary linear combination exists, the nonstationary time series should be cointegrated. The stationary linear combination is called the cointegrating equation and may be interpreted as a long-run equilibrium relationship among the variables. The purpose of the cointegration test is to determine whether a group of nonstationary variables is cointegrated or not. We carried out Johansen cointegration test. This test reports the so-called trace statistics and the maximum eigenvalue statistics. The output of Johansen test also provides estimates of the cointegrating relations β and the adjustment parameters α_{ec}. We applied Johansen test between WTI and each exogenous variable. The null hypothesis of the test denotes that there is no cointegration relationship between variables. If the probability of Trace statistic eventually maximum Eigenvalue statistic is lower than 5%, we accept the alternative hypothesis of cointegration between variables. And therefore we can further examine the longterm relationship among these variables. The results are shown in tables below:

Trace test	No. of cointegration equations	Trace statistics	Probability
	None	6.678723	0.6153
	At most 1	1.39672	0.2373
Maximum eigenvalue test	No. of cointegration equations	Max Eigen statistics	Probability
	None	5.282002	0.7059
	At most 1	1.39672	0.2373

Table 4. The results of trace and the maximum eigenvalue statistics (WTI, USD/EUR).

Trace test	No. of cointegration equations	Trace statistics	Probability
	None	22.95825	0.0031
	At most 1	1.162411	0.281
Maximum eigenvalue test	No. of cointegration equations	Max Eigen statistics	Probability
	None	21.79584	0.0027
	At most 1	1.162411	0.281

Table 5. The results of trace and the maximum eigenvalue statistics (WTI, USD/JPY).

Trace and maximum eigenvalue test results (Table 4) indicate that there is no cointegration among variables at the 0.05 significance level. So between WTI price and exchange rate USD/EUR there is no long-run equilibrium relationship. It means that in the long term point of view there is no proved relation among these variables. It may be because in Europe more often Brent crude oil, and other types of crude oil from Russia and OPEC are traded.

Trace and maximum eigenvalue test (Table 5) indicates 1 cointegrating eqn(s) at the 0.05 level. So between WTI price and exchange rate USD/JPY there is a long term equilibrium relationship. After the confirmation of a cointegration relation between variables we can establish an Error Correction Model (ECM). The ECM has cointegration relations built into the specification so that it restricts the long-run behavior of the endogenous variables to converge to their cointegrating relationships while allowing for short-run adjustment dynamics. The cointegration term and it is known as the error correction term since the deviation from long-run equilibrium is corrected gradually through a series of partial short-run adjustments.

The basic specification of EC model looks as follows:

$$\Delta y_t = \alpha_0 + \alpha_1 \Delta x_{t-1} + \alpha_{ec}(y_{t-1} - \beta_0 - \beta_1 x_{t-1}) + \varepsilon_t \tag{4}$$

where coefficient a represents short dynamic and β long term relationship between variables y and x. Coefficient α_{ec} is an error correction element, which describes the speed of correction of the deviation from long-term relationships.

Coefficients of EC model between WTI and USD/JPY are stated below:

Adjustment Coefficients			Cointegrating Coefficients	
α_0	α_1	α_{ec}	β_0	β_1
-0.0185	-0.01343	-0.0008	0.026586	0.059388

From these results we clearly see the difference between the impact of Exchange rate from short and long term view. In the short term view the Exchange rate of the US dollar and the Japanese yen has a negative impact on WTI oil price. However in the long term view this impact is inverse It could be because in the long term, prices have tended to adapt to the situation of decreasing or increasing of the exchange rates. The error correction element value is 0.0008 which means that every time period deviation from long term equilibrium decreases by -0.08%. In our case this period is one day, which means that this correction is appropriate.

The estimated model is as follow:

$$\Delta WTI_t = -0.0185 - 0.1343\Delta \mathrm{USDJPY}_{t-1} - 0.0008(WTI_{t-1} - 0.026586 - 0.059388\mathrm{USD}JPY_{t-1}) + \varepsilon_t \tag{5}$$

Johansen Cointegration test for WTI price and Exchange rate between US dollar and Swiss franc:

Trace test	No. of cointegration vectors	Trace statistics	Probability
	None	15.49471	0.0213
	At most 1	3.841466	0.3892
Maximum eigenvalue test	No. of cointegration vectors	Max Eigen statistics	Probability
	None	17.16444	0.0169
	At most 1	0.741539	0.3892

Table 6. The results of trace and the maximum eigenvalue statistics (WTI, USD/CHF).

Trace and maximum eigenvalue test (Table 6) indicates 1 cointegrating equations at the 0.05 level. So between variables there is a long term relation. We investigate this relation using ECM.

Coefficients of EC model are stated below:

Adjustment Coefficients			Cointegrating Coefficients	
a_0	a_1	a_{ec}	β_0	β_1
-0.0275	-0.00971	-0.06069	0.044751	6.250402

Also in this relation we see differences between short and long term dynamics. In the short term period the increasing of Exchange rate causes a decrease of oil price and vice versa in the long term period. In the long term view, the impact of change of value USD/CHF is higher than USD/JPY. But in the short term the oil price is more responding to change in USD/JPY value. The error correction element is higher in relation between WTI and USD/CHF, and its value is -0.06, which means that each day the deviation from long term equilibrium decreases by 6%.

$$\Delta WTI_t = -0.0275 - 0.00971\Delta USDJPY_{t-1} - 0.06069(WTI_{t-1} - 0.044751 - 6.250402USDJPY_{t-1}) + \varepsilon_t \quad (6)$$

Structural model of oil price development

In this section, we have created an econometric model based on monthly data that would describe the oil price development. For this purpose we have used monthly data for the period from period January 1994 to September 2010, with 201 observations.

The model is as follow:

$$WTI = \beta_0 + \beta_1 OECD + \beta_2 US + \beta_3 CHINA + \beta_4 OECDSTOCK + \beta_5 USSTOCKS + \beta_6 USCOM + \beta_7 WP\ \beta_8 QUOTAS + \beta_9 CHEAT + \beta_{10} GDPGROWTH + \beta_{11} OECDDAYS + \beta_{12} USEXCHANGE + \beta_{13} DUMMY + \varepsilon_t \quad (7)$$

Where:

Oil price of **WTI** (West Texas Intermediate) as dependant variable. The value of the variable WTI represents the average spot price of WTI crude oil with Incoterm FOB (Free on Board), expressed in U.S. dollars for barrel.

The exogenous variables

Demand side:

One of the important fundamental variables, which explains the situation in the oil markets is demand for this commodity, hence its consumption. With increasing consumption, demand is growing, and causing that the demand curve shifts to the right and consequently increases the price of oil. Thus, we assume that the increase of consumption and in the presence of inelastic supply curve, oil price will increase. We chose consumption of OECD countries (**OECD**) and United States (**US**), as the largest consumer of this commodity. Both variables are reported in thousands of barrels per day. We can see relatively stable growth of these variables. However, we can notice that in the dynamically growing countries such as India, Brazil and China, this growth is much higher.

Therefore, we incorporate into the model the Chinese demand for oil (**CHINA**). These data were quarterly, so we adjusted them using linear interpolation. It must be noted that these data are based on an estimation of the sum of domestic oil production and import of this commodity, because of the lack of official data on consumption.

Stocks:

After the first oil crises in the 70s and 80s of 20th century, OECD decided to set minimum stocks of oil in order to avoid unforeseen supply disruptions of this energy commodity. As explanatory variables we chose the stocks of countries belonging to the OECD (**OECDSTOCK**) and USA (**USSTOCKS**) - the stocks of crude oil including strategic reserves and refined products. U.S. oil stocks account for about 25 percent of OECD stocks, and because of their significant share of total OECD stocks, they are reported separately. Both variables are expressed in thousand barrels. We assume that if the stocks level rises, it will cause a decrease in demand for oil and then decrease its price. The commercial oil stocks in USA (**USCOM**) were included as another variable to the model. This is the amount of reserves after deduction of Strategic Petroleum Reserves, which the U.S. federal government usually holds in case of long interruption of oil supplies. By this regressor we expect a negative impact on oil price. Furthermore, we have included a variable that has been established by Kaufmann in his study - we called it **OECDDAYS** and it is calculated as a proportion of OECD stocks and OECD demand. This variable can be interpreted as a degree of independence from the OECD and OPEC price shocks.

Supply side:

In this category we include a variable that indicates the global production of oil **(WP)**. It is measured in thousands of barrels per day. We assume that with increasing production, the oil price declines. OPEC, the world's largest producer as a cartel, declares raises or cuts of their production quotas (**QUOTAS)** and this information is a very important sign for the oil markets and has an impact on the oil prices. These quotas define for each member of OPEC the amount, which can be produced per day by the particular member. This variable is stated in thousands barrels per day for OPEC countries as a whole as well. Production quotas are changing at the Board meetings of OPEC as a response to current prices and demand for crude oil. We expect a negative relationship between oil prices and production quotas, thus raising the value of production quotas causes oil price to decrease. Using these restrictions, OPEC tries to control price development and maintain the stability in the oil markets. But it has been observed that their efforts were often violated by several members of the cartel (In May 2009, the OPEC´s compliance with a series of cuts agreed in the second half of 2008 has reached about 80 %, and it was an unprecedented figure in OPEC´s history). Therefore, we think that a violation **(CHEAT)** could be a further variable, because of its role in the supply side. It is a difference between OPEC production and their quotas and will be expressed in thousand barrels per day. We expect that an increase of violations of production quotas, will cause the oil prices to decline because of overproduction.

Also we incorporated variables that describe economic activity **(GDP growth)** of selected countries, especially, the largest consumption countries, such as USA, Japan, Germany as the largest economy among European countries and China as one of the fastest growing economy in the world and the second biggest economy in the world. Because the data for this variable are available only in annual frequency, we applied the statistical method of interpolation to get monthly values. The last variable included in this model is US dollar exchange rate against euro (**USDEUR**). To capture the unforeseen factors such as geopolitical factors etc. we established a **dummy** variable that will indicate the political or climate events that might affect the development of oil prices. In such situation, the dummy variable has a value 1 and value 0 otherwise.

The included events are following:

- 11th September 2001 - the terrorist attack on the United States of America
- December 2002, January and February 2003 - General strike in the oil industry in Venezuela
- 20th March 2003 - Invasion of Iraq
- 14th September 2004 - Hurricane Ivan - the damage of oil drills in the Gulf of Mexico
- 29th August 2005 - Hurricane Katrina, which damaged oil drills in the Gulf of Mexico
- July to December 2008 – the Global financial crisis.

5.2 Results

Before creating the model, we tested stationarity of each variable, using ADF test, and to identify number of lags included, we have chosen Akaike information criteria. The result of test is stated in table below.

Variable	Intercept/Trend	Lag length	ADF statistic	Significance[1]
WTI	None	2	-0.566609	
D(WTI)	None	1	-6.407248	***
LOG(WTI)	Intercept, Trend	2	-3.282916	*
OECD	None	14	0.215167	
D(OECD)	None	13	-2.450925	***
US	None	5	0.900117	
D(US)	None	4	-11.09349	***
CHINA	None	3	4.535421	
D(CHINA)	None	2	-11.15506	***
OECDSTOCK	None	0	0.848878	
D(OECDSTOCK)	None	0	-13.36523	***
USSTOCKS	None	0	1.645177	
D(USSTOCKS)	None	0	-12.55420	***
USCOM	None	0	0.155353	
D(USCOM)	None	0	-13.59005	***
WP	None	2	2.210754	
D(WP)	None	1	-14.35961	***
QUOTAS	None	0	-0.113869	
D(QUOTAS)	None	0	-13.29784	***
CHEAT	None	3	0.476998	
D(CHEAT)	None	2	-9.948083	***
GDP_US	Intercept	12	-3.093607	**
GDP_GER	Intercept	12	-3.180358	**
GDP_JAP	None	13	-2.286409	**
GDP_CHINA	None	13	-4.046799	***
OECDDAYS	Intercept	14	-3.098642	**

[1] Rejecting of null hypothesis on significance level: 10% - * ; 5% - ** ; 1% - ***

Table 7. ADF test.

From the results of the test we see that almost each variable is nonstationary. We also tested stationarity of logarithm of WTI prices. The result of ADF test rejected the null hypothesis of nonstationarity. Because of that, we have applied a modeling logarithm of WTI prices with 198 observations after adjustments, and could interpret it as a relative change of oil price. The list of variables that came out as significant for describing the development of oil prices are stated in the table below:

Variable	Coefficient	Std. Error	t-Statistic	Prob.
C	3.680507	0.336197	10.94749	0.0000
CHINA	0.000293	9.62E-06	30.46082	0.0000
GDP_US	-0.086903	0.015054	-5.772764	0.0000
GDP_US(-3)	0.034622	0.014790	2.340933	0.0203
DUSCOM(-1)	-4.04E-06	2.05E-06	-1.971315	0.0501
D(QUOTAS)	5.57E-05	2.27E-05	2.455246	0.0150
OECDDAYS	-0.014453	0.003477	-4.156537	0.0000
USDEUR	-0.564405	0.127172	-4.438121	0.0000
DUMMY	0.135698	0.064702	2.097258	0.0373

Table 8. The results of structural model.

From the results (Table 8), we can clearly see that all these variables are significant on 5% level. The R-squared of this model is 0.89, which means that the model described the oil prices for 89%. To test whether residuals are white noise, we use Breusch-Godfrey test. The null hypothesis of this test means that there is no correlation between residuals. Thus, we accept the null hypothesis on 5% significance level.

The final equation is as follows:

$$LOG(WTI) = 3.6805 + 0.0003*CHINA_t - 0.0869*GDP\_US_t + 0.0346*GDP\_US_{t-3} - 4.0938*10^{-06}*DUSCOM_{t-1} + 5.5681*10^{-05}*DQUOTAS_t - 0.0145*OECDDAYS_t - 0.5644*USDEUR_t + 0.1357*DUMMY \quad (8)$$

The results of the model show that variables describing the supply side of oil are not so significant for oil price setting as much as the variables describing the demad side. In spite of that, the OPEC qoatas (DQOATAS) seem to be statistically significant, but there is a positive correlation with oil price, which does not comply with the link of our assumption. The interpretation of this result could be that the change of OPEC quotas has not an impact on oil price, because of the often low level of OPEC´s compliance (Cheats). On the other hand, US commercial stocks, OECD stocks as well as the OECD strategic reserve as a whole seem to be significant and have a negative correlation with oil price. The increase of stocks of the key oil consumption countries leads to a decrease of the global demand and drop of the oil prices. In case of the GDP growth indicators they showed to be significant only for China and USA – the two largest countries in oil consumption. While the GDP growth of China has a positive impact on oil prices immediately, the GDP growth of USA has a positive impact on oil prices, but with a three months lag. However, the GDP growth of

USA at time *t* has a negative impact. This does not correspond with our assumption, since USA is the world´s largest oil consumer. However, this result would be in line with the results of other studies that have been done in relation to this issue. The negative impact of the US Dollar on oil price is marks clear evidence in the results of model that the increase of value of US Dollar against euro leads to the oil prices to decrease, or, the decrease of US Dollar exchange rate causes the oil prices to raise. Finally, the dummy variable indicates a positive correlation with the oil prices. The events leading to oil supply interruptions or to damages in the downstream sector have an impact on the oil prices.

6. Conclusion

In this chapter we have tried to answer the question, what is the impact of movement of the US dollar exchange rate against world currencies on the oil prices? In fact US dollar devaluation creates several problems for the world oil industry. The US dollar is the currency of invoicing in global crude oil trade while oil producing countries use other currencies to buy goods and services from different nations. US dollar devaluation leads to a decrease in drilling activity and then oil supply interruption and an increase of the demand of oil in the countries using world currencies other than US Dollar. Furthermore, US Dollar devaluation decreases the purchasing power parity of oil exporting countries, especially if their currencies are tied to the US Dollar. In other words, the devaluation of US Dollar affects the global supply and demand.

Overall, our empirical results find that there is a high negative correlation between the US Dollar exchange rate against EUR and oil price and statistically significant - *P-value* (0.0000) and the coefficient is relatively high (-0.56). Therefore, our assumption has been confirmed and the results of this study confirm also what the previous studies have shown that the impact runs from US Dollar exchange rate to oil price. In addition to the US Dollar exchange rate, there are other variables that were examined in our structural model which have impact on oil prices. The finding being that the demand side factors have a higher significance in the model than the supply side factors. However, in spite of that, the OPEC quotas (DQOATAS) seem to be statistically significant, but there is a positive correlation with oil price, which is not in line with our assumption. This result could be interpreted so that the change of OPEC quotas has not the impact on oil price, because of the often low level of OPEC´s compliance (Cheats). On the other hand, US commercial stocks, OECD stocks as well as the OECD strategic reserve as a whole seem to be significant and have a negative correlation with oil price. The increase of stocks of the key oil consumption countries leads to a decrease of the global demand and a drop in the oil prices. In case of GDP growth indicators, they showed to be significant only for China and USA - the two largest countries in oil consumption. While the GDP growth of China has a positive impact on oil prices in time t, the GDP growth of USA has a positive impact on oil prices, but with three months lag (t-3). However, the GDP growth of USA at time *t* has a negative impact. This does not correspond with our assumption, since USA is the world´s largest oil consumer.

However, this result would be in line with the results of other studies that have been done in relation to this issue. The negative impact of the US Dollar on oil price is marks clear evidence in the results of model that the increase of value of US Dollar against euro leads to the oil prices to decrease, or, the decrease of US Dollar exchange rate causes the oil prices to raise. Finally, the dummy variable indicates a positive correlation with the oil prices. The events leading to oil supply interruptions or to damages in the downstream sector have an impact on the oil prices.

7. Acknowledgements

This chapter is a part of research project VEGA No. 2/0009/12.

8. References

Alhajji, A. F. (2004): The Impact Of Dollar Devaluation On The World Oil Industry: Do Exchange Rates Matter? Middle East Economic Survey, Vol. XLVII, No. 33.

Amano, R. A. & van Norden, S. (1998). Oil prices and the rise and fall of the US real exchange rate. Journal of International Money and Finance. 17, 299-316.

Baláž, P. (2008): Energy- Key factor of EU'S Economic Policy. Ekonomicky casopis (Journal of Economics), vol: 56, number: 3, pages: 274-295

Barsky, R. - LUTZ, K. (2000): Center for Economic Policy Research. [Discussion.] In: W. Bernard (2005): The Downsizing of the Dollar in Oil Prices. <www.theglobalist.com/>

Barsky, R. – LUTZ, K. (2000): A Monetary Explanation of the Great Stagflation of the 1970's. [Discussion paper, No. 2389.] Center for Economic Policy Research. www.cepr.org/pubs/dps/DP2389.asp

Bright, E. Okogu (2003): The Middle East and North Africa in a Changing Oil Market. [Working paper.] Washington: IMF.

Bénassy-Quéré, A., Mignon, V. & Penot, A.(2005). China and the relationship between the oil price and the dollar. CEPII Working Paper.

Cheng, K. C. (2008). Dollar depreciation and commodity prices. In: IMF, (Ed.),2008 World Economic Outlook. International Monetary Fund, Washington D.C., pp. 72-75.

C K Liu, H. (2002). US Dollar Hegemony Has Got to Go. www.atimes.com/global-econ/

C K Liu, H. (2005). Dollar Hegemony Against Sovereign Credit. www.atimes.com/global-econ/

Coudert, V. Mihnon, V. & Penot, A. (2008). Oil Price and the Dollar. *Energy Studies Review,* Vol. 15:2, 2008.

Economic time series page. <http://www.economagic.com/em-cgi/data.exe/fedstl/currns+1>

EEDEN, P. (2000): Understanding Gold. <http://www.USagold.com/gildedopinion/>

Heinberg, R. (2004). The Endangered US Dollar. Museletter No. 149 - August 2004. http://old.globalpublicmedia.com/the_endangered_us_dollar

Fama, E. F. (1965). The Behavior of Stock-Market Prices. The Journal of Business, Vol. 38, No. 1. (Jan., 1965), pp. 34-105.

Historial Exchange Rate Regime of United states. <intl.econ.cuhk.edu.hk/exchange_rate_regime/index.php?cid=3 - 19k –>

Hoontrakul, P. (1999): Exchange Rate Theory: A Review. Sasin-GIBA, Chulalongkorn University, Thailand. <http://www.library.ucla.edu/libraries/>

Howard, J. (2005): The Silent Oil Crisis. Powerswitch. www.powerswitch.org.uk/portal/ index.php?

Jalali-Naini, A.R. & Manesh, M. K. (2006). Price volatility, hedging and variable risk premium in the crude oil market. *Organization of the Petroleum Exporting Countries*, OPEC Review.

Keleher, R. E. (1998): US Dollar Policy: A Need for Clarification. Joint Economic Com-mittee Study, United Stats Congres. www.hoUSe.gov/jec/fed/fed/dollar.htm#endnotes #endnotes

Krichene, N. (2005). A simultaneous equations model for world crude oil and natural gas markets. IMF Working Paper.

Lee, K., Ni, S., Ratti, R.A., (1995), " Oil shocks and the macro economy: the role of price variability". The Energy Journal 16(4), 39-56

Leeming, D. (2005): The End Of Cheap Oil. Ontario Planning Journal. www.ontarioplanners.on.ca/ content/journal/

Mork, Knut A. (1989), .Oil and the Macroeconomy When Prices Go Up and Down: An Extension of Hamilton.s Results,. Journal of Political Economy, 91, pp. 740-744.

Márquez, H. (2006): Oil Market Analysts Issue Dire Warnings. www.ipsnews.net/

Monbiot, G. (2005): Are Global Oil Supplies About To Peak? The Guardian. http://www.guardian.co.uk/

Mundell, R. (2003): Commodity Prices, Exchange Rates and the International Monetary System. [Paper presented at the conference *Consultation on Agricultural Commodity Price Problems*.] Rome: FAO, March 25 - 26, 2002.

Obadi, S. M. (1999): The impact of the oil sector on economic development of the countries of Middle East and North Africa with a focus on member countries of OAPEC. Ekonomický časopis/Journal of Economics, *47*, 1999, č. 4, s. 545 - 567, Bratislava, Slovakia.

Obadi S. M. (2006). Do Oil Prices depend on the Value of US dollar?. Journal of Economics No: 3, 53/2006, p. 319-329, The Institute of Economic Research, Slovak Academy of Sciences, Bratislava, Slovakia.

OECD (2004): The OPEC Annual Statistical Bulletin.

Putland, G. R. (2003): The War To Save The US Dollar. <www.trinicenter.com/oops/iraqeuro.html>

Rahn, R.W.(2003). How Far Will the Dollar Fall? Cato Institute. http://www.cato.org/pub_display.php?pub_id=2483

Reap, S. (2004): The Impact of Higher oil Prices on the Global Economy with focUS on Developing Economy. IEA. <www.iea.org/>

Rifkin, J. (2004): A Perfect Storm About To Hit. The Guardian. h ttp://www.guardian.co.uk/

Spero, J. E. & Hart, J. A. (1997). Oil, Commodity Cartels, and Power. Routledge: St. Martin's Press, Chapter 9, pp. 276 - 301.

Spiro, D. E. (1999). The Hidden Hand of American Hegemony: Petrodollar Recycling and International Markets, Cornell University Press.

The Daily Oakland Press (2005): Decline in Value of Dollar Helps Fuel Rising Gas Prices. http://theoaklandpress.com/stories/041405/opi_20050414004.shtml

Weller, C. E. & Lilly, S. (2004): Oil Prices Up, Dollar Down - Coincidence? http://www.americanprogress.org/site/pp.asp?

Yousefi, A., Wirjanto, T. S.(2004). The empirical role of the exchange rate on the crude-oil price formation. Energy Economics. 26, 783-99.

Zamani, M. (2004). An Econometrics Forecasting Model of Short Term Oil Spot Price, Paper presented at the 6th IAEE European Conference, Zurich, 2–3 September.

11

A State-of-the-Art Review of Finance Research in Physical and Financial Trading Markets in Crude Oil

Andrew C. Worthington
Griffith University
Australia

1. Introduction

It goes without saying that crude oil is important for both developed and emerging economies, with crude oil-related expenditures representing a significant proportion of total economic inputs, and an even more substantial share of total energy consumption. For example, energy expenditures of about $1,233 billion in the US currently account for some 8.8% of its GDP (from a high of 13.7% in 1981), with crude oil products (including distillate, jet fuel, liquefied petroleum gas, and gasoline) representing some 35.27% of total end-use energy expenditure, ahead of 23.56% for natural gas and 19.76% for coal. Of this, the residential, commercial, and industrial sectors respectively account for 11.4%, 5.9% and 47.1% of end-use expenditure and the transportation sector the remainder. It is then clear to see economic activity positively depends on crude oil requirements and expenditures. It also explains why Hamilton (1983) very early concluded that a dramatic increase in the price of oil has preceded almost every post-World War II recession in the US. Accordingly, as a key factor input, crude oil price changes have the potential to alter dramatically the financial performance of economies at large and the performance and behavior of the firms that operate within them. Hence, understanding the relationship between oil prices and various financial markets is paramount, and there is a role for finance research in this regard.

For the most part, the finance discipline has addressed two main bodies of research in the physical and financial trading markets in crude oil. The first body of research has concerned the pricing dynamics of crude oil in both spot and derivative markets. The purpose here, among other things, is to provide market participants with accurate price and volatility forecasts to facilitate capital budgeting plans, help develop corporate strategy, and better manage revenue streams and commodity risk. Important research substreams include the relationships between crude oil prices and downstream products, such as refined petroleum and heating oil, and their relationship with financial instruments, such as crude oil futures, forwards, and options (including convenience yield and the role of inventories) as a tool for improving risk management practices and price forecasts. They also include the relationship between crude oil prices and linked real assets, such as exploration lease contracts and oil reserves, to facilitate the valuation of projects and firms in the crude oil industry. A particularly well developed stream of finance research within this body of work concerns the market efficiency of crude oil prices. In contrast, a less developed but emerging area of interest revolves around behavioral finance and the actions and impacts of heterogeneous agents in crude oil markets.

The second body of research relates to the role of crude oil as a factor input and thence its impact upon financial assets, especially equity. For the most part, this work examines whether oil prices constitute a source of systematic asset price risk and whether exposure to this risk varies across industries and even countries. This serves as a suitable complement to the macroeconomics literature with its particular focus on the impact of oil prices shocks at the economy level. The underlying rationale provided for the inclusion of oil prices within asset pricing models typically stems from the importance of crude oil to the economy. Crude oil price changes have the potential to alter significantly future cash flows of firms through their influence on the costs of factor inputs. However, the significance of an oil price factor is potentially industry dependent. For instance, the transportation industry is highly vulnerable to oil price increases owing to its heavy reliance on oil-based factor inputs, while revenues in the oil and gas industry are likely to benefit from increases in oil prices. Similarly, the ability of individual firms to pass on rising factor costs to customers and by differing degrees of oil price hedging complicates the detection of any indirect or direct impact of oil price changes on stock returns.

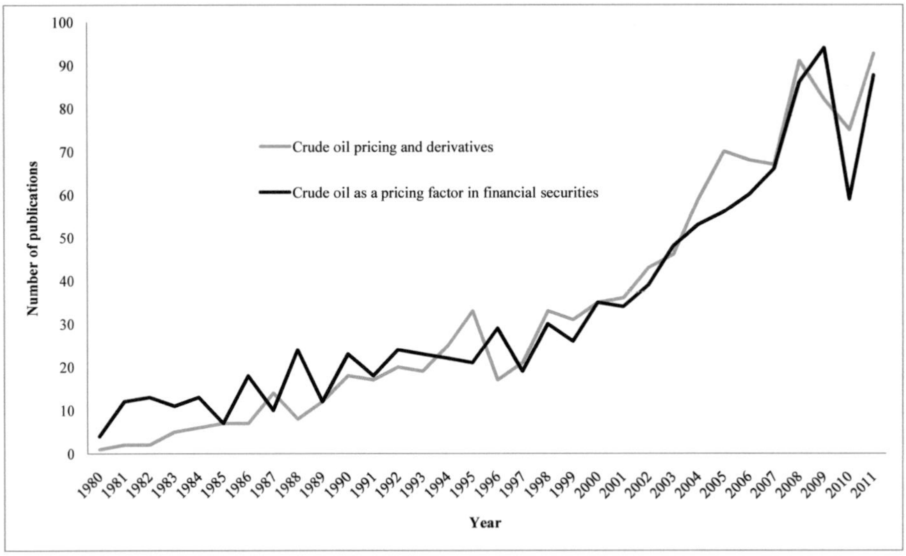

Fig. 1. Crude oil-related finance research, 1980–2011.

As shown in Fig. 1, interest in both of these areas of research has grown substantially in recent decades. The purpose of this chapter is then to survey comprehensively the burgeoning finance research as it relates to physical and financial trading markets in crude oil. Apart from discussing the objectives and purpose of crude oil-related finance research, the chapter also assesses the methods used by existing academic and industry researchers. This highlights the empirical problems that have received attention in the literature, and the efforts to overcome these problems. It therefore provides guidance to those conducting finance-related research in crude oil and an aid for crude oil industry analysts, regulators, policymakers, practitioners and other stakeholders interpreting the findings of these studies.

Overall, the chapter serves to complement the other chapters in this volume concerning crude oil and encompassing the disciplines of engineering, chemistry, geology, geophysics, economics and management.

The remainder of the chapter comprises six sections. Section 2 details the scope of the review. Section 3 outlines the theory and empirical work on intertemporal price relationships in the physical and financial markets for crude oil. Section 4 discusses the theory and evidence on financial market efficiency in crude oil markets while Section 5 undertakes a similar process in considering crude oil as a pricing factor in financial markets. Section 6 provides a conclusion to the chapter and Section 7 lists the references used.

2. Scope and limitations of review

This chapter is concerned with finance research on the physical and financial trading markets in crude oil. While closely related, it is not the purpose of this review to address the sizeable macroeconomic literature concerning the fundamental determinants of crude oil prices and their subsequent impact on the prices of refined products (such as gasoline or distillate) or the impact of crude oil prices and expenditures on macroeconomic activity, growth, and development. Instead, we take crude oil prices as given and focus on their impact on the prices and other behavior of crude oil-related financial securities and derivatives.

Where possible, we concentrate on the most-recent finance research in crude oil markets published since 2000. We searched EconLit, the *Journal of Economic Literature* electronic database, to identify journal articles that were representative of the field. References therein helped identify other relevant articles. We also used Google Scholar to locate references not included in EconLit. Nonetheless, as it was not possible to include all of the studies discovered using this search procedure, we focus on the most significant and/or most readily accessible. Of the 73 references in the chapter, all are refereed journal articles with 19 (26%) published in *Energy Economics*, 4 (5%) in the *Energy Journal*, 3 (4%) each in the *Journal of Futures Markets* and the *Journal of Banking and Finance*. The remainder appears across a range of generalist economic, econometric, and finance journals. Only eight (11%) are published before the 2000s.

3. Intertemporal price relationships in crude oil markets

This first area of research concerns the relationships between the prices for crude oil in its physical (or spot) markets as the underlying consumption asset and the prices of crude oil commodity derivatives whose prices and conditions derive from these markets, as represented by crude oil futures and their associated options markets. Crude oil futures are standardized, exchange-traded contracts in which the contract buyer agrees to take delivery, from the seller, a specific quantity of crude oil at a predetermined price on a future delivery date. Crude oil options are contracts in which the underlying asset is a crude oil futures contract. The holder of a crude oil option possesses the right (but not the obligation) to assume a long (or bought) position (with a call option) or a short (or sold) position (with a put option) in the underlying crude oil futures at the specified strike (or exercise) price. This right ceases to exist when the option expires after the expiration date.

Currently, crude oil future and options contracts can be traded on a number of product-specific exchanges, of which the most important are the New York Mercantile Exchange (NYMEX) (part of the CME Group), the Intercontinental Exchange (ICE) (formerly the International Petroleum Exchange) and the Tokyo Commodity Exchange (TOCOM). The quotation of NYMEX Light Sweet Crude and Brent Crude oil options is in US dollars and cents per barrel and traded in lots of 1,000 barrels (42,000 gallons) of crude oil, as are the ICE Brent and West Texas Intermediate (WTI) crude oil options and futures. The TOCOM futures contract specifications for Middle East (monthly average of Dubai and Oman prices) crude oil are per 50 kiloliters (approximately 314.5 barrels) and are in Japanese yen.

As with other commodity futures and options, there are an array of market participants engaged in the trading of crude oil futures and options. These include commercial enterprises with a direct stake in the price of oil where these contracts serve as a hedging instrument for future production. For example, to protect themselves against falling cash (spot) market prices, producers, traders, and marketers can sell futures to lock in prices for future delivery, protecting the value of future crude oil sales. Futures can also serve as a hedging instrument to shift exposure to price risk. For example, refiners, traders, and marketers will buy futures to protect against their exposure to rising cash market prices. Similarly, professional energy traders trading refined crude oil product contracts, such as heating oil and unleaded gasoline, in tandem with crude oil futures can lock in the crack spread, or the theoretical cost of the refining margin between crude oil and its refined products. Finally, investors and speculators with no intention or required purpose for buying or selling crude oil can attempt to profit by buying or selling futures contracts.

As derivatives on a consumption asset (an asset held primarily for consumption purposes), crude oil and futures and options written on crude oil exhibit different price relationships to investment assets (assets held primarily but not solely for investment purposes), including precious metals and financial securities. For any asset in the absence of income and storage costs (including any physical outlay costs, interest charges and possibly a risk premium) the forward price will equal the spot price, otherwise arbitrageurs can buy (sell) the asset and enter short (long) forward contracts on the asset. Commodities that are consumption assets (like crude oil) provide no income but are subject to significant storage costs (as negative income). This would suggest that the futures price is equal to the spot price plus storage costs; otherwise, arbitrageurs would sell the commodity, save the storage costs, invest the proceeds at the risk-free interest rate, and take a long position in the futures contract.

However, users of crude oil as a consumption commodity are not indifferent between holding the physical commodity and futures contracts written on that commodity. For example, an oil refiner is unlikely to regard a futures contract on crude oil in the same way as crude oil held in inventory and therefore available for processing, now or later; benefits sometimes referred to as the convenience yield provided by the commodity. The intertemporal price relationship for crude oil will then reflect the market's expectation concerning its future availability such that the greater the possibility that shortages will occur, the higher the convenience yield. Conversely, if users of crude oil have large inventories, there is little chance of shortages in the future, and the convenience yield tends to be low.

Ultimately, the cost of carry reflects these factors. For crude oil, this is most determined by storage costs, such that the futures prices will differ from the spot price by an amount reflecting the cost of carry net of the convenience yield. If we assume fixed storage costs, the negative cost of the convenience yield will determine the costs of carry that positively relates to the level of inventory, approaching zero when inventory levels are very high. Higher convenience yields are then associated with futures prices less than the spot price, as a price premium exist for earlier, including immediate (spot), delivery.

Regardless of these circumstances, the price of crude oil futures will move to equal (and thus become an unbiased predictor of) the future spot price as time progresses. Otherwise if the futures price is higher (lower) than the spot price, arbitrageurs will sell (buy) the futures contract, buy (sell) crude oil and then make delivery, locking in a profit because the price of the contract sold (bought) is already higher (lower) than the amount spent buying (selling) the underlying asset to cover the position. The former situation, when the futures price is above the expected future spot price, is known as contango, the latter, when the futures price is below the expected future spot price, is known as normal backwardation.

An expanding number of studies have empirically examined aspects of these basic theoretical relationships, primarily with respect to crude oil futures and, to a lesser extent, crude oil options. One basic dimension of this research aims to examine prices, risk and convenience yields in futures prices. Lin and Duan (2007), for example, defined business cycle of crude oil as a seasonal commodity with demand/supply shocks and found that the convenience yield for crude oil indeed exhibited seasonal behavior, with the convenience yield for WTI crude oil highest in the summer, while that for Brent crude oil is the highest in the winter. In line with expectations, they also found that crude oil convenience yields are negatively related to inventory levels and positively related to interest rates. Lastly, Lin and Duan (2007) verified that physical oil prices are more volatile than futures prices at low inventory. Bhar and Lee (2011) also considered crude oil futures but with the aim of assessing market risk in futures markets. They found that the risk premium in the crude oil market is time varying and driven by the same sorts of risk factors as equity and bond markets.

Several other studies have also examined these fundamental relationships between the spot and futures markets in crude oil. Alquist and Kilian (2010), for instance, argued that despite their widespread use as predictors of the spot price of oil, oil futures prices tend to be less accurate as predictors than a default forecast of no change. Alquist and Kilian (2010) further argued that the relatively poor performance of futures prices in predicting spot prices was driven by strong variability of the futures price about the spot price, as captured by the oil futures spread, and in turn, the marginal convenience yield of oil inventories. Using a two-country, multi-period general equilibrium model of the spot and futures markets, Alquist and Kilian (2010) concluded that increased uncertainty about future oil supply shortfalls caused the futures spread to decline and the precautionary demand for oil to increase, thereby resulting in an immediate increase in the spot price. Thus, we may consider the negative impact of the oil futures spread as an indicator of fluctuations in the price of crude oil driven by precautionary demand.

Moshiri and Foroutan (2006) likewise argue that while movements in physical oil prices are complex, nonlinear models could provide useful forecasts of future price changes using

NYMEX daily crude oil futures prices from 1983 to 2003. Murat and Tokat (2009) likewise used the ability of futures prices to predict spot prices, but instead focused on the crack spread, justifying their choice on the basis that this was the most relevant signal to refiners and consumers primarily concerned with their exposure to the crack spread, but also by hedge funds speculating on the crack spread. Their results indicated the causal impact of crack spread futures on the spot oil market in both the long- and short-run. Further, crack spread futures were almost as good as the crude oil futures in predicting spot oil markets. In other work, Ye et al. (2010) investigated the changing relationship between the price and volume traded of short- and long-maturity NYMEX light sweet crude oil futures contracts and major changes in the physical crude oil market. They found, among other things, that excess production capacity in the physical market and electronic trading on the NYMEX futures markets played a substantial role in providing an early indication of market regime shifts in the other market and thus could potentially improve short-run crude oil spot price forecast models.

The question naturally arises as to the sources of prices in changes in futures outside their relationship with the physical market. Not unexpectedly, the most basic argument is that the macroeconomic fundamentals that drive the spot market in crude also drive the futures markets. For example, Zagaglia (2010) used factor-augmented vector autoregression models to examine the dynamics of NYMEX oil futures prices using a large panel dataset that includes global macroeconomic indicators and financial market indices, along with the quantities and prices of energy products. The results showed a strong linkage between crude oil futures markets and other financial markets. In other work, Casassus et al. (2010) used an equilibrium model to consider the correlation between inflation and oil futures returns. As a consumption good, Casassus et al. (2010) realistically found the positive correlation found in most earlier empirical studies, however other important sources of correlation related to monetary and output shocks. In fact, the reaction of central banks to these shocks was at least equally important in determining futures price changes.

Kaufmann and Ullman (2009), for one argued that the relationship between between spot and futures markets was relatively weak, arguing that long-term increase in oil prices be exacerbated by speculators, with innovations first appearing in spot prices for Middle Eastern crude oil prices and then spreading to other spot and futures prices. Yet other innovations first appear in the far month contract for WTI and likewise spread to other exchanges and contracts. Buyuksahin and Harris (2011) also considered the role of speculators in crude oil futures markets. Using a set of unique data from the US Commodity Futures Trading Commission, Buyuksahin and Harris (2011) tested the Granger causal relation between crude oil prices and the trading positions of various types of traders in the crude oil futures market. In contrast to Kaufmann and Ullman (2009), they found little evidence that hedge funds and other non-commercial position involved Granger-caused price changes. Lastly, it is understandable that the one-off macroeconomic and political shocks so important in driving prices and risk in physical markets in crude oil also drive futures prices. As just one example Chiu and Shieh (2009) found great success with modeling futures with autoregressive conditional heteroscedasticity models and found that high-volatility regimes in futures markets corresponded with the Gulf War, the Asian Financial Crisis, and September-11.

Apart from futures market behavior itself and its relationship with spot markets, a small body of work has examined the interrelationships between crude oil and other futures markets. One part of this exploits the obvious links between crude oil futures and related energy markets. For example, Westgaard et al. (2011) considered the relationship between gas oil and Brent crude oil using ICE futures. They found that in normal periods, energy futures prices, both crude oil-related and otherwise, are strongly correlated, but that during volatile periods corresponding to financial and environmental crises, the spread between gas and crude oil is likely to deviate, taking some time to revert to its original equilibrium value. In contrast to Murat and Tokat (2009), this was taken to suggest that energy traders and hedgers should treat the crack hedge with some care in volatile market environments.

Even farther afield, Cortazar and Eterovic (2010) used the prices of crude oil futures contracts to estimate copper and silver future prices. They found that a simple multicommodity model able to incorporate the nonstationary long-term process of oil was useful in accurately estimating long-term copper and silver futures prices, in fact achieving a much better fit than the available individual or multicommodity models. Du et al. (2011) similarly considered the relationships between crude oil futures markets and those of other (agricultural) commodities. They found that speculation and inventories were important in explaining the volatility of crude oil prices and that like many energy markets, crude oil futures prices displayed mean-reversion, an asymmetric relationship between returns and volatility, volatility clustering, and infrequent compound jumps. As to the relationships with agricultural futures markets like corn and wheat, Du et al. (2011) concluded these had tightened because of the role of ethanol production.

A final area relevant in the examination of crude oil futures markets concerns hedging. For the most part, empirical research here focuses on identifying an optimal hedge and determining it effectiveness, i.e. the reduction in the variability of a portfolio of spot assets and futures positions relative to an unhedged position. As a recent example, Yun and Hyun (2010) analyzed the hedging effectiveness over different hedge types and periods by Korean oil traders. As there, and in the wider commodity futures literature, while there are benefits to hedges, but the magnitude of the optimal hedges and their effectiveness can vary significantly. Numerous other studies have examined additional dimensions of crude oil spot and futures pricing (Agnolucci, 2009; Anderluh & Borovkova, 2008; Barros Luis, 2001; Crosby, 2008; Deaves & Krinsky 1992; Deryabin, 2011; Doran & Ronn, 2008; Fleming & Ostdiek, 1999; Hikspoors & Jaimungal, 2007; Horan et al., 2004; Hughen, 2010; Meade, 2010; Moosa & Al-Loughani, 1994; Moosa & Silvapulle, 2000; Paschke & Prokopczuk, 2010; Tokic, 2011; Trolle & Schwartz, 2009).

This final area offers one of the more useful directions for further research on crude oil futures. While there seems to be some empirical consensus that crude oil futures markets exhibit the basis relationships we expected, including with spot markets, hedging and other risk management strategies remain of interest. In particular, a greater academic understanding is required of the hedging behavior of both crude oil producer and consumer firms. This should permit market participants a more accurate insight into their own situation and possibly more effective solutions as well as facilitating the design of new and innovative derivative products on established and yet to be established exchanges.

4. Price behavior in crude oil markets

The second main area of finance research in crude oil markets concerns the efficient market hypothesis (EMH). While there has been strident criticism of the veracity of the EMH in recent years, particularly by the behavioral finance literature, it remains (along with asset pricing models) one of the original and main areas of finance research. In essence, the EMH assumes that market participants have rational expectations, that on average the market is correct (even if no individual is), and that whenever new and relevant information appears, participants update their expectations appropriately. Importantly, there is no requirement that participants behave rationally in response to this information, beyond the usual rational utility maximizing assumptions made in most orthodox finance models. As such, when faced with new information, some market participants may overreact or underreact to this new information. This is also consistent with the EMH as long as the reactions are statistically random such that no market participants can reliably employ this information to make an abnormal profit. Thus, any or all participants can be mistaken about market conditions, but the market as a whole is always right, and therefore market assets, including commodity assets such as crude oil, are priced appropriately at some equilibrium level.

Within this, convention now states that there are three subforms of the EMH—weak-form efficiency, semi-strong-form efficiency and strong-form efficiency—each of which subsumes weaker forms and has different implications for how markets behave. Weak-form efficiency states participants cannot earn excess returns using strategies employing historical market information, including past prices, volume and price volatility. Accordingly, current market prices fully incorporate all available historical information. Semi-strong form efficiency argues prices adjust within an arbitrarily small but finite amount of time and in an unbiased fashion to publicly available new information, so there are no excess returns from market behavior trading on that information, but excess returns are available for trading based on non-historical information. Consequently, market prices include all currently available public information. Lastly, strong-form efficiency asserts that prices reflect all information and excess returns are unavailable to anyone. Conventionally, this is that market prices incorporate all publicly and privately held information, including insider information.

Clearly, the concept of market efficiency has important implications for crude oil markets and their participants. First, market efficiency is associated with appropriate equilibrium spot prices. Second, the level of market efficiency thought to exist will determine the trading and other strategies of market participants. Third, where prices are set in markets that are inefficient, profitable (risk-adjusted) opportunities may be available. Finally, and drawing on the discussion in the previous section, if futures markets are efficient, hedging efficacy is enhanced. This is because if futures prices reflect all available information, then they provide the best forecast of spot prices. Thus, hedgers do not underpay or overpay for the service of risk transfer.

Consider first tests of weak-form efficiency in crude oil spot markets. Elsewhere, these typically range from single tests of the autocorrelation of price changes and unit root tests to establish a random walk to more sophisticated single and multiple variance ratio tests and cointegration analysis and error correction models. For the most part, these and other tests of weak-form market efficiency effectively condenses down to the degree of randomness in historical price changes and the ability of lagged prices to forecast futures prices. As one example, Tabak and Cajueiro (2007) analyzed the efficiency of crude oil markets (Brent and WTI) by estimating the fractal structure of the time series, concluding that the crude oil spot market has become more efficient over time. Wang and Liu (2010) later extended this analysis (only to WTI) through observing the dynamics of local Hurst exponents based on multiscale detrended fluctuation analysis. The empirical results there showed that short-, medium- and long-term market behaviors in crude oil were generally exhibiting behavior that was more efficient over time. Further, small fluctuations in the WTI crude oil market were persistent. However, larger fluctuations had higher instability, both in the short- and long-term. In other work, Alvarez-Ramirez et al. (2010) studied the efficiency of crude oil markets using lagged detrended fluctuation analysis (DFA) to detect delay effects in price autocorrelations quantified in terms of a multiscaling Hurst exponent (i.e., autocorrelations are dependent of the time scale). The results, based on spot price data over the period 1986-2009, indicated important deviations from efficiency associated with lagged autocorrelations. Moreover, Alvarez-Ramirez et al. (2010) argued that any evidence found in favor of price reversion to a continuously evolving mean underscored the importance of adequately incorporating delay effects and multiscaling behavior in the modeling of crude oil price dynamics.

A common analysis in the finance discipline in relation to the weak-form efficient market hypothesis concerns tests of observed market anomalies, that is, specific and persistent instances of market inefficiency explained by observable fundamental or technical characteristics. In the only known study of an anomaly of this type in the crude oil market, Yu and Shih (2011) examined the presence of a calendar effect in the form of the weekend effect. In stock markets, this typically takes the form of higher returns (log or discrete price changes) on Fridays and lower returns on Mondays. Yu and Shih (2011) found using a probability distribution approach that oil markets exhibited instead a Thursday effect, with a shortening of the holding period from Wednesday to Friday and a leftward return distribution. Naturally enough, the authors concluded that this had some implications for risk management by participants trading in crude oil markets on a frequent or long-term basis.

A small number of additional studies also consider weak-form market efficiency in crude oil markets, but in crude oil futures markets. To start with, Lean et al. (2010) examined the market efficiency of crude oil spot and futures prices using both the mean-variance and stochastic dominance approaches. Using WTI crude oil data from 1989 to 2008, they found no evidence of any of the tested relationships, thereby inferring that there was no arbitrage opportunity between crude oil spot and futures markets, that investors are indifferent to investing spot or futures, and that the spot and futures crude oil markets were efficient and rational. In contrast, Wang and Yang (2010) used high-frequency (intraday) data on crude oil (along with heating oil, gasoline, and natural gas) futures markets to find some evidence for weak-form market inefficiency, but only in heating oil and natural gas, not crude oil. Lastly, Kawamoto and Hamori (2011) considered the weak-form market efficiency in crude oil futures markets but with futures of different maturities. WTI futures were yet again employed, though with cointegration analysis and error correction models. The results showed that WTI futures were consistently efficient within an 8-month maturity range and consistently efficient and unbiased within a 2-month maturity range.

Unlike analyses of market efficiency in many other commodity markets, very few concern high-order measures of market efficiency, namely the semi-strong and strong forms. In the only known recent study of semi-strong form market efficiency in crude oil markets, Demirer and Kutan 2010 examined the informational efficiency of crude oil spot and futures markets with respect to OPEC conference and US Strategic Petroleum Reserve (SPR) announcements and an event study methodology to detect abnormal returns between 1983 and 2008. Typically, event studies examine the statistical significance of particular information releases or market actions on price changes, with the expectation that in efficient markets, any new information should be quickly and accurately incorporated in current prices. For example, long delays in the incorporation of new information in prices may be reflective of informational or behavioral limitations in at least some market participants. The results indicated that only OPEC production cut announcements yield a statistically significant long-term impact with SPR announcement invoking a short-term market reaction following the announcement date.

Fundamentally, tests of semi-strong and strong form market efficiency in crude oil markets, both spot and futures, remain problematic. One problem concerns the identification of a fundamental equilibrium or benchmark-pricing model against which we can gauge price change as being efficient or inefficient. With strong form market efficiency, the more serious problem is the identification of trading behavior by market specialists and government and corporate insiders using superior pricing models and/or information to trade in crude oil markets. Nonetheless, the consensus in most crude oil markets, especially futures, is that market efficiency prevails in the long run, but that in the short run transitory and moderate inefficiencies may arise.

5. Crude oil as a financial pricing factor

Since the development of the Sharpe and Lintner capital asset pricing model (CAPM) in the 1960s, numerous studies have attempted to identify the determinants of security prices and correspondingly their returns. While both Sharpe and Lintner separately proposed that the pricing of assets only accords to their covariance with a market portfolio, they did not explicitly acknowledge other factors, such as the macroeconomy, in the pricing relationship. The subsequent development of Ross's arbitrage pricing theory and other multifactor models provided an opportunity to incorporate such factors. With this multifactor specification in mind, numerous studies, have sought to investigate whether individual macroeconomic variables constitute a source of systematic asset price risk at the aggregate market and industry level.

Crude oil is, of course, one key macroeconomic factor of obvious empirical interest. While it may be reasonable to expect that oil price changes influence stock markets, determining whether oil prices constitute a source of systematic asset price risk is difficult to ascertain and hindered by scant research. As Hammoudeh et al. (2004 p. 428) note: "There has been a large volume of work investigating the links among international financial markets, and some work has also been devoted to the relationships among petroleum spot and futures prices. In contrast, little work has been done on the relationship between oil spot/futures prices and stock indices, particularly the ones related to the oil industry". Similar sentiment is echoed in the earlier work of Sardorsky (1999, p. 450): "In sharp contrast to the volume of work investigating the link between oil price shocks and macroeconomic variables, there has been relatively little work done on the relationship between oil price shocks and financial markets".

The rationale provided for the inclusion of oil prices in asset pricing models of typically stems from the importance of oil to the economy in question [see, for example, Hamao (1988) and Jones and Kaul (1996)]. Oil price changes have the potential to alter significantly future cash flows of firms through their influence on the costs of factor inputs (Faff and Brailsford, 1999). However, the significance of an oil price factor is potentially industry dependent. For instance, the transportation industry is highly vulnerable to oil price increases due to its heavy reliance on oil-based factor inputs, while the oil and gas industry is likely to benefit from such increases in oil prices [see, for example, Sadorsky (2001) and El-Sharif et al. (2005)]. Furthermore, Faff and Brailsford (1999) note that the detection of any indirect or direct impact of oil price changes on stock returns is complicated by the ability of individual firms to pass on rising factor costs to customers and by differing degrees of oil price hedging.

Additionally, identifying oil price effects is complicated by potential indirect relationships. For example, fluctuations in oil prices have the capability to indirectly influence stock price returns through its impact on inflation, and hence discount rates, and through its potential to dampen demand for goods (i.e. through its impact on discretionary spending). Obviously, knowing the extent of the impact of crude oil prices on financial assets like stock is of some practitioner interest. For instance, if an investor believes that the price of oil will surge above market expectations, then that investor could estimate an oil price factor to enable the construction of a portfolio that has the highest (positive) sensitivity to oil price increases. Likewise, the manager of a managed portfolio that is concerned with rising oil prices could construct their portfolio so that it has the lowest sensitivity to an oil factor.

For the early literature that does examine the significance of crude oil as a source of systematic asset price risk, the results have been inconsistent across studies and countries. Chen et al. (1986) and Hamao (1988) found no evidence of an oil price factor for the US and Japan, respectively. In contrast, Sardorsky (1999) and Kaneko and Lee (1995) conclude that oil prices are a significant factor in the US and Japan, respectively. However, of most promise are the early studies of Faff and Brailsford (1999), Sardorksy (2001), Hammoudeh et al. (2004) and El-Sharif et al. (2005) which attempted to examine the significance of an oil price factor at the industry level. These studies indicated that oil prices do constitute a source of systematic asset price risk and that the exposure to this risk varies across industries.

As just one example, Faff and Brailsford (1999) investigated the sensitivity of Australian industry equity returns to an oil price factor. Continuously compounded monthly returns over the period July 1983 through March 1996 are employed. Five of the twenty-four industries displayed a significant relationship with the price of oil. Industries with a significantly positive relationship with the price of oil included, as expected, the oil and gas and diversified resources industries. Those with a significantly negative relationship included paper and packaging, transport and banking. In a similar manner, Sadorsky (2001) sought to identify risk factors in the stock returns of Canadian oil and gas companies. Utilising an augmented market model [similar to Faff and Brailsford (1999)], Sadorsky (2001) regressed the excess monthly returns of the oil and gas stock index on the excess return of the market portfolio and the oil price return. An exchange rate factor and a term premium factor were added to subsequent regressions. The results indicated a strong positive relationship with oil prices, the market portfolio, and an inverse relationship with the term premium and the exchange rate.

El-Sharif et al. (2005) provided a complementary analysis concerning U.K. data to that of Sadorsky (2001) (Canadian data) and Faff and Brailsford (1999) (Australian data). Daily data covering the period January 1989 to June 2001 was used. As with much of the previous research in this area, a multifactor model was employed to investigate the relationship between excess returns to the oil and gas sector and oil price returns, excess returns on the market portfolio, the exchange rate, and the term premium. Due to inter-temporal variability in the relationship between natural resource and equity prices [as highlighted in the work of Sadorsky (2001) and Faff and Brailsford (1999)], El-Sharif et al. (2005) divided the sample into twenty-five six-month periods, however, it is suggested that shorter sub-periods may yield superior results due to the increasing oil price instability. A consistently positive, and sometimes highly significant, relationship was found between oil prices and the excess stock returns of oil and gas companies. Four additional industries were tested (mining, transport, banking and software and computer services) with no significant relationship with oil price changes indicated. However, the significance of oil price changes on stock returns cannot be underestimated, with Huang et al (2005) concluding that oil price volatility explains stock returns better than changes in industrial output.

While the aforementioned studies concerned aggregate broad-based stock market indices, very few papers have attempted to relate conditional macroeconomic volatility to the conditional stock market volatility of a particular industry. In one such study, Sadorsky (2003) examines the determinants of US technology stock price volatility over the period July 1986 to December 2000. Utilizing both monthly and daily data, Sadorsky (2003) tests whether oil price volatility (WTI crude oil futures) and a series of other macroeconomic variables have any impact on technology stock price volatility (using the Pacific Stock Exchange Technology 100 Index as a surrogate). Conditional volatility was estimated using a generalization of a 12-month rolling standard deviation estimator, similar to that of ARCH estimates. The econometric results indicate, inter alia, a significant link between lagged conditional oil price volatility and conditional technology stock price volatility. However, industrial production and the consumer price index were found to have a larger impact on technology stock price volatility than oil. Hence, Sadorsky (2003) provides one of the few papers that attempts to investigate the association between conditional oil price volatility and stock price volatility.

A number of studies have since developed these approaches in a variety of contexts. In Australia, McSweeney and Worthington (2008) examined the impact of crude oil prices on Australian industry stock returns using multifactor static and dynamic models containing crude oil and other macroeconomic factors from January 1980 to August 2006. The macroeconomic factors used comprised the market portfolio, oil prices, exchange rates, and the term premium. The industries consist of banking, diversified financials, energy, insurance, media, property trusts, materials, retailing, and transportation. McSweeney and Worthington (2008) found crude oil prices are an important and persistent determinant of returns in the banking, energy, materials, retailing, and transportation industries. The findings also suggest oil price movements are persistent. Nonetheless, the proportion of variation in excess returns explained by contemporaneous and lagged oil prices appeared to have declined during the sample period.

In the US, Odusami (2009) also reconsidered the impact of crude oil shocks on stock market, concluding a nonlinear effect. There was also some evidence of the impact of OPEC meetings, with the suggestion of prospective and important new information release, on US stock returns. Killian and Park (2009) also examined the role of crude oil shocks in the US, finding stronger, more significant and asymmetric impacts at the industry level to crude oil demand shocks than crude oil supply shocks. Importantly, Killian and Park (2009) also suggested that the impact of crude oil shock on stock markets was not so much through domestic cost or productivity shocks, rather through shifts in the final demand for goods and services, with some of the strongest responses outside the energy industry in transportation, consumer goods, and tourism services. Similar findings are echoed elsewhere in work on the effect of crude oil prices shocks on stock markets in Turkey (Eryiğit, 2009), Brazil, China, India and Russia (Ono, 2011), Greece, the US, the UK and Germany (Lake & Katrakilidis, 2009), Nigeria (Somoye et al., 2009), Gulf Cooperation Countries (GCC) (Arouri et al. 2010), Russia (Hayo and Kutan, 2005, Bhar & Nikolova 2010). Lastly, there is also evidence garnered from broader international studies (Nandha & Brooks, 2009; Jawadi et al., 2010) and from the experience of global industries, including oil companies (Sardorsky, 2008) and shipping (El-Masry et al. 2010). The literature includes a large number of similar applications (Chan et al., 2011; Ghoilpour, 2011; Hammoudeh & Choi, 2006; Hammoudeh et al., 2010; Nandha & Faff, 2006).

Overall, the evidence is now becoming increasingly consistent that crude oil markets exert a significant but varying impact on stock and other financial markets through the asset pricing mechanism. This effectively complements a broader and longer-established literature on the impact of macroeconomic activities on financial markets and the influence of crude oil on the aspects of the macroeconomy, including GDP, (un)employment and inflation. It is also clear that the impact of crude oil on asset pricing varies across industries, though the exact transmission mechanism through which this works is not especially clear and therefore demands further attention. Lastly, it is evident that the impact of crude oil markets on financial asset pricing and hence returns generally reflects its contribution to economic activity, with clear evidence of a long-term decline associated with the stage of economic development and efficiency gains and diversification in energy input markets.

6. Conclusion

This chapter reviewed the finance literature as it relates to physical (spot) and financial (futures and options on futures) markets in crude oil. As discussed, this builds on an older and more established economics literature in terms of the determinants of crude oil prices and the interrelationships between crude oil markets and the macroeconomy, particularly output, employment, and inflation. However, by its nature this literature also extends this general work with a focus on behavior in and between crude oil spot and derivative markets and between crude oil markets and those for financial securities. As argued here, for the most part, existing research on the finance dimensions of crude oil markets have focused on three core areas, namely, intertemporal price relationships in the physical and financial markets for crude oil, evidence on financial market efficiency in crude oil markets, and the role of crude oil as a pricing factor in asset pricing financial markets.

In terms of intertemporal relationships, there seems to be some empirical consensus from a relatively voluminous literature that crude oil futures markets exhibit the basic relationships expected in these markets. However, much remains unknown. For example, can we gather further insights on the risk management strategies of producers and users and the role futures and options can potentially play? Similarly, what impact do alternative energy spot and derivative markets have on those for crude oil and vice versa in a world increasingly characterized by energy substitution and diversification? In contrast, there is much less work on market efficiency concepts in the physical and financial crude oil markets. Moreover, most of this concerns weak-form efficiency and there is generally little attention given to tests of semi-strong or strong form efficiency, largely because of perceived difficulties in specifying equilibrium or benchmark pricing models and the limited availability of suitable data.

Finally, notwithstanding a sizeable and consistent literature on the role of crude oil as a pricing factor in financial markets, including both debt and equity, and seemingly despite efforts to expand this beyond aggregate market studies to those focusing on industries, much also remains to be done. For example, we know much less about the exact transmission mechanism through which changes in crude oil prices impact upon financial markets. Logic would dictate that this is through costs, such that industries and firms relatively more dependent on crude oil (and its processed forms) would be more affected. This would clearly explain crude oil as a pricing factor in, say, the transportation sector, but how exactly do we explain crude oil appearing as a priced factor in other sectors? It also remains a challenge to disentangle this direct effect from a more general indirect impact on consumer expenditure. Likewise, it also ignores the efforts by firms to hedge themselves prices increases in key factor inputs, like crude oil.

7. References

Agnolucci, P. (2009). Volatility in Crude Oil Futures: A Comparison of the Predictive Ability of GARCH and Implied Volatility Models. *Energy Economics,* Vol.31, No.2, (March 2009), pp. 316-321.

Alquist, R. & Kilian, L. (2010). What Do We Learn from the Price of Crude Oil Futures? *Journal of Applied Econometrics,* Vol.25, No.4, (June-July 2010), pp. 539-573.

Alvarez-Ramirez, J., Alvarez, J. & Solis, R. (2010). Crude Oil Market Efficiency and Modeling: Insights from the Multiscaling Autocorrelation Pattern. *Energy Economics,* Vol.32, No.5, (September 2010), pp. 993-1000.

Anderluh, J. & Borovkova, S. (2008). Commodity Volatility Modelling and Option Pricing with a Potential Function Approach. *European Journal of Finance,* Vol.14, No.1-2, (January-February 2008), pp. 91-113.

Arouri, M.E., Lahiani, A. & Bellalah, M. (2010). Oil Price Shocks and Stock Market Returns in Oil-Exporting Countries: The Case of GCC Countries. *International Journal of Economics and Finance,* Vol.2, No.5, (November 2010), pp. 132-139.

Arouri, M.E.L., Bellalah, M. & Nguyen, D.K. (2011). Further Evidence on the Responses of Stock Prices in GCC Countries to Oil Price Shocks. *International Journal of Business,* Vol.16, No.1, (January 2011), pp. 89-102.

Barros Luis, J. (2001). The Estimation of Risk-Premium Implicit in Oil Prices. *OPEC Review,* Vol.25, No.3, (September 2001), pp. 221-259.

Bhar, R. & Lee, D. (2011). Time-Varying Market Price of Risk in the Crude Oil Futures Market. *Journal of Futures Markets,* Vol.31, No.8, (August 2011), pp. 779-807.

Bhar, R. & Nikolova, B. (2010). Global Oil Prices, Oil Industry and Equity Returns: Russian Experience. *Scottish Journal of Political Economy,* Vol.57, No.2, (May 2010), pp. 169-186.

Buyuksahin, B. & Harris, J.H. (2001). Do Speculators Drive Crude Oil Futures Prices? *International Journal of Advanced Robotic Systems,* Vol.6, No.4, (December 2009), pp. 12-16.

Casassus, J., Ceballos, D. & Higuera, F. (2010). Correlation Structure between Inflation and Oil Futures Returns: An Equilibrium Approach. *Resources Policy,* Vol.35, No.4, (December 2010), pp. 301-310.

Chan, K.F., Treepongkaruna, S., Brooks, R. & Gray, S. (2011). Asset Market Linkages: Evidence from Financial, Commodity and Real Estate Assets. *Journal of Banking and Finance,* Vol.35, No.6, (June 2011), pp. 1415-1426.

Chen, N-F., Roll, R. & Ross, S.A. (1986). Economic Forces and the Stock Market. *Journal of Business*, Vol.59, No.3, (July 1986), pp. 383-403.

Chiu, T-Y, & Shieh, S-J. (2009). Regime-Switched Volatility of Brent Crude Oil Futures with Markov-Switching ARCH Model. *International Journal of Theoretical and Applied Finance,* Vol.12, No.2, (March 2009), pp. 113-124.

Cortazar, G. & Eterovic, F. (2010). Can Oil Prices Help Estimate Commodity Futures Prices? The Cases of Copper and Silver. *Resources Policy,* Vol.35, No.4, (December 2010), pp. 283-291.

Crosby, J. (2008). A Multi-factor Jump-Diffusion Model for Commodities. *Quantitative Finance,* Vol.8, No.2, pp. 181-200.

Deaves, R. & Krinsky, I. (1992). Risk Premiums and Efficiency in the Market for Crude Oil Futures. *Energy Journal,* Vol.13, No.2, (June 1992), pp. 93-117.

Demirer, R. & Kutan, A.M. (2010). The Behavior of Crude Oil Spot and Futures Prices around OPEC and SPR Announcements: An Event Study Perspective. *Energy Economics,* Vol.32, No.6, (November 2010), pp. 1467-1476.

Deryabin, M.V. (2011). Implied Volatility Surface Reconstruction for Energy Markets: Spot Price Modeling versus Surface Parametrization. *Journal of Energy Markets,* Vol.4, No.2, (Summer 2011), pp. 67-85.

Doran, J.S. & Ronn, E.I. (2008). Computing the Market Price of Volatility Risk in the Energy Commodity Markets. *Journal of Banking and Finance,* Vol.6, No.4, (December 2008), pp. 2541-2552.

Du, X., Yu, C.L. & Hayes, D.J. (2011). Speculation and Volatility Spillover in the Crude Oil and Agricultural Commodity Markets: A Bayesian Analysis. *Energy Economics,* Vol.33, No.3, (May 2011), pp. 497-503.

El-Masry, A.A., Olugbode, M. & Pointon, J. (2010). The Exposure of Shipping Firms' Stock Returns to Financial Risks and Oil Prices: A Global Perspective. *Maritime Policy Management*, Vol.37, No.5, (September 2010), pp. 453-473.

El-Sharif, I., Brown, D., Burton, B., Nixon, B. & Russell, A. (2005). Evidence on the Nature and Extent of the Relationship between Oil Prices and Equity Values in the UK. *Energy Economics,* Vol.27, No.6, (November 2005), pp. 819-830.

Eryiğit, M. (2009). Effects of Oil Price Changes on the Sector Indices of Istanbul Stock Exchange. *International Research Journal of Finance and Economics*, No.25, No.1, (September 2009), pp. 209-216.

Faff, R. & Brailsford, T. (1999). Oil Price Risk and the Australian Stock Market. *Journal of Energy Finance and Development*, Vol.4, No.1, (June 1999), pp. 69-87.

Fleming, J. & Ostdiek, B. (1999). The Impact of Energy Derivatives on the Crude Oil Market. *Energy Economics*, Vol.21, No.2, (April 1999), pp. 135-167.

Ghoilpour, H.F. (2011). The Effect of Energy Prices on Iranian Industry Stock Returns. *Review of Middle East Economics and Finance*, Vol.7, No.1, (April 2011), Article 3.

Hamao, Y. (1988). An Empirical Examination of the Arbitrage Pricing Theory Using Japanese Data. *Japan and the World Economy*, Vol.1, No.1, (October 1988), pp. 45-61.

Hamilton, J.D. (1983). Oil and the Macroeconomy since World War II. *Journal of Political Economy*, Vol.91, No.2, (April 1983), pp. 228-248.

Hammoudeh, S. & Choi, K. (2006). Behavior of GCC Stock Markets and Impacts of US Oil and Financial Markets. *Research in International Business and Finance*, Vol.20, No.1, (March 2006), pp. 22-44.

Hammoudeh, S., Dibooglu, S. & Aleisa, E. (2004). Relationships among US Oil prices and Oil Industry Equity Indices. *International Review of Economics and Finance*, Vol.13, No.4, (December 2004), pp. 427-453.

Hammoudeh, S., Yuan, Y., Chiang, T. & Nandha, M. (2010). Symmetric and Asymmetric US Sector Return Volatilities in Presence of Oil, Financial and Economic Risks. *Energy Policy*, Vol.38, No.8, (August 2010), pp. 3922-3932.

Hayo, B. & Kutan, A.M. (2005). The Impact of News, Oil Prices, and Global Market Developments on Russian Financial Markets. *The Economics of Transition*, Vol.13, No.2, (February 2005), pp. 373-393.

Hikspoors, S. & Jaimungal, S. (2007). Energy Spot Price Models and Spread Options Pricing. *International Journal of Theoretical and Applied Finance*, Vol.10, No.7, (November 2007), pp. 1111-1135.

Horan, S.M., Peterson, J.H. & Mahar, J. (2004). Implied Volatility of Oil Futures Options Surrounding OPEC Meetings. *Energy Journal*, Vol.25, No.3, (September 2003), pp. 103-125.

Huang, B-N., Hwang, M.J. & Peng, H-P. (2005). The Asymmetry of the Impact of Oil Price Shocks on Economic Activities: An Application of the Multivariate Threshold Model. *Energy Economics*, Vol.27, No.3, (May 2005), pp. 455-476.

Hughen, W.K. (2010). A Maximal Affine Stochastic Volatility Model of Oil Prices. *Journal of Futures Markets*, Vol.30, No.2, (February 2010), pp. 101-133.

Jawadi, F., Arouri, M.E. & Bellah, M. (2010). Nonlinear Linkages between Oil and Stock Markets in Developed and Emerging Countries. *International Journal of Business*, Vol.15, No.1, (January 2010), pp. 19-31.

Jones, C. M. & Kaul, G. (1996). Oil and the Stock Markets. *Journal of Finance*, Vol.51, No.2, (June 1996), pp. 463-491.

Kaneko, T. & Lee, B-S. (1995). Relative Importance of Economic Factors in the U.S. and Japanese Stock Markets. *Journal of the Japanese and International Economies*, Vol.9, No.3, (September 2005), pp. 290-307.

Kaufmann, R.K. & Ullman, B. (2009). Oil Prices, Speculation, and Fundamentals: Interpreting Causal Relations among Spot and Futures Prices. *Energy Economics*, Vol.31, No.4, (July 2009), pp. 550-558.

Kawamoto, K. & Hamori, S. (2011). Market Efficiency among Futures with Different Maturities: Evidence from the Crude Oil Futures Market. *Journal of Futures Markets*, Vol.31, No.5, (May 2011), pp. 487-501.

Killian, L. & Park, C. (2009). The Impact of Oil Price Shocks on the U.S. Stock Market. Interntaional Economic Review, Vol.50, No.4, (November 2009), pp. 1267-1288.

Lake, A. & Katrakilidis, C. (2009). The Effects of Increasing Oil Price Returns and its Volatility on Four Emerged Stock Markets. *European Research Studies*, Vol.12, No.1, (March 2009), pp. 149-161.

Lean, H.H., McAleer, M. & Wong, W-K. (2010). Market Efficiency of Oil Spot and Futures: A Mean-Variance and Stochastic Dominance Approach. *Energy Economics*, Vol.32, No.5, (September 2010), pp. 979-986.

Lin, W.T. & Duan, C-W. (2007). Oil Convenience Yields Estimated under Demand/Supply Shock. *Review of Quantitative Finance and Accounting*, Vol.28, No.2, (February 2007), pp. 203-225.

McSweeney, E.J. and Worthington, A.C. (2008). A Comparative Analysis of Oil as a Risk Factor in Australian Industry Stock Returns, 1980-2006. *Studies in Economics and Finance*, Vol.25, No.2, (June 2008), pp. 131-145.

Meade, N. (2010). Oil Prices—Brownian Motion or Mean Reversion? A Study Using a One-Year Ahead Density Forecast Criterion. *Energy Economics*, Vol.32, No.6, (November 2010), pp. 1485-1498.

Moosa, I.A. & Al-Loughani, N.E. (1994). Unbiasedness and Time Varying Risk Premia in the Crude Oil Futures Market. *Energy Economics*, Vol.16, No.2, (April 1994), pp. 99-105.

Moosa, I.A. & Silvapulle, P. (2000). The Price-Volume Relationship in the Crude Oil Futures Market: Some Results Based on Linear and Nonlinear Causality Testing. *International Review of Economics and Finance*, Vol.9, No.1, (February 2000), pp. 11-30.

Moshiri, S. & Foroutan, F. (2006). Forecasting Nonlinear Crude Oil Futures Prices. *Energy Journal*, Vol.27, No.4, (October 2006), pp. 81-95.

Murat, A. & Tokat, E. (2009). Forecasting Oil Price Movements with Crack Spread Futures. *Energy Economics*, Vol.31, No.1, (January 2009), pp. 85-90.

Nandha, M. & Brooks, R. (2009). Oil Prices and Transport Sector Returns: An International Analysis. *Review of Quantitative Finance and Accounting*, Vol.33, No.4, (November 2009), pp. 392-409.

Nandha, M. & Brooks, R. (2009). Oil Prices and Transport Sector Returns: An International Analysis. *Review of Quantitative Financial Analysis*, Vol.33, No.4, (November 2009), pp. 393-409.

Nandha, M. & Faff, R. (2006). Short-Run and Long-Run Oil Price Sensitivity of Equity Returns: The South Asian Markets. *Review of Applied Economics*, Vol.2, No.2, (September 2006), pp. 229-244.

Odusami, B.O. (2009). Crude Oil Shocks and Stock Market Returns. *Applied Financial Economics*, Vol.19,No.2, (May 2009), pp. 291-303.

Ono, S. (2011). Oil Price Shocks and Stock Markets in BRICs. European Journal of Comparative Economics, Vol.8, No.1, (June 2011), pp. 29-45.

Paschke, R. & Prokopczuk, M. (2010). Commodity Derivatives Valuation with Autoregressive and Moving Average Components in the Price Dynamics. *Journal of Banking and Finance,* Vol.34, No.11, (November 2010), pp. 2742-2752.

Sadorsky, P. (1999). Oil Price Shocks and Stock Market Activity. *Energy Economics,* Vol.21, No.5, (October 1999), pp. 449-469.

Sadorsky, P. (2001). Risk Factors in Stock Returns of Canadian Oil and Gas Companies. *Energy Economics,* Vol.1, No.1, (January 2001), pp. 17-28.

Sadorsky, P. (2003). The Macroeconomic Determinants of Technology Stock Price Volatility. *Review of Financial Economics,* Vol.12, No.2, (March 2003), pp. 191-205.

Sadorsky, P. (2008). The Oil Price Exposure of Global Oil Companies. *Applied Financial Economics Letters*, Vol.4, No.1, (March 2008), pp. 93-96.

Somoye, R., Olukayode, C., Akintoye, I.R. & Oseni, J.E. (2009). Determinants of Equity Prices in the Stock Markets. *International Research Journal of Finance and Economics,* No.30, (August 2009), pp. 177-189.

Tabak, B.M. & Cajueiro, D.O. (2007). Are the Crude Oil Markets Becoming Weakly Efficient over Time? A Test for Time-Varying Long-Range Dependence in Prices and Volatility. *Energy Economics,* Vol.29, No.1, (March 2007), pp. 28-36.

Tokic, D. (2011). Rational Destabilizing Speculation, Positive Feedback Trading, and the Oil Bubble of 2008. *Energy Policy,* Vol.39, No.4, (April 2011), pp. 2051-2061.

Trolle, A.B. & Schwartz, E.S. (2009). Unspanned Stochastic Volatility and the Pricing of Commodity Derivatives. *Review of Financial Studies,* Vol.22, No.11, (November 2009), pp. 4423-4461.

Wang, T. & Yang, J. (2010). Nonlinearity and Intraday Efficiency Tests on Energy Futures Markets. *Energy Economics,* Vol.32, No.2, (March 2010), pp. 496-503.

Wang, Y & Liu, L. (2010). Is WTI Crude Oil Market Becoming Weakly Efficient over Time? New Evidence from Multiscale Analysis Based on Detrended Fluctuation Analysis. *Energy Economics,* Vol.32, No.5, (September 2010), pp. 987-992.

Westgaard, S., Estenstad, M., Seim, M., & Frydenberg, S. (2011). Co-integration of ICE Gas Oil and Crude Oil Futures. *Energy Economics,* Vol.33, No.2, (March 2011), pp. 311-320.

Ye, M., Zyren, J., Shore, J. & Lee, T. (2011). Crude Oil Futures as an Indicator of Market Changes: A Graphical Analysis. *Energy Journal,* Vol.32, No.2, (August 2011), pp. 167-202.

Yu, H-C. & Shih, T-L. (2011). Gold, Crude Oil and the Weekend Effect: A Probability Distribution Approach. *Investment Management and Financial Innovations,* Vol.8, No.2, (June 2011), pp. 39-51.

Yun, W-C. & Hyun, J.K. (2010). Hedging Strategy for Crude Oil Trading and the Factors Influencing Hedging Effectiveness. *Energy Policy,* Vol.38, No.5, (May 2010), pp. 2404-2408.

Zagaglia, P. (2010). Macroeconomic Factors and Oil Futures Prices: A Data-Rich Model. *Energy Economics,* Vol.32, No.2, (March 2010), pp. 409-417.